BARBARA J. NISS AND
ARTHUR H. AUFSES, JR.

TEACHING TOMORROW'S MEDICINE TODAY

The Mount Sinai School of Medicine,
1963–2003

New York University Press • *New York and London*

NEW YORK UNIVERSITY PRESS
New York and London
www.nyupress.org

Library of Congress Cataloging-in-Publication Data
Niss, Barbara.
Teaching tomorrow's medicine today :
the Mount Sinai School of Medicine, 1963–2003 /
Barbara J. Niss and Arthur H. Aufses Jr.
p. ; cm.
Includes bibliographical references and index.
ISBN 0–8147–0706–8 (cloth : alk. paper)
1. Mount Sinai School of Medicine—History.
2. Medical colleges—New York (State)—New York—History.
[DNLM: 1. Mount Sinai School of Medicine.
2. Schools, Medical—history—New York. W 19 N726t 2004]
I. Aufses, Arthur H.   II. Title.
R747.M88N55   2004
610'.71'17471—dc22   2004017283

New York University Press books are printed on acid-free paper,
and their binding materials are chosen for strength and durability.

Manufactured in the United States of America

10 9 8 7 6 5 4 3 2 1

Contents

Preface and Acknowledgments

WHEN STARTING TO READ A BOOK, it is important to know what to expect. What story are the authors telling? What is their viewpoint or bias; what ax do they have to grind? That is the purpose of this Preface, to provide the information on what we set out to do and why, and to thank the many people who have helped us along the way.

This book relates the story of the founding and the subsequent forty years of maturation of the Mount Sinai School of Medicine. We have placed the emphasis on an analysis of the early years of the School, along with a more general report of the current state of the institution. We recognize that we do not have the distance of time or objectivity to place today's events in context, and so we relate the current shape of things and rely on others in the future to assess the meaning.

This leads us to the inevitable question of why we are writing this book now (2003), on the fortieth anniversary of the School. There are many reasons why we have chosen to write about the Mount Sinai School of Medicine at this time. The most obvious reason is that this is the time when we could write about it; in ten years, for the fiftieth anniversary, it is not certain that we will have the desire or time to devote to such a project. Another compelling reason for pursuing this work now is that many of the founders and early leaders of the School are still active and available to us to provide insights and information that are not otherwise recorded. Finally, the publication of the School's history now completes the wider telling of the Mount Sinai story that was begun in *This House of Noble Deeds: The Mount Sinai Hospital, 1852–2002*[1] and *The Sinai Nurse: A History of Nursing at The Mount Sinai Hospital, New York, New York, 1852–2000.*[2]

Clearly, as people who have worked at Mount Sinai since 1954 (Aufses) and 1986 (Niss), we are not unbiased reporters of past events. Aufses had the experience of being actively involved in many of the events recounted here; he was in fact a representative of the Department of Surgery on the original Curriculum Committee. Still, we have striven mightily to ensure that nothing untrue appears in these pages, and we have verified our memories, and others', against official documents from the period.

We would like to thank the many people who have read (and reread) sections of the text, drafted notes and chapters about their departments, sat for interviews, and replied to innumerable questioning e-mails. We would mention in particular (in alphabetical order): Stuart Aaronson, F. Carter Bancroft, Tibor Barka, Mark Chassin, Robert Desnick, Patrick Eggena, Ravi Iyengar, Panayotis Katsoyannis, Terry Krulwich, Sherman Kupfer, Jeffrey Laitman, Robert Lazzarini, John Morrison, Peter Palese, Irving Schwartz, Paul Wassarman, Harel Weinstein, and Savio Woo. We would especially like to thank Milton Sisselman and Nathan Kase for their extensive assistance in helping us to get the facts, and tone, of the times just right. We are also most grateful to Emily Falk for her invaluable work during the summer of 2002. In addition to her research on the early years of the basic science departments, she traveled the Northeast, interviewing not only current undergraduate and graduate students but also a number of alumni. The results of her efforts form the last, and perhaps the most enlightening, chapter. Harriet Aufses read, and reread, every revision of every section; her comments were invaluable. We must also thank Drs. Mark Chassin and Larry Hollier, Chairmen of the Departments of Health Policy and Surgery, respectively, and Lynn Kasner Morgan, Associate Dean for Information Resources and Systems and Director of the Levy Library, the chiefs of our respective departments, for their assistance and understanding while we pursued this project.

We also recognize the support and trust the Mount Sinai leadership and its Boards of Trustees have placed in us by encouraging us to write this history. In addition, we want to thank Andrew Heineman, a longtime member of the Board of Trustees, for his generosity. We are again indebted to New York University Press for its patience and assistance, in particular to Eric Zinner, Editor-in-Chief; Emily Park, Editorial Assistant; and Despina Papazoglou Gimbel, Managing Editor. And, while we acknowledge the assistance of these many people, we must note that any errors of omission or fact are solely our responsibility.

Finally, we would most especially like to thank our families for their continued support and patience, which allow us to carry on this work. They encourage us in our obsessive efforts when needed and forgive us when necessary.

PART I

I

The History of the School

IN THE MID-1950S, The Mount Sinai Hospital was at the pinnacle of hospitals in the United States. It had a reputation for attracting exemplary physicians who provided high-quality patient care and performed the clinical research that had made Mount Sinai's name known around the world. It had state-of-the-art obstetric and research buildings, which had opened in 1952: the Klingenstein Pavilion and the Berg and Atran Laboratories. It was already looking forward to the creation of a new clinical building on Madison Avenue that not only would house the large psychiatry program but also would replace the outmoded ward beds in the older buildings. There were more than 200 research programs ongoing in the Hospital, funded, in part, by the largest amount of federal research dollars to go to any stand-alone hospital in the country. The Mount Sinai Trustees were very involved in the day-to-day operations of the institution and kept a careful eye on the quality of the care provided and the reputation of the Hospital.

But when the medical leaders of The Mount Sinai Hospital surveyed the scene around them, it was apparent to them that the landscape was changing. This led to the beginning of a process whereby the leaders of the medical staff set out to educate and persuade the lay leadership that if Mount Sinai did not respond to these changes, the Hospital's ability to provide high-quality patient care and research programs would be impaired. The solution proffered by the medical staff centered on the creation of an undergraduate medical school. The medical leaders most actively involved were the full-time Hospital department directors: Drs. Hans Popper, Chief of Pathology; Horace Hodes, Chief of Pediatrics; Alexander Gutman, Chief of Medicine; Alan Guttmacher, Chief of Obstetrics and Gynecology; M. Ralph Kaufman, Chief of Psychiatry; Allan Kark, Chief of Surgery; and Martin Steinberg, Director of the Hospital.

3

The campus of The Mount Sinai Hospital, mid-1950s. The area shown is looking southeast, with Fifth Avenue in the right foreground, 101st Street on the left, Madison Avenue along the top left, and 98th Street at the top right.

The physicians put forth many explanations for what had changed and why action was needed now. One was that hospitals were being eclipsed as elite institutions by academic medical centers that combined hospitals and university-affiliated medical schools on one campus. These large centers seemed to more easily attract young physicians because they could offer an academic title and environment. Also, the physicians feared that faculty members from these institutions were filling up the grant review agencies and specialty board panels and might be inclined to favor people at academic medical centers.

One factor leading to the elevation of these university-based medical centers was the change in the nature of medical research. Mount

Sinai Hospital had thrived in the era of clinical research, where data was received at the bedside, analyzed in the laboratory, with the results coming in the form of a disease description or a treatment taken back to the bedside. After World War II, the basic sciences of chemistry, molecular biology, and physiology had assumed a much larger role in elucidating the underlying workings of disease. Biomedical research became based in scientific labs, often on a medical school or university campus. It took trained scientists to run these laboratories and to teach the techniques and science involved. It was becoming beyond the ability of practicing physicians to pursue molecular research because of the nature of their training and their limited access to modern laboratory facilities. Mount Sinai was becoming out of step with the evolution of medical research.

There was another major change that had affected Mount Sinai's ability to attract and retain house staff after their training, but this was not given much public airing. From the beginning, in 1872, when Mount Sinai first created an internship, the Hospital had benefited from the anti-Semitism that existed in the graduate training programs at other hospitals.[1] While it was difficult for young Jewish men to gain acceptance to medical school, it was even harder to find a place in a high-quality postgraduate training program. This applied to many hospitals in New York City, making Mount Sinai Hospital the obvious choice for young Jewish doctors. Among these candidates were some of the brightest and the best, and the quality of Mount Sinai's interns reflected this. Ironically, after World War II, when anti-Semitism lessened in other graduate training programs, Mount Sinai suffered as Jewish trainees now could, and did, select other institutions.

All of these factors culminated in the most persuasive argument that the proponents of the school used: Mount Sinai's leadership position in the world of medicine would surely slip as all of these issues combined to make Mount Sinai less desirable to the best physicians and so less able to provide the best care and to perform pioneering research. As Dr. Martin Steinberg, the Director of the Hospital, put it:

> this concert of full-time people and I began to feel that we were either going to be a university hospital or we would be left behind. We began to see for us the kind of a role that hospitals like, let's say, Lenox Hill had, or maybe even Roosevelt, and that if we were going to be the

Mount Sinai we were so fiercely proud of, that the only way to accomplish it was to be a university hospital.[2]

The response to these arguments varied. Leo Gottlieb, the Trustee who chaired a joint committee of physicians and Trustees looking into the future of medical education at Mount Sinai, felt that the idea of founding a medical school was "so impractical and so contrary to the direction that I felt the Hospital should be moving" that he wrote a twenty-one-page memo opposing the concept of a school.[3] Others did not see what good it would do the Hospital to have a medical school. There were already medical students from Columbia University's College of Physicians and Surgeons (P&S) on the wards, along with the residents, to provide an academic atmosphere. Some faculty had academic titles from Columbia thanks to their roles in teaching the undergraduate students, as well as offering extensive postgraduate training courses under the auspices of Columbia. There were real questions also about what the influx of large numbers of full-time clinical faculty would mean in terms of the voluntary staff's ability to admit patients. Would there be fewer—or any—beds left for them?

Still, the physician leaders pushed to have a responsible agency come to Mount Sinai and study the possibility of founding a school, as well as review the current affiliation with P&S to see whether more could be made of that academic tie. Enough Trustees were convinced by their arguments that the plan was approved. At the same time, May 1958, it was also agreed that the Hospital would hire more geographic full-time physicians who would become "career men" at Mount Sinai. These doctors would receive salary supplements, research support, and office space from the Hospital, in return for which they would concentrate their efforts at Mount Sinai.[4] So, while there was clearly opposition to the concept, there was not enough to stop the momentum of studying the idea and at least finding out more.

By the end of 1958, Popper and Gutman took another step by going to Chicago to speak with representatives of the American Medical Association (AMA) and the Association of American Medical Colleges (AAMC). They described the Hospital's desire to found a medical school and asked what the position of these organizations would be if they proceeded. Both groups assured them that they believed Mount Sinai capable of founding a school, but they noted that no hospital had founded a medical school on its own since the

Flexner Report on medical education in 1910 and that this was not a precedent they wished to encourage. They did not discourage the Hospital from proceeding but urged the leaders to find a university affiliation.

Time passed, and Mount Sinai committees continued to study the issue, as well as deal with the other urgent business of the Hospital. Leaders were struggling to finalize plans for the project to build what became known as the Klingenstein Clinical Center amid a flurry of decisions about which plan would best meet the needs of the future and still be affordable in the present. Eventually a plan for a twelve-story building to house psychiatry, medical/surgical beds, a new boiler plant, and engineering space was chosen. Also, a nurses' residence was begun on 101st Street to accommodate a growing professional staff. Work began on both projects in 1959.

Another significant event that year was the publication of the Bane Report.[5] This national survey reported that there was a shortage of physicians and that an additional twenty-five medical schools were needed to train a sufficient number of doctors for the United States. This was quickly added to the arguments for the school, although it was not as important as the internal factors.

In 1960, the pace of the planning for the school quickened. In January, nineteen members of the staff asked the Board of Trustees to approve a field survey by the AMA's Committee on Medical Education and Hospitals to determine the financial and other requirements for founding a school. This was approved. In the spring, Popper and Steinberg made a visit to Chicago to again sound out the AMA and the AAMC on their chances for approval if they were to create a school. Again, they were somewhat encouraged to proceed but also were urged to make efforts to find a university affiliation.

This had been much on the minds of the leaders at Mount Sinai. They could see the value in having ties to a university and, indeed, had approached many universities about becoming their medical school. It was felt that enlarging the current relationship with Columbia University would not be possible. Mount Sinai, although noted as a major academic affiliate of the university, filled a very circumscribed role, primarily in teaching postgraduate courses and undergraduate electives. Few medical students were required to rotate through Mount Sinai. Still, talks proceeded in a slow fashion as Mount Sinai continued to look elsewhere.[6]

Mount Sinai officials approached several universities around the metropolitan area and even went as far afield as Princeton, Dartmouth, and Brandeis. They spoke to New York State officials about the possibility of becoming a third state-sponsored medical school, but the state declined. In some ways, what may have made Mount Sinai's offer attractive were also the aspects that were troubling to these academic institutions: Mount Sinai wanted to have the school on its own campus, to finance it on its own, and to run it itself. The Hospital was, of course, willing to follow the academic guidelines and policies of the university, but it was clear that this was to be a Mount Sinai venture, requiring minimal commitment from the university. Having the hospital be dominant was not the model for university medical centers, but even with these terms, some colleges were interested, and talks were, in fact, moving ahead with The City College of New York.

Spring of 1961 saw the first formal visit from an official medical group, the joint AMA-AAMC sponsored organization, the Liaison Committee on Medical Education (LCME). Four members visited both Mount Sinai and City College to investigate their readiness to start a school. The report was positive, and the talks with City College continued. By July 1961, the planning committee for the school approved moving forward with an affiliation.

But it was not to be, and the reason for this places Mount Sinai within the broader setting of the Jewish community in New York. While Mount Sinai was the largest Jewish hospital in New York City, with 1,200 beds, it was still only one of several Jewish teaching hospitals in the city. The Montefiore Hospital had outgrown its roots as a chronic-care facility and also supported large residency and research programs. To the leaders of Montefiore, the question was this: if there was to be another Jewish medical school in the city (the Albert Einstein College of Medicine already existed), why should Montefiore not be included? For all of the reasons a school made sense at Mount Sinai, it would enhance Montefiore's position, as well. And so Montefiore entered the negotiations with City College. The two hospitals attempted to reconcile their plans and to work together, but this was not possible. City College withdrew its interest.

This was just one aspect of the dialogue that existed in the New York Jewish community about the founding of a medical school at Mount Sinai or elsewhere. The arguments were debated within the organization of the Federation of Jewish Philanthropies. In December

1961, Leonard Block, a prominent member of the Jewish philanthropic community and a staunch supporter of Federation, privately published a pamphlet entitled *Let us not sponsor a Medical School*, aimed at leaders of Federation and the Jewish community.[7] In it, Block marshaled data to rebut the major arguments for the founding of a school. He showed that Mount Sinai had matched 100 percent of its residency positions for the past ten years, so officials could not argue that there had been slippage in the desirability of the residency programs. He also noted that two Federation hospitals with close medical school ties had not matched as well. Further, he stated that even if the new school could attract quality medical students who stayed on as house staff, since the Hospital already attracted good residents, there would be no added benefit. Block also provided statistics that supported his claim that more research had been "done at Mount Sinai in 1960 than at any other hospital in Greater New York,"[8] including at Presbyterian Hospital, which was larger. Again, what benefit could a school bring?

To answer the question of being able to attract high-quality career physicians and scientists, Block cited the Mayo Clinic to show that "there is no indication from any of the data reviewed over the last ten years that teaching the undergraduate is necessary to obtain the type of teaching environment that stimulates and excites the interest of brilliant men."[9] He also noted that the high quality of the current full-time chairmen of departments at Mount Sinai belied that requirement.

In a general vein, Block asserted that the Bane Report was not universally accepted as true, and, even if it were, it would not be good to sponsor another sectarian institution when the Jewish community had always fought against sectarianism as a hindrance to itself and to others. He also argued that a strong university affiliation really was a necessity: "There are a number of examples of medical schools that are hospital controlled, but none of them are top grade medical schools."[10] In closing, he asked whether it was worthwhile to start a new school as "insurance" that a hospital's preeminence would continue. The cost of building such a school would have to be taken from other "needful projects," while "even the most pessimistic forecasters concede that good hospitals, rendering excellent patient care, can be run and maintained without medical school affiliation."[11]

In the end, Block's well-reasoned arguments did not matter: the momentum was too great, the desire for a school too strong, the proponents too dominating. The Federation ultimately supported Mount

Sinai's plan for a school. When Mount Sinai dedicated the new Esther and Joseph Klingenstein Clinical Center, in November 1962, it was clear by the wording of the program that the Hospital was committed to this new direction:

If you have built castles in the air, your work need not be lost; that is where they should be. Now put the foundation under them. —Thoreau

By this philosophy The Mount Sinai Hospital has lived and grown. For a century and a decade, farsighted, dynamic leaders have attempted to forecast the future and to outline in broad terms how the hospital can keep abreast of the accelerating tempo of the years ahead. One by one, their "castles in the air" have been anchored on solid foundations, keeping the name of Mount Sinai synonymous with the history of medical progress.

The task is a Herculean one—a never-ending one. In this rapidly changing world, the medical sciences advance so swiftly that a teaching and research hospital which lags behind is soon obsolete. Extensive research programs must be continued to develop new drugs, new techniques, new equipment. A broad, ever-expanding teaching program is required to communicate this new knowledge and these new techniques to the professional men and women who administer them for the benefit of patients. We at Mount Sinai are thus committed to change . . . to meet the growing needs of our community, to move ever forward.[12]

Building a medical school had become a mission, and the cast had been assembled that could bring it to fruition. Chief among these on the medical side was Hans Popper, M.D., Ph.D. Popper had been recruited to Mount Sinai in 1957 to serve as Director of the Department of Pathology. He was born in Vienna in 1903 and came to America in 1938 to escape arrest by Nazi officials. He settled first at Cook County Hospital, in Chicago, and began fashioning a body of work that would lead to his being called the "Father of Hepatology." His research in liver disease continued throughout his life, regardless of the many other duties he assumed.[13] Throughout his life, Popper believed in the need for basic medical research to propel medicine forward. While at Cook County, he was involved in the creation there of the Hektoen Institute for Medical Research.

Hans Popper, M.D., Ph.D., in his office in the Atran Building. Dr. Popper kept this office throughout his career at Mount Sinai, from Chairman of Pathology, to Dean of Academic Affairs and then Dean and President. Over his right shoulder is his beloved microscope.

Popper emerged as the leading proponent of the new school among the medical staff. He drafted the earliest statements of philosophy, was named Acting Dean in 1963, and became Dean for Academic Affairs when a permanent Dean was identified. The most likely reason that Popper himself was not named Dean was his age and inclination; Popper was already past the age of sixty, and he did not want to give up the necessary time from his first love, the study of the liver. He was a scientist of international standing, a prodigious author, a sought-after lecturer and visiting professor. In his ultimate role as Dean for Academic Affairs, he retained a great deal of power over the direction of the school and yet escaped many of the administrative burdens.

The dream of founding a medical school would have died aborning without the active support of the lay leaders of the Hospital. While not all Trustees agreed with the need for a school (e.g., Leo Gottlieb, Chairman of the Joint Committee on Medical Education, resigned his

Gustave L. Levy, President of The Mount Sinai Hospital and then Chairman of the Mount Sinai Boards of Trustees, 1962–1976. Photo credit: Fabian Bachrach.

position on the Board), most were willing to be convinced. Gustave L. Levy, active in Federation activities, had joined the Board of Trustees of The Mount Sinai Hospital in 1960. In 1962, he became President of the Board and spearheaded the effort to create a school at Mount Sinai. In January 1963, he created the Joint Committee on Medical School Planning, a joint trustee/physician committee. Without Levy's decisiveness, fund-raising ability, and political skills, developing a school would not have been possible within the same time frame or scale.[14] These skills had been honed in his roles as Senior Partner and Chairman of the Management Committee at Goldman, Sachs, as well as Chairman of the New York Stock Exchange.

One of Levy's early actions as President was to request William Golden, Trustee chairman of the Research Administrative Committee (RAC), to have an outside group come to study and report on the research program at the Hospital. This became another opportunity to examine the idea of having a school. A site visit to assess the research capabilities of Mount Sinai was conducted in November 1961 by Drs.

William B. Castle, from Harvard, and W. Barry Wood Jr., from Johns Hopkins. The visitors heard a round of introductory speeches by Golden and some of the full-time clinical chairmen and investigators on staff. Tours of the laboratories and lunch were followed by fourteen scientific presentations by researchers, including one by young Nathan Kase, M.D., then a resident in OB/GYN and later Dean of the School of Medicine. An RAC meeting to discuss the current round of grant applications followed.

In their report, dated December 14, 1961, Castle and Wood made some suggestions about running the research programs, told the Hospital that a closer relationship with a medical school was advisable, and recommended that an advisory planning board composed of medical educators be established to begin work on a school. They noted that "we have no doubt that the Mount Sinai Hospital with its magnificent staff of clinicians and research scientists already possesses the professional talent to become the nucleus of one of the nation's great medical schools. We are also of the opinion that now is the time that serious planning should be begun."[15]

A year passed before an advisory group was created, but during that time the Hospital took the important step of applying to New York State for a charter for The Mount Sinai Hospital School of Medicine. At this time, the Hospital Board also approved a $500,000 gift to the School "when and if" the charter was received. The provisional charter was approved on June 28, 1963; a final charter would be forthcoming with the opening of the School.

This was just one event in what was a busy year for the planners of the Mount Sinai School. In January 1963, the Hospital received word that the AMA had approved it as a site for a future medical school, again urging Mount Sinai to place the school "within the body of a graduate university." Two months later, a formal Advisory Committee on the formation of a school was formed. The members were Drs. William B. Castle, Bernard D. Davis, and Robert Glaser, from Harvard; Alvin C. Eurich, Litt.D., of the Fund for the Advancement of Education; Dr. Seymour S. Kety, National Institute of Mental Health; Dr. Vernon Lippard, of Yale University; Dr. William C. Rappleye, of the Josiah Macy Jr. Foundation; Dr. Paul Weiss, of Rockefeller Institute; Dr. W. Barry Wood Jr., of Johns Hopkins; and Drs. George Baehr and Paul Klemperer, retired former department chiefs at Mount Sinai and leaders in New York medicine.

This is the Provisional Charter of The Mount Sinai Hospital School of Medicine, granted in 1963. Subsequent charters changed the name to the current Mount Sinai School of Medicine.

Baehr especially was a well-known figure at the Hospital. He had risen through the ranks at Mount Sinai from his graduation from the house staff in 1910, eventually becoming Chief of the First Medical Service and Director of Medical Research. A clinician and scholar of note,[16] he led the Mount Sinai–staffed U.S. Army medical unit (Base Hospital No. 3) during World War I and served on advisory committees for several government agencies at the city, state and federal level. A leader in the field of public health, Baehr was one of the founders of the Health Insurance Plan (HIP) of New York. He had tremendous knowledge of the workings of New York medicine and politics and used his insight and influence to further Mount Sinai's cause in official circles.

The first meeting of this advisory group was convened in March 1963, along with the members of the Joint Committee on Medical Education. Gustave Levy chaired the session. At this meeting, Hans Popper made the first formal statement about what the philosophy guiding the development of the School would be. One month after this meeting, the joint committee of Trustees and medical staff formally approved this statement as a base from which to proceed while establishing the School. As noted later in a report to the Josiah Macy Jr. Foundation:

> Probably the most significant aspect of the philosophy was the apparent determination to make the most of the school's situation as a hospital-based institution while at the same time eliminating the inherent dangers such as . . . a possible tendency to underemphasize the basic sciences. Too, the basic concept underlying the philosophy, which in effect made humanism the cornerstone of medical education, was a distinct departure from the general practice in medical schools.[17]

Popper proposed that Mount Sinai create a biomedical center consisting of the Hospital, the Medical School, a Graduate School for Human Studies, a Graduate School for Biological Sciences, and an Institute for Environmental Medicine. To plan a school, it was necessary to look into the future and to try to see what the practices of medicine and research would be like, using:

> certain trends which are recognized and definite. One of these is the increasing specialization in medicine and almost reciprocal reduction of the attraction of the general practice of medicine. As more and more specific therapeutic and diagnostic techniques become available to the specialists, their ascendancy in attraction over general practice will increase. Preventive medicine . . . will assume greater importance.[18]

Popper felt that the curriculum would have to meet the following requirements:

1. Thorough knowledge of the clinical and biological problems confronting the physician of the future, combined with sufficient knowledge to permit specialization in the graduate years, without necessarily providing within the school curriculum

all the information which might be required of a general practitioner.

2. Dedication to the personal problems of the patient and his family, with thorough knowledge of the problems of a changing society. . . .

3. Continuing training in the physical sciences to enable the physician to achieve maximum proficiency in the techniques of medicine of the future and to deal with environmental changes, natural and manmade. . . .

4. Opportunity for the student to transfer from the medical curriculum to the paramedical sciences, permanently or for a period.[19]

At this advisory group meeting, there were extensive discussions about the philosophy and the curriculum and about what the School would need to get started: a permanent Dean, a basic science faculty, and a strong academic affiliation, although it was felt that it was more important to find the right fit than to rush into a relationship with just any university. It was also agreed that the School should try to attract lower-income students who had traditionally been passed over and to allow time for students to take elective classes early in their training.

Another important milestone in the creation of the School was the establishment of a separate Board of Trustees for the School of Medicine. The first meeting of this group was held on August 2, 1963. The original members of the Board were Gustave L. Levy, Joseph Klingenstein, Sheldon R. Coons, Henry A. Loeb, Ira A. Schur, Gladys Straus, and M. Ronald Brukenfeld. At this meeting, three more trustees—William T. Golden, Max Abramovitz, and Milton Steinbach—were added, bringing the total number of School Trustees to ten. Steinbach was elected the first President of the Mount Sinai School of Medicine at this meeting; Levy was named Chairman of the Board of the School. He was also President of the Board of the Hospital.

One of the issues that the new Board had to handle was the reality of the School as a physical space. In a press conference in the fall of 1963, Levy said that the first students would come to Mount Sinai in 1968, a very ambitious starting date. The Board was now faced with determining what that meant in terms of identifying space on the Hospital campus where the School would go. The School created the need for student housing, student laboratories, research laboratories for the faculty that would be recruited, classrooms, and lecture halls. It also required an expansion and significant upgrade of the Jacobi Library, which had served the Hospital for many years. These questions were being addressed by the firm of Joseph Blumenkranz and Associates, which had been hired in the spring of 1963 to look at the Hospital facilities and site and to devise a master plan for both the physical space and some of the more academic aspects of the School.

The initial plans very much reflected the influence of the Hospital's needs over the planning process, with new clinical areas dominant and the teaching aspects receiving less attention. As the School became more real and central to the process, this evolved into a much more School-focused project, resulting in what became the Annenberg Building, with no inpatient beds. Indeed, the stimulus for the School subtly changed from being what the Hospital needed to survive to being what was needed for the research and educational missions of Mount Sinai to thrive. This is witnessed by the text of a report submitted to the LCME in September 1964:

> Whereas the welfare of the hospital was a large factor in our beginning efforts, at this time in the mind of the founding group, both lay and medical, the newer goals and necessary experiments in medical education have become the overriding considerations and the hospital is now considered an important part of the experiment rather than its prime benefactor.[20]

When the discussion over the School began, the medical staff gave Levy an estimated cost of the project as $5 million. They felt that, since the expensive hospital component of the Medical Center already existed, it would be relatively less costly to build a school for Mount Sinai than for other organizations. This was a short-lived dream. Estimates for the amount needed rose quickly and continuously. In 1964, a grant application to the U.S. Public Health Service (USPHS) for funds to underwrite the construction on the building was submitted in the amount of $26 million. In the September 1964 LCME report, the total cost of the School was estimated at $54 million, with $30 million for construction of School facilities and $24 million for Hospital renovation and upgrades for teaching purposes.[21] This estimate was conservative, considering that the architectural consultants had presented three schemes to Mount Sinai in August that had ranged from $52 million to $67 million.

The physical needs were only one aspect of the financial planning for the School. The leaders also had to address plans for recruiting and supporting an entire basic sciences faculty for the Medical and Graduate Schools; greatly expanding and underwriting the full-time clinical faculty; creating scholarship and loan funds, only some of which already existed at the Hospital; and creating an endowment fund to ensure the future of the enterprise.

Parallel to this was the need to undertake the creation of an academic entity at the Hospital. The first step was the statement of philosophy that Hans Popper had submitted in the first meeting of the Advisory Committee in March 1963.[22] This evolved, in 1965, into an important publication called "The Mount Sinai Concept." The key contribution of this article is the articulation of the tripod upon which the School would be based:

> The developing school intends to rely on two of the legs of the usually referred to tripod which exist on the campus now, namely patient care and research, and to lengthen the third leg [education] from postgraduate to medical student teaching. For this another tripod is being depicted. One leg is exact biology in medicine. The second is counteraction of the depersonalization of the organ specialized physician by the broadening influence of a Graduate School for Human Studies as part of The Mount Sinai Biomedical Center. The third leg is Community Medicine, which by experimental patterns strives to give in a setting of specialists good care to every patient and every disease including presymptomatic stages.[23]

The idea that the School would incorporate this new concept of community medicine was unusual for Mount Sinai, both generally and specifically. On the whole, it was a new idea in medical education. Certainly the need to place a patient in the context of his family and community was not unheard of; what was new was that now the techniques of how best to do this would be taught and emphasized in the medical curriculum. It was also unusual for Mount Sinai itself as an institution. While Mount Sinai had long served as a primary healthcare provider for the community of East Harlem, the concept of proactively studying the health needs of the community or trying to ensure health, as opposed to treating disease, was not considered feasible or appropriate.[24]

These are some of the physicians who helped get the School going: from left to right, Hans Popper, Chairman of the Department of Pathology; George Baehr, Director of the First Medical Service, 1928–50; George James, Dean of the School; and Horace Hodes, Chairman of the Department of Pediatrics, 1948–76. This image was taken at the dedication banquet on October 20, 1968.

Other academic areas that needed to be created before the School could begin were the recruitment of a scientific faculty; the development of an administrative structure, including policies and procedures; and, most important, the appointment of a full-time Dean to lead the School. A meeting of the Board of Trustees of the new School, in March 1964, laid the groundwork for all of these areas. At this meeting, the Board itself expanded to twenty-two members. This larger group then formalized the first official departments and faculty of the School when it accepted the departments of Medicine; Obstetrics and Gynecology; Pathology; Pediatrics; Psychiatry; and Surgery. The Directors of these areas became Acting Chairmen in the School. Drs. Alexander Gutman, Saul Gusberg, Hans Popper, Horace Hodes, M. Ralph Kaufman, and Allan Kark, respectively, also became Professors of the School. The

Trustees at this meeting also heard another exposition on the philosophy of the School and the building plans to date. And, importantly, at this same meeting, Gustave Levy appointed a new committee, with himself as Chairman, to mount a search for a Dean.

Hans Popper had been named Acting Dean in 1963 when accelerating events had forced Mount Sinai to identify an interim chief academic officer. Popper's leadership role in conceptualizing the School made him the natural choice for this post, and he worked tirelessly for the establishment of the School. Still, it was a continuing theme of Advisory Committee and site visit meetings: who would be the Dean? Many suggestions were solicited and offered about what kind of qualities should be sought in the Dean. Advisory Board members noted he should be an academic of substantial stature, since as Dean he would be the leader of the faculty, charged with representing the faculty needs to the Board of Trustees, while also implementing the Board's policy decisions with the faculty. Most suggested that the Dean should be a basic scientist, to show the Hospital's commitment to developing the sciences at the new School. Others thought the Dean should be a clinician so that he might have standing among the strong group of physicians already at Mount Sinai. Many felt that the Dean should have experience administering a large program; beyond that, all agreed that the new Dean had to be a dynamic leader who could "sell" the School to prospective faculty, students, and donors. One noted that "He needs vision and a hard-boiled attitude."[25] It was also suggested the new Dean be less than fifty-five years of age.

Lists of names were compiled and many candidates were contacted, but no positive progress was made. The need was becoming more urgent as the various official medical bodies placed great importance on the naming of a Dean to help indicate the quality of the School and the direction in which it would go. Finally, at a special meeting called on January 4, 1965, the Board of Trustees agreed that the deanship would be offered to Irving Schwartz, M.D., and that he would be Dean and Chief Executive Officer of the Medical School, with other titles, such as Vice President of the combined organizations, to be negotiated between Levy and Schwartz. Although not on the search committee's initial list of potential candidates, Schwartz had an academic background and a strong interest in medical education and was known for his scientific work in neurohypophyseal hormones and peptides.[26] His appointment was followed within days by the arrival of a four-man team from the LCME to perform a site visit so that the School could be granted a reasonable assurance of being accredited when the first students arrived. This "reasonable assurance" was viewed as necessary to receiving the $26 million Public Health Service grant that had been submitted. The site visit went well, and the Hospital was told that a favorable report would be written.

Schwartz's presence as the Dean at the site visit was a critical element necessary for LCME approval. However, both he and the Trustees recognized immediately that his great strengths in teaching and research would not be utilized to the fullest as Dean of the Medical School.[27] At its January 18, 1965, meeting, the Board heard a report about the continuing negotiations with Schwartz but now approved his appointment as Dean of the Graduate School of Biomedical Sciences and Chairman of the Department of Physiology in both the Medical School and the Graduate School.[28] These roles, which Schwartz filled with distinction for sixteen and fourteen years, respectively, were unquestionably better suited to his talents.

By early March, Hans Popper was gathering information on George James, M.D., the Commissioner of Health for New York City, as a possibility to become the Dean of the School of Medicine. Again, he was not on any candidate lists and was, indeed, very different from others who had been mentioned. His name may have been submitted by Milton Steinbach, a Trustee and President of the School,[29] or by Schwartz and Popper,[30] but certainly he was a well-known figure to the people at Mount Sinai. A graduate of Columbia University, with an M.D. degree from Yale, he had spent many years in the field of public health, serving in the New York State and New York City Departments of Health. He was named Commissioner of the New York City office in 1962. During his tenure there, the city's water supply was fluoridated, and family planning services were instituted throughout the city. He was an early and frequent advocate of the value of understanding health and disease in the context of the family and community. He had held faculty appointments at Yale, Columbia, and Johns Hopkins over the years and was a productive author and frequent speaker at professional meetings and governmental hearings on health care issues.

In April 1965, James submitted a letter outlining his understanding of the position: he would be Dean of the Medical School, and all other Deans would report to him. Also, he would serve as "Chancellor" and

Chief Executive Officer of the Mount Sinai institutions. In addition, he would undertake the creation of a Department of Community Medicine, an area of importance in the "Mount Sinai Concept," and a task for which James was uniquely qualified with his many years in urban public health.[31] On June 14, 1965, George James was approved as Vice President and Dean of the Medical School, Chairman of the new Department of Community Medicine, and Professor of Community Medicine.

In July 1965, Mount Sinai received word that its grant application to the USPHS had been approved but not funded. Mount Sinai changed the application to reflect the current building plans and re-submitted it in February of 1966. The grant process was a massive undertaking, but only one aspect of the growing administrative burden that the formation of the School generated. Throughout the developing years of the School, it had fallen primarily to the Office of the Director of the Hospital to coordinate the work that needed to be done to make things happen. Along with the Director, Martin Steinberg, there were also four Associate Directors. All were involved in the founding of the School, but two assumed larger roles. S. David Pomrinse, M.D., an Associate Director who operationally was in charge of "front of the house" activities, such as nursing, medical staff relations, and government relations, was very active in the planning for the School. His efforts involved working on the organizational structure of the proposed Medical Center, curriculum, and faculty issues, including the early drafts of the faculty practice plan. He helped develop the first budgets for the School, trying to codify exactly what the needs would be: how many anatomy professors, how much laboratory space, how many square feet. He helped with programming the space around the campus, trying to define the future needs of the Hospital and the School and to identify appropriate areas for them. In this vein, he suggested a plan to wrap a new hospital building around the structure of the planned Medical School tower in the hope that it would, in the long run, be the most cost-efficient way to build new hospital facilities. After much discussion, it was determined that this was not to be the case and that the Hospital would have to wait for its new beds. In 1969, Pomrinse became Director of the Hospital and Vice President for Hospital Operations, Mount Sinai Medical Center. He also served as the Chairman of the Administrative Medicine Department in the School.

He remained a voice decisively attuned to the needs of the Hospital in what was a rapidly changing healthcare environment.

Milton (Mike) Sisselman, the Hospital Associate Director in charge of operations in the "back of the house,"—housekeeping, laundry, engineering, and so on—handled a great deal of the planning work and the grant application paperwork for the School. He dealt extensively with the consultants Mount Sinai had brought in to advise on program and facilities. He later oversaw much of the building process of the Medical School building, including planning for the destruction of six Hospital buildings in the middle of the campus, and the resultant need for programmatic changes in the Hospital to accommodate the moves. He was involved with the financing of the building through the federal grant and state loans and worked with the contractors during the many years of construction. With the establishment of a formal President's Office, Sisselman assumed the Medical Center title of Vice President for Coordination and Planning. He later became the Vice President for Professional Services, serving as the Executive Director of the Medical Service Plan (later the Faculty Practice Plan), leading to its full launch in 1976.[32]

A third Hospital staff member who worked assiduously on the creation of the School was Sherman Kupfer, M.D. He was not in the Hospital administrative structure but was a physician and scientist with strong academic credentials. He had been on the Research Administrative Committee (RAC) at Mount Sinai since 1955 and had served as its Executive Secretary for many years. This group reviewed all the grant applications drafted at Mount Sinai before they were submitted to granting agencies. This trustee/physician committee ensured that the projects were appropriate and credible and that issues of patient and animal safety in research were addressed. Mount Sinai was something of a pioneer in this area, establishing the committee in the 1930s, long before there were federal requirements for this type of oversight. Kupfer himself was a successful grant applicant, having secured funding from the National Institutes of Health (NIH) for Mount Sinai to establish its Clinical Research Center in 1962, where he served as the Center Director until 1985. Kupfer clearly had a strong grasp of the research efforts of the Hospital. As an Assistant Professor of Physiology at Cornell University Medical College, he also had an understanding of the academic enterprise. By 1966, he was appointed an Assistant

Dean, rising to Associate Dean for Faculty Appointments and Research Administration with the opening of the School in 1968.

Many aspects of the School were now in place or in process. The two pieces missing were the $26 million federal grant, which would make a tremendous difference in the Hospital's ability to build the Medical School building, and a university affiliation. In fact, it was becoming clear that the two were connected. In a September 1965 letter to Gustave Levy, George James wrote about a recent visit to Washington, D.C., where he spoke with the Assistant Surgeon General, Dr. Harald Graning, about the Mount Sinai grant application:

> He [Graning] told me that our request for the medical education portion of the school caused a tremendous amount of discussion in the Advisory Council. The only reason we did not receive the highest priority was because of our lack of a university affiliation. The group was disturbed because of the precedent which would be set, making it more feasible for many other professional schools (i.e., optometry, podiatry, etc.) to be established as trade schools and not true academic institutions.[33]

Still, the leaders felt that the fact that their grant rating had been fairly high allowed them time to be selective in arranging an affiliation. By 1966, this feeling was changing. In August of that year, Mount Sinai hired the firm of Heald, Hobson and Associates to help find a university affiliation for the School. In November, the preliminary report recommended Brandeis University, the Polytechnic Institute in Brooklyn, and the new City University of New York (CUNY) as possibilities. Brandeis was quickly eliminated, and meetings were held with the other two. By May 1967, The City University was the lead candidate. The points in CUNY's favor were that it was financially secure, its faculty was of "high quality," and it would provide Mount Sinai with the funding for ten professorships, primarily in the basic sciences. An important drawback was that it was a large public institution, and there was the fear that the School of Medicine would be lost in the larger organization and would not be perceived as a private entity.[34] The Polytechnic Institute was less financially sound but also had a good faculty, and the new School was interested in biomedical engineering. It would also be willing to consider a move to the Mount Sinai campus.

The decision was eventually made to affiliate with CUNY. An agreement was signed on July 31, 1967, and on October 26, the New York State Board of Regents amended the School's charter to make it the Mount Sinai School of Medicine of The City University of New York. Much excitement greeted the affiliation. Dean James was confident that the young university and Mount Sinai could work together to improve healthcare service and education throughout the City. There were many plans for collaborative projects between the two institutions in the areas of training healthcare workers, from nurses, nurse's aides, and technicians to MBAs in healthcare administration and Ph.D.s in biomedical engineering. It was also hoped that the members of the graduate programs at the two institutions would work closely together.

The other event to greet the affiliation was a letter dated September 8, 1967, informing Mount Sinai that the federal grant application had been funded in the amount of $26 million, the largest grant of its kind up to that time. This was followed in October with the kickoff of a fund-raising campaign with a goal of $107 million. The level was raised, in 1969, to $137 million and then, in 1970, to $152 million.

Mount Sinai still was working out the details of what the organizational structure of the combined institutions—School and Hospital—would be. It had initially been decided to have two separate corporations with the CEOs, the Dean and the Director, reporting to their respective Boards and working out overlapping areas between themselves. The management firm of Cresap, McCormick and Paget was brought in to study the issue and to make a report. The report advised and Mount Sinai agreed to the creation of an administrative structure that would include three corporations at Mount Sinai: the Hospital, the School of Medicine, and the Medical Center. The Hospital and the School were operational units, while the Medical Center was an umbrella structure with senior administration and a few overarching departments such as Public Affairs and Development. Each corporation would be led by a board of lay trustees. These three boards would be overlapping in membership, and all three would have the same person serving as Chairman. In the early years, this was Gustave Levy. The President of the Medical Center would be the chief executive officer of the corporations, with the Dean and the Director of the Hospital, the chief operating officers, reporting to him. However, it was decided initially to fill the positions of both President and Dean

Gustave L. Levy, Chair-
man of the Mount Sinai
Boards of Trustees, puts
the Presidential necklace
around George James'
neck during James' inau-
guration as the first President of the Mount Sinai
Medical Center, October
20, 1968. This event was
held at The City College of
New York.

with the same person, which served to further the interest of the young
School. It was also agreed that the formal activation of the Medical
Center structure would wait until the then Hospital Director, Martin
Steinberg, retired in 1969. After this time, the new Director would
report to the President and not directly to the Board of Trustees, as had
been the case previously.

By the end of 1967, all the components for the School were in place:
the Dean, the university affiliation, and the federal funding. Indeed, the
first course under the auspices of the Mount Sinai School of Medicine
was offered that year by the Page and William Black Post-Graduate
School. Even the facilities were coming along. Because of the delay in
developing the tower building for the School, the Trustees, in 1966, had
authorized the purchase of an old bus garage at 10 East 102nd Street. At
a cost of $7 million, the Trustees bought the building and during
1967–68 renovated it to house faculty laboratories, offices for the basic
science and Graduate School departments, and lecture rooms. It also
housed multidisciplinary laboratories for eighty first- and second-year

students. Each of these laboratory units held a desk, laboratory bench, and storage space for the student. The building was wired for closed-circuit television, a recent innovation in educational methods. The Basic Sciences Building, named for Trustee Nathan Cummings in 1971, also housed a branch library containing a basic science collection, funded in part by a National Library of Medicine grant of $19,000.

As witnessed by the library grant, fund-raising from all sources was proceeding apace. In 1967, Trustee Alfred Stern assumed the chairmanship of the Medical School Fund-Raising Committee, later the Development Committee. He set about the creation of a formal fund-raising office at Mount Sinai and hired the first development officer. This allowed Mount Sinai to stop using the services of a consulting firm, and released the Dean's Office and other administrators from some grant-writing and fund-raising duties. While most of the private and federal money raised went to the construction of the Medical School building, efforts were made to begin an endowment fund for

Presidents all. From left to right: Joseph Klingenstein, President of The Mount Sinai Hospital, 1956–62; Milton Steinbach, President of the Mount Sinai School of Medicine, 1962–68; George James, M.D., President of the Mount Sinai Medical Center, 1968–72; Gustave L. Levy, President, then Chairman of The Mount Sinai Hospital, 1962–76.

the School, much of which was in the form of endowed chairs. The Mount Sinai *Annual Report for 1967* proudly noted the establishment of eleven such chairs by the end of that year.[35] The School was busy recruiting chairmen to invest in those chairs and faculty members to support them. By the time of the first Faculty Assembly of the School, in September of 1967, there were 1,200 faculty, including five hundred at the rank of Professor, Associate Professor, and Assistant Professor.[36]

This is not to say that all was smooth sailing for the young school; the amount of work to be done was overwhelming. At a September

This is the old bus garage on 102nd Street that Mount Sinai bought from New York City to serve as the first home of the School while the Annenberg Building was being completed. This view, seen from Madison Avenue, was taken in September 1967. The sign on the building says: "For A Healthier City: This structure is being converted into a basic sciences building which will be the first completed facility of the Mount Sinai School of Medicine of The City University of New York. It will receive its first class of tomorrow's doctors September, 1968."

The multidisciplinary laboratories in the Nathan Cummings Basic Sciences Building when it opened in September 1968. These were designed to be the same as those that would appear in the Annenberg Building when it opened five years later.

1967 meeting of the Faculty Council, which consisted of all faculty of the rank of Associate Professor or above, the Dean organized the group into fifteen committees to work out the details of the educational objectives and policies of the School. The curriculum was a continuing work in progress, with subcommittees working and reworking syllabi and course schedules until the last minute.[37] The Dean's Office worked closely with representatives from The City University to establish rules for tenure and ranking that were consistent with overall University policies.

The Building Committee, led by Trustee Max Abramovitz and staffed by Mike Sisselman, was fortunate to have as members Saul Horowitz Jr., of HRH Construction, and Alfred A. Baum, of Jaros, Baum

Alfred Stern, Chairman of the Fund Raising Committee of the Board of Trustees, at the lectern at the dedication dinner, October 20, 1968. On his left are Gustave L. Levy, Chairman of the Mount Sinai Boards of Trustees, and The Honorable Nelson Rockefeller, Governor of New York State.

and Bolles, specialists in HVAC systems. The committee met continually with architects from Skidmore, Owings & Merrill and with Morse Diesel, Inc., the construction managers. The delay in receiving the federal funds had slowed the process, but the plans for the Medical School building were nearing final form. The design had coalesced into a tower building with a ground floor housing a lobby and an auditorium, twenty-five floors for offices, laboratories, and teaching space, and two multilevel mechanical areas, resulting in a building equivalent to thirty-one stories. In June 1965, it had been announced that the eight children of Mrs. Moses Annenberg had each pledged $1 million toward the funding of the Medical School building. In recognition of this generosity, the building was to be named the Annenberg Building. When completed, Annenberg would be the largest single structure housing a medical school in the world. It would have a library, lecture halls, classrooms, and individual study and research units that could accommodate an entering class of more than one hundred students. Some hospital units were also included, with an expanded Outpatient Department, a cardiopulmonary center, operating rooms, and enlarged radiology facilities, but no inpatient beds were included in the final design. As the building took shape, there began the difficult task of allocating the interior space among the competing departments, attempting to balance their current and future needs. At Mount Sinai, as with all urban medical centers, space was ever the issue.

OPENING DAY

On May 24, 1968, the absolute charter for the Mount Sinai School of Medicine was received. On September 6, the first students arrived to the newly readied facilities in the Basic Sciences Building on 102nd Street. They were greeted by the Trustees and 1,500 faculty members, including the administration—George James, Dean; Hans Popper, Dean for Academic Affairs; David Koffler, Assistant Dean; George Christakis, Associate Dean for Admissions and Student Affairs; and Sherman

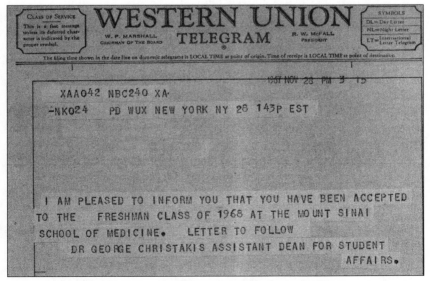

The telegram sent to all the students in the first first-year class at the Mount Sinai School of Medicine.

Kupfer, Associate Dean for Appointments and Research Administration. There were thirty-six first-year students, twenty-three transfer students in a third-year class, and nineteen graduate students. Mount Sinai had been allowed to recruit a third-year class, unusual for new schools, because of its long history of clinical teaching with medical students from other schools. There had been more than five hundred applicants for the first-year class. Because of the extensive interest and high quality of the applicants, Mount Sinai was allowed to expand the class to thirty-six. Mount Sinai had promised the accrediting agencies to increase class size to one hundred as soon as was practical, which meant with the opening of the Annenberg building.

But why would these students want to take a chance on a new medical school? The answer seems to be a combination of factors. Many students knew the reputation of The Mount Sinai Hospital and were willing to accept that the School would be of an equally high calibre. Others had heard that Mount Sinai's School would pay attention to the social concerns of patients and students, and, in the socially charged era of the 1960s, this was important to many. As one member of the first-year class noted years later, "I've always been a bit of a maverick. It was sort of part of the times as well, and the idea of going to a brand new medical school certainly appealed."[38] From the Class of 1973: "It was a different kind of student who came to Mount Sinai. . . . Those were people who were risk takers . . . [it] was a very political time in which people thought that medicine had a larger responsibility to the larger public good. Medicine had to be a part of these fights."[39] And, finally, "the people who were involved in forming it and active, seemed to me to be very idealistic and very approachable, very excited and enthusiastic. . . . I loved the small class; I loved the sense that we were actually part of developing the program."[40]

And, to a significant degree, the students were involved, in part because of their willingness to speak out and in part because the administration, Dean James in particular, was willing to listen. He kept an open-door policy, inviting groups of students in for discussions about the School.[41] When they requested that students be placed on the major committees of the Medical School, Dean James allowed them access to every committee except the Faculty Promotions Committee. When the students urged that the letter grading system be abandoned because they felt it engendered an overly competitive spirit, the system was changed to pass/fail.

Dean George James talking to students after a class.

The fact that the opening of the School went as smoothly as it did was a testament to the efforts of the many people who had worked on the various components of the School: the facilities, the curriculum, the recruitment of faculty, the establishment of policies and practices regarding admissions and student life. Of particular note are the efforts of Edra Spilman, Ph.D., who was appointed in 1967 to serve as the Director of the Department of Laboratory Education. In this role, he was responsible for planning the educational space for the students, in particular the structuring of the multidisciplinary laboratories. He had been in a similar position at Case Western Reserve University, which had pioneered such units. In 1968, as the students arrived and the plans for the Annenberg Building were more concrete, the Department of Laboratory Education was dissolved, and Spilman took the title of Associate Dean for Special Services. In this role, he worked on developing the new building, including helping assign and design space throughout the building. He had the highly charged duty of overseeing the institutional space bank, deciding among the competing requests for additional space. He retired from Mount Sinai in 1977, two years after the full opening of the Annenberg Building.

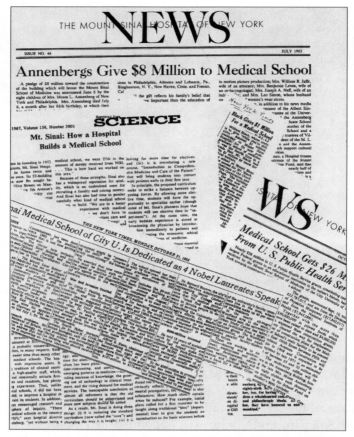

Some of the headlines about the birth of the School, 1965–68

One of the most important tasks that had faced the leaders of the School was the recruitment of prominent chairmen for the new basic science departments that were added. When the School opened in September 1968, the following appointments to chairs were in place: Tibor Barka, M.D., in Anatomy; Panayotis G. Katsoyannis, Ph.D., in Biochemistry; Kurt Deuschle, M.D., in Community Medicine; Edwin D. Kilbourne, M.D., in Microbiology; Jack Peter Green, M.D., Ph.D., in Pharmacology; and Irving Schwartz, M.D., in Physiology.[42] That same year also saw the successful recruitment of Solomon Berson, M.D., as the Chairman of the Department of Medicine, a key appointment.[43]

The faculty of the School grew swiftly over the years as the class size grew and as more academic affiliations were established. When the School opened in 1968, there were 1,500 faculty members at Mount Sinai and its affiliates: the Beth Israel Medical Center, City Hospital Center at Elmhurst, The Hospital for Joint Diseases and Medical Center, and the Bronx Veterans Administration (VA) Hospital. The faculty governance structure of the School included a Faculty Council, which encompassed all faculty of the rank of Associate Professor or above, with subcommittees covering important academic areas such as Appointments and Promotions, Curriculum, and Student Affairs, as well as an Executive Faculty. Later, a Basic Sciences Faculty Assembly and a Clinical Sciences Faculty Assembly were also formed, but these were not a formal part of the structure and disappeared in 1971.

The early years of the Mount Sinai School of Medicine witnessed a series of "firsts": the first faculty meeting in 1967, the first day of classes, and then exams. The School received its first official accreditation visit from the Liaison Committee on Medical Education and was

The Nobel Laureate speakers at Dedication Day, October 20, 1968. From left to right: Linus C. Pauling, Ph.D., George W. Beadle, Ph.D., Dean George James, Francis H. C. Crick, Ph.D., Sir Peter Medawar.

The Annenberg Building under construction, taken from Central Park, 1972. This photo was taken by the Mount Sinai Hospital engineer Sydney Hechinger, who helped oversee the Annenberg construction process.

accredited for the maximum seven-year term, an event that pleased Mount Sinai greatly. Everything was new, and the students and faculty were excited by the events. As students in the early classes later noted: "The faculty, it was as if they had been waiting for twenty years to have these babies come and then we show up and we were long haired . . . and so we were both their darling babies but we were also their worst nightmare."[44] And also: "it was clear once we got here that we were a new school. We were the pride and joy of this new baby, and that great things were expected of us—not in a demanding way, but just in a reassuring way."[45]

One of the most important events of these early years was the first commencement of the Mount Sinai School of Medicine. The initial third-year class graduated on May 27, 1970. Prior to this ceremony came the activation on campus of the New York Lambda chapter of Alpha Omega Alpha (AOA), the national medical honor society. This was celebrated with a large dinner attended by 150 members of the faculty, all AOA members, at which time five members of the Class of 1970 were inducted into the Chapter. With the election of the students from the Class of 2003, a total of 585 MSSM graduates have now been inducted into AOA. In addition, thirty-four faculty, thirty-eight house staff, and eight alumni have been elected to the Chapter by the student members.

When the School opened in 1968, it had two departments that were somewhat unusual for a medical school. These were the Department of Administrative Medicine and the Department of Library Science. The former was organized to help teach medical students about the healthcare system, as well as to work with Baruch College of The City University on the implementation of an MBA in Health Care Administration. The Department also served as a School "home" for Hospital administrators, providing them with academic titles and teaching duties. The Department was chaired initially by Martin Steinberg, M.D., then Director of the Hospital. With his retirement in 1969, the chairmanship passed to S. David Pomrinse, M.D., the next Director of the Hospital, and the chairmanship remained vested in the Director's position for many years. In 1973, an endowed chair was created in Health Care Management, the Edmond A. Guggenheim Professor, and

Members of the Class of 1973 sitting in a lecture in 1970. All classes were held in the Cummings Basic Sciences Building at this time.

Hans Popper, M.D., Ph.D., at the podium at the first commencement of the Mount Sinai School of Medicine of The City University of New York, in May 1970. Dean George James is on the left, with Trustees and faculty on the stage.

leaders of this Department held the chair in succession. In 1980, with the arrival of Samuel Davis to head the Department, the name was changed to Health Care Management. After his departure in 1984, the chairmanship was never again filled on a permanent basis, but the Guggenheim Chair continued with Norman Metzger as the incumbent. Metzger was a pioneer in the field of labor relations and human resources in healthcare organizations and had headed Mount Sinai's Department of Human Resources for years. In 1991, the administration of the Mount Sinai/Baruch MBA moved to the Department of Community and Preventive Medicine. In 1995, when Metzger retired from the Guggenheim Chair, the health services research programs and remaining faculty were transferred to the new Department of Health Policy, along with the named chair.

The Department of Library Science also had a short-lived formal presence. The Department was officially activated in November 1968 with Alfred Brandon as the first Professor and Chairman. He was also Director of the Jacobi Library, as well as the branch library that had been created in the Basic Sciences Building. Mr. Brandon was very well known in the field of medical library science and originally served as a consultant to Mount Sinai in regard to the expansion of its medical library for the School of Medicine. He was then hired full time to oversee both the expansion of the collection and the physical development of first the branch library and then the new facility in the Annenberg Building.

Unfortunately, Brandon left before the completion of the new library. It was Rachael K. Goldstein, as Library Director, who presided over the opening of the Gustave L. and Janet W. Levy Library in the Annenberg Building. This new entity was created by the merging of the collections of the old Abraham Jacobi Library, created in 1883 and named for Jacobi in 1910, and the branch library that had been created in 1968. It also included many volumes from The Mount Sinai Hospital School of Nursing Library, which had closed with the School in 1971. This combined collection held more than 70,000 bound volumes and approximately 1,750 periodical or serial titles.

On November 22, 1974, a dedication celebration was held to honor Mr. and Mrs. Gustave Levy for their generosity in naming the library and to show off the facilities on the tenth and eleventh floors of the Annenberg Building. The equipment included computer terminals for the new automated literature searches that had recently become possible, foreshadowing the Library's continuing role as a leader in utilizing the latest information technology in support of research and education.

In 1975, the Department of Library Science in the School of Medicine was dissolved, and the Library became a part of the Department of Medical Education. The Library's role in the School has grown with the advent of computer technology and the digitalization of information. The Library's professional staff has also continued to contribute to the teaching of medical students over the years through electives and curriculum time in required courses.

As the School began to function, many adjustments had to be made as Mount Sinai's leaders struggled to run two very different organizations. One area was in the relationship between the voluntary and the growing full-time clinical faculty.[46] From the beginning, the new School faced a long process of trying to define and control faculty practice, as well as to assure the voluntary faculty that their contributions were essential and valued. This aspect grew more contentious as the class size grew and the need for more clinical faculty rose with it. It also became clear that the functioning of the Faculty Council was not all

An administrative meeting in the office of the Dean and President. From left to right, facing the camera, George Christakis, M.D., Associate Dean; Hans Popper, M.D., Ph.D., Dean of Academic Affairs; George James, M.D., Dean. Horace Hodes, M.D., Chairman of Pediatrics, is on the right, in a white coat.

that it could be. In 1971, a process started to review the governing structure of the school and to suggest changes. This group, led by Edwin Kilbourne, Ph.D., Chairman of Microbiology, submitted a document in November 1971 that became the basis for the Academic Council that was established in 1972. This new plan provided more equitable representation for all the constituencies of the growing School.

There was also a rocky process as the School sought its own identity in relation to a Hospital that had been central for so long. In addition, there was the fear that the Hospital's needs would be forgotten in the process of getting the School going, in particular the replacement of the aging Hospital facilities. This uneasiness also played out in the management structure of the institutions. It had been planned that with the retirement of Martin Steinberg and the formal activation of the Mount Sinai Medical Center in 1969, the Director of the Hospital would report to George James as the President. In fact, this did not occur. David Pomrinse, Steinberg's successor as Director, was also appointed Executive Vice President in the Medical Center structure and retained his authority to act unilaterally and to have direct access to the Trustees. While James and Pomrinse met frequently and there was appropriate interaction when necessary, James grew to feel that this had not been the intention of the planners and that the situation must change. Tragically, George James died unexpectedly in March 1972 after suffering a heart attack, his second in two years.

It is hard to analyze the impact of James's premature death on the young School, but the effect his life had on Mount Sinai was undeniable. Certainly he embodied much of the energy that drove and brought to fruition the establishment of the School and its linkages to the community. He had an excellent working relationship with the leaders of CUNY, facilitating many cooperative ventures in health education. Perhaps more would have been done along these lines if he had lived longer, but that is unsure given the fiscal climate of the 1970s. James was also at ease in the public arena, familiar to and sought out by medical organizations and the press. He was able to make the new School real to many people and to reassure them that this was a solid academic and scientific venture. He was decisive when necessary and flexible when needed, invaluable traits when creating a school from scratch.

Mount Sinai deeply mourned the loss of George James. Hans Popper was quickly named Acting Dean and President. (He was later fully appointed to these offices.) Popper understood the need for him to serve, but he was dismayed by the impact this "promotion" had on his plans. He had just retired from the chairmanship of Pathology and other administrative posts so that he might pursue his liver research on a Fogarty Fellowship at the NIH. A few months later, the institution was shaken again by the untimely death of Solomon Berson, Chairman of the Department of Medicine. This came at a time when the chair in Surgery, a major power position in the Medical Center, was already vacant. Fenton Schaffner, M.D., stepped in to run the Department of Medicine while a search was mounted. It was a difficult time for Mount Sinai.

Popper urged the early formation of a Dean's search committee and hurried the process along by suggesting candidates. While speaking to Thomas C. Chalmers, M.D., at the National Institutes of Health, trying to explain his delay in serving the Fogarty Fellowship, Popper discussed his situation. A few weeks later, Chalmers accepted the Trustees' invitation to become the third Dean and President of Mount Sinai.

THE END OF THE BEGINNING, 1973–1983

Thomas Clark Chalmers was born in Queens in 1917. He attended Phillips Exeter Academy and Yale University and, in 1943, became a third-generation graduate of Columbia University's College of Physicians and Surgeons. He had plans to become a family physician but grew to feel that it was no longer possible for one doctor to master every field needed for this. Instead, he became an internist and established a reputation as a leading liver disease specialist involved in the study of hepatitis. Along with this, he analyzed the research of others to see what worked. He did valuable research on the proper way to perform clinical trials and was instrumental in developing the area of meta-analysis, the process of aggregating data from many studies to provide a statistical look at the efficacy of a particular treatment or practice. As Chalmers said, in a 1990 article in *Science*, "[Meta-analysis] is going to revolutionize how the sciences, especially medicine, handle data. And it's going to be the way many arguments will be ended."[47] Chalmers's commitment to finding and utilizing only the treatments proven to be effective was evidenced by his own resistance to receiving a hip replacement until clinical trials had shown that to be the best course for arthritic hips. As the *New York Times* said, "He lived by his example."[48]

Prior to his coming to Mount Sinai, Chalmers was on the faculty of Tufts University School of Medicine and Harvard Medical School, and he served as Chief of Medical Services at the Lemuel Shattuck Hospital in Boston. In 1968, he moved to Washington, D.C., to become the Assistant Chief Medical Director for Research and Education at the Veterans Administration. Two years later, he was appointed Associate Director for Clinical Care and Director of the Clinical Center at the NIH, also assuming the title of Professor of Medicine at George Washington University School of Medicine.

Chalmers was a man who thrived on work and had an enthusiasm for medicine and an interest in the world around him. He had a sign on his desk that read, "My God—It's Friday!" He planted tomato plants on the rooftop outside his office. He was very involved with the organization International Physicians for the Prevention of Nuclear War, a group that won the Nobel Peace Prize in 1986; Chalmers attended the award ceremony. From this interest grew a course that Chalmers offered to the medical students at Mount Sinai, entitled "Preventing

Thomas Clark Chalmers, M.D., Dean and President of Mount Sinai, 1973–1983. Due to chronic back problems, Chalmers preferred to work standing up and had this special desk built for him.

Nuclear War." He believed that the study of medicine was not confined to the years of medical school and postgraduate training. He considered it a lifelong pursuit and wanted to ensure that the curriculum at Mount Sinai provided the students the skills to do so, a theme that had been sounded earlier by James. (As recounted later by a medical student, Dean James was asked what the core of medicine was, and he responded, "It was all well and good to think about what the core is, but remember that the core of today's apple is tomorrow's garbage and things do change. You need to adapt to your learning and not just be spoon-fed or be told what it is you need to know."[49] Chalmers would have likely agreed.)

When Chalmers assumed his position at Mount Sinai late in 1973, it was already clear that the 1970s would be very different from the 1960s. As James's term had progressed, the rosy financial picture of government support had begun to change. The interest derived from the Annenberg Building fund helped mitigate some of the impact of the reduced federal funding, but in 1971 that source of funds was gone, and the Medical Center posted a deficit. Cutbacks were made in programs, many hoped-for projects were not pursued, and purchases of equipment were delayed. During the 1970s, faculty relations were also a pervasive concern, with the development of town/gown (voluntary faculty/full-time faculty) difficulties. Indeed, medical practice by full-time salaried faculty was an issue facing all medical schools during these years.

As the federal money dried up, inflation also took its toll and the competition for grant funds increased. Research projects were slowed, and equipment grew older as the institution lacked the expected government funds to replace it; space in the newly opened Annenberg Building quickly filled. At the same time, the aging Hospital facilities needed refurbishing and rebuilding. As the class size grew, the affiliated hospitals and staff took on a heavier part of the teaching load. The affiliates of this time—the Bronx VA Medical Center, Beth Israel Medical Center, the City Hospital Center at Elmhurst, the Jewish Home and Hospital for Aged, and North General Hospital—became even more vital components of the teaching program of the School. This flowering of the multisite teaching programs was an important element in the 1970s and 1980s.

Perhaps the biggest bright spot for the School was the long-awaited opening of the Annenberg Building starting in 1973. In its funding requests, the School had promised the government to admit one hundred students in the fall of 1973, and having the Annenberg facilities was essential to that effort. The student areas of the building, up to the thirteenth floor, were readied first to allow for that expansion.

It was not until May 26, 1974, that the Annenberg Building was officially dedicated, with Vice President Gerald Ford as the principal speaker. The fact sheet issued to reporters at the dedication noted that excavation had begun in 1970 and that 100,000 tons of rock and earth had been carted away. The completed building had a total gross square footage of 1,036,897 and was the height equivalent of a thirty-one story building. The cost ultimately exceeded $117 million. Along with the formal program and speeches, Governor Malcolm Wilson issued a proclamation proclaiming May 26 Mount Sinai Medical Center Day in New York. The *New York Times* published a special supplement about the building and the young School. Press releases were written and

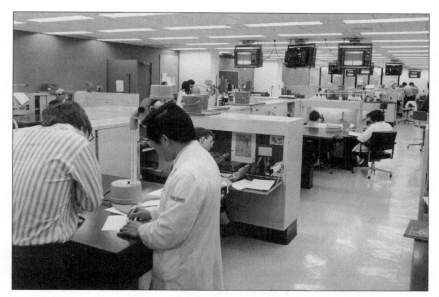

The multidisciplinary laboratories in the Annenberg Building, 1974. Closed-circuit television, an innovation in medical education at the time, was installed throughout the new building.

many articles printed about the day, many noting that the Annenberg Building was believed to be the largest structure devoted to medical education in the world. The architectural press was less enthusiastic about the design of the brown Cor-Ten steel building, and, indeed, the outer coating of the building did experience problems, prompting lawsuits and more construction.

The Annenberg dedication was a most important day in the history of Mount Sinai, the culmination of endless days of work and planning and disruption (the building was not fully opened for nine more months). From the days when the initial plans for the School started to take shape in the federal grant application in 1962, Trustees and staff had been consumed with funding and building the Medical School building. For these twelve years, people like Mike Sisselman, Sidney Hechinger (who reported to Sisselman and was the former head of Hospital Engineering and the Mount Sinai liaison person for the contractors), Edra Spilman, and a staff of planners, engineers, and administrators had concentrated on getting the building built and equipped on time and, as closely as possible, on budget. As Spilman said in a memo to Chalmers on February 14, 1975:

> Today, Valentine's Day, 1975, is a momentous day for me, for today I accepted the last floor to be completed above the tenth floor of Annenberg. . . . When I walk through the floors, I relive the many hours and days of planning with the future occupants and the architects—the planning of room and wall locations, planning offices, laboratories, cold rooms and even toilets, planning casework with high benches, low benches, and no benches, planning furniture and color accent walls. I recall vividly the difficult task of cost cutting without which we would not have had a building. . . .
>
> Today, the steel and concrete, the dry wall and paint, have all been put into place to form the most complex building ever built, to form a citadel of learning, research, and patient care. . . . As we marched relentlssly up the building floor by floor, all of those features came to life and created an environment beyond our expectations. . . . I point with justifiable pride at Annenberg and say "We did it!"[50]

Another tragedy struck Mount Sinai in November of 1976 when Gustave Levy, Chairman of the Boards of Trustees, died unexpectedly from a stroke. His legacy to Mount Sinai was the School of Medicine, and it would be impossible to overstate his contributions to establishing the School. He led the institutions through difficult, changing fiscal times and provided the School with a firm foundation. He was intelligent and decisive, demanding the best from everyone but especially from himself. With his passing, Alfred Stern assumed the leadership of the institutions. Stern had just successfully completed the long fund-raising campaign that raised the $152 million that had built the Medical School. He was knowledgeable about Mount Sinai and the changing healthcare environment and was well respected by his fellow Trustees.

Curriculum matters at the School continued to evolve. One curricular area that troubled Chalmers, and every Dean since, was the effort to devise instructional methods other than the large-group lecture, the easiest method for a busy faculty. Also, problems had developed with the multidisciplinary teaching program, making it hard to maintain the early uniqueness of the courses. With each adjustment of the curriculum, Mount Sinai moved toward more traditional teaching and curricular models. Still, some of the innovations that made Mount Sinai

First-year students in Anatomy class, Fall 1984.

distinctive were retained, including its emphasis on the humanities and the value of community medicine.

Although the progress was sporadic, again and again Mount Sinai affirmed its commitment to human values in medicine. In the mid-1970s, the students themselves began a project called HUMED to provide a stronger humanities component. The student project grew into a larger effort by the Department of Medicine for undergraduate students and graduate physicians, with teaching by psychiatrists and an ethicist, one of the first ethicists in a department of medicine. Richard Gorlin, M.D., Chairman of Medicine, and Howard D. Zucker, M.D., Professor of Psychiatry, spearheaded this effort. A forty-five-minute film made for the program, *A Complicating Factor: Doctors' Feelings as a Factor in Medical Care*, was distributed widely around the country. From this beginning, an ethicist was formally added to the School faculty in 1988.

With all of these concerns, much was still accomplished during Chalmers's ten-year tenure. Credit for this must be shared with Sam Davis, the Director of the Hospital, as well as the hard-working Associate and Assistant Deans and Chairmen who formed the administrative

Richard Gorlin, M.D., Chairman of the Department of Medicine, on rounds with medical students, 1980s.

structure around Chalmers. First among these were Drs. Sherman Kupfer and Arthur H. Aufses, Jr. Chalmers had raised Kupfer to the position of Deputy Dean, and he really ran the day-to-day operations of the School. He concurrently held the title of Senior Vice President for Education and Research in the Medical Center. Aufses was Chairman of the Department of Surgery and Senior Vice President for Clinical Affairs in the Medical Center. These two men, along with Chalmers, were primarily responsible for the direction of the School and its relationship to the Hospital and the affiliated institutions. Chalmers also created committees that advised him on issues, new projects, and areas of opportunity, including operations at the affiliates, the development of new clinical facilities, departmental problems, and so on. These were the Small Dean's Advisory Committee and later the President's Advisory Group (PAG).

Another individual who rose to the top in the administration during Chalmers's tenure was Barry Stimmel, M.D., a cardiologist and researcher in the area of narcotic addiction and the leader of Mount Sinai's Narcotic Rehabilitation Center. He served as the Dean for Academic Affairs, as well as the Dean for Admissions and Student Affairs.[51] There were also the Deans of the other Mount Sinai Schools: Terry Krulwich, Ph.D., in the Graduate School of Biological Sciences, and Samuel K. Elster, M.D., in the Page and William Black Post-Graduate School of Medicine.

During Chalmers's years, a remarkable thirteen new appointments were made at the department chairman level. These vacancies were viewed as opportunities, as well as challenges. Both Medicine and Surgery were without leaders when Chalmers arrived in 1973. These positions were filled the next year by Richard Gorlin, M.D., and Arthur H. Aufses Jr., M.D., respectively. Nine other chairs were later recruited: in Pediatrics, Kurt Hirschhorn, M.D. (1976); in Dermatology, Harry Shatin, M.D., and Raul Fleischmajer, M.D. (1975 and 1979); in Ophthalmology, Steven Podos, M.D. (1975); in Pathology, Jerome Kleinerman,

Cynthia Gruber, Registrar (left), and Joyce Shriver, Ph.D., Assistant Dean for Student Affairs, select names from a black doctor's bag to determine clinical rotations for the Class of 1979.

M.D. (1978); in Radiology, Jack Rabinowitz, M.D. (1978); in Anesthesiology, David C. C. Stark, M.D. (1979); in Administrative Medicine, Samuel Davis (1979); and in Obstetrics and Gynecology, Nathan Kase, M.D. (1981). Physiology became vacant in 1979, without a full-time chairman until 1984. Urology was without a permanent director from 1982 until after Chalmers left in 1983.

Under Chalmers's direction, two new departments were also created: the Department of Biostatistics, in 1975, now called Biomathematical Sciences, and the Department of Geriatrics, in 1982. Both were innovations. The Department of Geriatrics was the first in a medical school in the United States. Within this department, Mount Sinai established the first, and still the only, mandatory rotation of medical students in a nursing home. For this new department, Mount Sinai recruited the prominent geriatrician Robert Butler, M.D., an important appointment that got the Department off to an excellent start.[52]

The Department of Biostatistics grew from Chalmers's belief that physicians needed to be trained in statistical methods in order to analyze their own research data, as well as to successfully evaluate the research of others. He sought support from the leaders of CUNY to begin this program with the hope that it would ultimately lead to a universitywide doctoral program in Statistics, with a major in Biostatistics. Recruitment began in 1974, and Harry Smith Jr., Ph.D., was appointed. He joined the staff at Mount Sinai full time in 1975. At this time Mount Sinai already had an Institute for Computer Science, led by Aron Safir, M.D., an ophthalmologist with great interest in using computers in medicine and biostatistics. This Institute maintained the computing facilities needed to allow Mount Sinai faculty access to the mainframe computers available from The City University, as well as CUNY e-mail systems. Mount Sinai did not then own its own computers for data storage or manipulation.

Smith and the Biostatistics faculty worked with several departments and programs, providing support for their research efforts. They also offered a small class in biostatistics for medical and graduate students, although this never developed to the scope Chalmers desired. In 1980, Biostatistics was combined with the Computer Institute in an attempt to rationalize Mount Sinai's use of computing resources. The resulting Department of Biomathematical Sciences was led by Smith and had three divisions. The Research Computing Unit (RCU), directed by Dr. Ford Calhoun, continued to oversee the hardware and software needed to maintain the School's connection to the CUNY machines, to provide user education, and to facilitate computer usage in research and education. It worked on creating student self-testing programs and made them available on computers in the Medical School. The Biostatistics unit consulted with Medical School departments on their research endeavors, and the Biomathematics group undertook original and collaborative research projects. In 1983, the Department was involved in 388 projects with 320 investigators, mostly in clinical research.

Smith retired in 1985, at which point Chalmers, himself retired from the Deanship, assumed direction of the Department as Acting Chairman. Two years later, Charles DeLisi, Ph.D., was named Chairman. In the 1989 Mount Sinai *Annual Report*, DeLisi described the Department's efforts: "Departmental research uses computer power and mathematical modeling to explore fundamental and applied problems in biology and medicine, such as epidemiology, membrane biophysics, physiology and protein and nucleic acid structure and function. The Department also provides extensive support in the planning and analysis of clinical trials."[53]

In 1990, DeLisi left, and Craig Benham, Ph.D., from the Department, was appointed Acting Chairman. In the early 1990s, many of the RCU functions were given over to the Levy Library as it established an Academic Computing division. Benham continued the Department's efforts to support the biomathematical needs of the Mount Sinai faculty, as well as continuing his own research program. The Department upgraded the computing facilities to offer researchers an enhanced computational platform and a database for genetic sequence data. Benham remained as Acting Chairman until 2001, at which time the Department became leaderless. Two years later, Larry Sirovich, Ph.D., was named Chairman of the Department.

A hallmark of Chalmers's tenure was the continuous emphasis on striving for excellence. This took many forms. One obvious venue was the quality of the student education and the School's desire to create physician-scientists. It was during Chalmers's tenure that the Saul Horowitz Jr. Memorial Award was created. This award, named in honor of the memory of Saul Horowitz Jr., a longtime Trustee of The Mount Sinai Hospital and the School of Medicine, was established in June 1978. This important award is still given annually to an alumnus/a of the Mount Sinai School of Medicine who has graduated not less than five years previously and who, by virtue of outstanding achievements since graduation, has made or gives promise of making significant contributions as a teacher, investigator, and/or practitioner in the field of medicine. (See Appendix A.)

Planning and evaluation were other components of this drive for excellence, and, indeed, the Hospital had established a Planning Office back in the early 1960s. By the time Chalmers arrived at Mount Sinai, the Office of Planning had become an established part of the Medical Center structure under Mike Sisselman's direction. Chalmers believed that to plan effectively, information "must be based on data gathered in an objective manner and presented for discussion and decision to the Board of Trustees."[54] In 1974, Chalmers requested that a special committee of Trustees and faculty be created and divided into three task forces: "One to define the goals and objectives of the academic center; another to analyze alternative patterns for financing academic appointments; the third to define methods for evaluation of competence of clinical staff in the areas of teaching, research and service responsibilities."[55] The combined entity became known as the Clinical Excellence Committee, which operated from 1974 to 1976 and exhaustively reviewed the programs and activities of the Medical Center. The endeavor was coordinated by Paul J. Anderson, M.D., and staffed by Carole Stapleton. Its report, *Planning for Change at Mount Sinai* (June 1977), led to the establishment of a program of departmental reviews by outside experts, using self-study reports. It also led to changes in the way full-time faculty practiced at Mount Sinai, as detailed in chapter 10. The Clinical Excellence Committee report then served as the basis for the voluminous self-study performed by the School for the Liaison Committee on Medical Education (LCME) site visit of February 1977. Mount Sinai became the first medical school in the country to receive a ten-year accreditation using the self-study model, an important achievement.

As this planning effort ended, others were begun. In 1977, Samuel Davis was moved up to become the Director of The Mount Sinai Hospital, the first Director since 1904 who was not a physician. He had a management background and worked with Chalmers on a major planning effort involving Trustees, administration, and staff. (In 1975, Davis's office had taken over the planning function from the President's Office and had brought in Raymond Cornbill to run it.) This long-term effort, including nine months devoted to defining the Medical Center's mission statement, helped stimulate thinking on the roles and responsibilities of the Hospital and the School. Another study resulted in the publication of *The Most Distinguished and Least Preferred Private Academic Medical Centers: A Comparative Report*, in 1980. This placed the Mount Sinai Medical Center in juxtaposition to the ten "best" medical schools and ten of the "least preferred" schools to assess its strengths and weaknesses.[56]

The uncertainty of government support gave urgency to the planning efforts. Scrutiny was brought to bear on the aging clinical facilities and cramped research quarters. As some had feared when the School was founded, having the educational enterprise on campus had indeed delayed needed work in the Hospital, as had the turbulent fiscal climate. It was clear that something had to be done soon, as the faculty and Trustees were deeply troubled by the physical condition of the Hospital. From this concern grew a Hospital rebuilding program aimed at improving patient care areas and freeing and modernizing some research space.

The Trustees were fortunate in that they already had on hand a bequest of $22 million from Edmond Guggenheim to help Mount Sinai build a new private patient pavilion.[57] Mr. Guggenheim was a descendant of the family that had given the Hospital the funds needed to build its first separate private pavilion on Fifth Avenue in 1904, and he had a deep interest in continuing his family's legacy. To complete Mr. Guggenheim's vision, a new fund-raising campaign was initiated to support the rebuilding, as well as to increase the endowment funds of the School. The 1904 Hospital buildings north of 100th Street were torn down in 1986, and ground was broken for the new Guggenheim Pavilion, designed by the internationally known architect I. M. Pei.

By 1981, Mount Sinai appeared to be at the end of the beginning. With the rebuilding plans, the changed regulatory environment, and the now established maturity of the School, Mount Sinai seemed poised for a new start toward the recently defined objectives. But the question of the balance of power in the Medical Center's administrative structure—Hospital versus School—continued to plague the institution. This led to the appointment of a Trustee committee, led by Marvin Asnes, to study the management structure of Mount Sinai, to compare it with ten other major medical centers, and determine the strengths and weaknesses of the Medical Center structure and the key issues to be resolved.[58] This group recommended that Mount Sinai commit to the principle of being academically driven, and to implement the "Center concept—providing

Drs. Chalmers and Glenn perform what has become a ritual in the passing of power at Mount Sinai. The incoming President and the outgoing President shake hands in front of a photographic portrait of their predecessors shaking hands, with the oil portrait of George James in the far background.

for distinct but organizationally integrated medical school and hospital components" by separating the offices of Dean and President, with the President becoming the chief executive officer (CEO) of all three Mount Sinai institutions.[59]

To implement these recommendations, it was felt that a change in leadership needed to be made, with new blood coming in from outside. The Trustees were also concerned that Chalmers would be eligible to retire over the next several years, just when the Medical Center hoped to be in the midst of a major rebuilding effort. To forestall this, Chalmers was asked to retire early from his administrative duties, and he was named a Distinguished Service Professor of the School. At this time, he set up the Clinical Trials Unit in the School, later named in his honor.

After a nationwide search, in June 1983, James Glenn, M.D., became President of the three institutions and Acting Dean of the School. Glenn, an internationally known urologist, was recruited from the Emory University School of Medicine, where he had been serving as Dean. The Board of Trustees passed a resolution confirming that the Director of the Hospital and the Dean of the School would report directly to him as President. After the transition, Sam Davis, who had received the title of President of the Hospital in 1981, left Mount Sinai. Glenn was confirmed as CEO of the three institutions. Barry Freedman remained as Director of the Hospital. In July 1984, Lester Salans, M.D., was named Dean, but Glenn did not feel that they would work well together, and Salans's tenure did not last into the fall. Mount Sinai again began the search for a leader for the School, appointing Nathan Kase, M.D., then Chairman of Mount Sinai's Department of Obstetrics, Gynecology, and Reproductive Science, as the new Dean, effective January 1985. Kase had ample opportunity to ponder the future of the School in the months prior to assuming the Deanship. As a Chairman, he had been a member of the Search Committee that had chosen Lester Salans as Dean in 1984. When Salans left prematurely, Glenn, the President of the Medical Center, again assumed the title of Acting Dean but in fact turned the role over to Kase until the latter's appointment became official.

In June 1987, Glenn submitted his resignation as President, and Kase was named Acting President and CEO. Mount Sinai began another long search for its second chief executive officer and fourth high-level appointment in as many years. On December 6, 1988, John Wallis Rowe, M.D., was inaugurated as the fifth President and CEO of the Mount Sinai Medical Center.

THE KASE YEARS, 1985–1997

Nathan Kase, M.D., was a native New Yorker and a graduate of Columbia University's College of Physicians and Surgeons (1955). His older brother, Jonah, had trained on the Mount Sinai house staff and served many years on the medical staff, and he himself had been a rotating intern and then resident at Mount Sinai in Obstetrics and Gynecology. A clinician well known for his work in gynecologic and reproductive endocrinology, Kase had risen to be Chairman of Obstetrics and Gynecology at Yale University School of Medicine and was recruited to become the Chairman of Mount Sinai's OB/GYN Department in 1981.[60] He returned with the mandate to revitalize that Department,

Nathan Kase, M.D., Dean of the Mount Sinai School of Medicine, 1985–1997.

and it was in that same spirit of revitalization that he undertook the job as Dean of the School.

With Kase's appointment in 1985, the Mount Sinai School of Medicine was marked by a new feeling of growth, achievement, and looking ahead. Many internal problems had been resolved under Chalmers, allowing the arrival of Glenn and then Dean Kase to usher in a period of planning and movement into new areas. The rebuilding of the Hospital and an extensive marketing campaign signaled Mount Sinai's more aggressive stance. The fact that the Dean's office now stood alone also allowed Kase to concentrate on School issues, without the additional concern of Hospital operations.

A look at the graduating classes that left Mount Sinai between 1970 and 1983 provides an interesting view of what type of physician Mount Sinai had turned out up to this point. Tables 1.1 and 1.2 are taken from a 1989 article by Barry Stimmel, then Dean for Academic Affairs,

Table 1.1.

Specialty Practice and Certification, MSSM Classes, 1970–1983

	Number	Percentage	Board Certified Percentage
PRIMARY CARE			
Family practice	50	(4)	(70)
Internal medicine			(100)
General	75	(7)	
Subspecialty	298	(26)	
Pediatrics	131	(12)	(82)
SURGERY AND SURGICAL SPECIALTIES			
General surgery	57	(5)	(52)[+]
Ophthalmology	51	(5)	(86)
Orthopaedics	48	(4)	(58)[+]
Otolaryngology	30	(3)	(77)[+]
Neurosurgery	6	(<1)	(50)[+]
Urology	21	(2)	(52)[+]
OTHER SPECIALTIES			
Anesthesiology	47	(4)	(79)
Critical care	11	(<1)	
Dermatology	39	(3)	(71)
Emergency Medicine	46	(4)	(46)
Neurology	24	(2)	*
Pathology	16	(1)	(100)
Preventive medicine	11	(1)	(100)
Physical medicine	9	(<1)	
Psychiatry	93	(8)	*
Radiology	66	(6)	(80)

*Psychiatry/neurology 76% certification.
+May not have met time requirements for certification

Table 1.2.

Major Professional Activity, MSSM Classes, 1970–1980

	No.	(%)
PATIENT CARE		
Office based	677	(60)
Hospital based Attending	128	(11)
Resident/fellow	176	(16)
FULL-TIME TEACHING	25	(2)
ADMINISTRATION	35	(3)
RESEARCH	62	(5)
NOT CLASSIFIED	28	(2)

Admissions, and Student Affairs, as well as Acting Chairman of the Department of Medical Education.[61]

Along these same lines, in 1988, for the twenty-fifth anniversary of the School, Dean Kase asked the staff of the Levy Library to undertake a study of the career paths of graduates from both the Hospital and the Medical School training programs. Using published directories and reference sources, the Library identified and researched 4,007 physicians who had graduated between 1935 and 1978. Current appointment data could not be found for 29 percent, leaving a total of 2,844 alumni. Of those, 23 percent (651) were listed as having only private or hospital practices, with no academic appointments. Of the 2,193 remaining, 11.2 percent (245) were listed as a chief of a service, department, or program, 3 percent were chairman (64), 0.6 percent (13) were Deans. Of those same 2,193 alumni, 21.3 percent were professors, 22.8 percent were associate professors, 34.5 percent were assistant professors, and 16.2 percent were lecturers or instructors.[62]

From the beginning, Dean Kase emphasized a commitment to strengthening the basic sciences and the research efforts of the Medical School. It was decided to shake things up and build for the future by recruiting several new basic science chairmen, a move that resulted in the turnover of three department chairs. Kase also began to develop multidisciplinary research programs focused in a few areas of biomedical knowledge that held promise, such as molecular biology, the neurosciences, and immunology. Although interdepartmental research cooperation already existed at Mount Sinai, these multidisciplinary centers pushed this trend further, as well as helped to define where these new disciplines would go. New departments and multidisciplinary centers created during this period include the Brookdale Center for Molecular Biology (1985), the Dr. Arthur A. Fishberg Center for Neurobiology (1986), the Derald H. Ruttenberg Cancer Center (1988), the Department of Cell Biology and Anatomy (formerly Anatomy only) (1989), the Department of Human Genetics (1992), the Department of Health Policy (1995), the Carl C. Icahn Institute for Gene Therapy and Molecular Medicine (1996), and the Center for Immunobiology (1997). From 1986 to 1996, the size of the basic science faculty doubled; an amazing twenty-seven new chairmen and program directors were recruited.

During the years under Dean Kase, the School was determined to create a more defined academic environment. This resulted in the re-

Dean Kase at a meeting with Department Chairmen. From left to right, Drs.
Kurt Hirschhorn (Pediatrics), Dean Kase, Arthur H. Aufses Jr. (Surgery),
Sherman Kupfer (Administration), and Richard Gorlin (Medicine).

structuring of some departments, new appointment and promotion
guidelines, the review of academic titles, and the revision of the faculty
and student handbooks. There was a great deal of faculty turnover
during these years as faculty members who did not meet the new criteria became subject to an "up-or-out" policy and new faculty were
actively recruited. Admission standards for students were raised over
the years, and steps were taken to reduce student indebtedness at
graduation. A student residence, the Jane B. Aron Residence Hall, was
opened in 1984, representing a significant boon to the students as well
as an aid in recruitment efforts. In 1988, the School began to require
that students pass Step I of the U.S. Medical Licensing Exams (USMLE)
in order to advance to the third year. Passing Step II became a requirement for graduation. In 1996, the first Annual Student Research Day
was inaugurated to give the students experience in presenting research
work and as recognition of that work. Earlier, the School had implemented a research track to encourage students to undertake research,
allowing them to graduate with distinction in research. Academic
convocations were instituted in 1986 to recognize professors appointed
to endowed chairs.

There were also changes in governance during Kase's deanship. In 1988, the Academic Council, the governing body of the faculty, was restructured to become the Faculty Council. This new group had no student representation on it, and there were updated formulas for representation by the faculty. Concurrent with this, a Student Council was formed to address student issues and to serve as a focus for student groups. These organizations remain today.

During Kase's years, the School was twice faced with accreditation reviews from the Association of American Medical Colleges. In 1986, a ten-year accreditation was granted. Beginning in 1994, another thorough self-study was initiated to prepare for the 1996 visit. This review led to additional changes in the curriculum, many of which had been begun previously: fewer large lectures, more small-group work, inclusion of modified problem-based learning, and more computer-assisted instruction. Again, the school received an outstanding review and a seven-year accreditation. The site visitors noted that in the previous ten-year period, there had been a 140 percent increase in peer-reviewed research grant funds and a 370 percent increase in support from faculty clinical practice.

There was also much attention given to the type of student Mount Sinai accepted and the type of doctor Mount Sinai created. Applications for enrollment increased tremendously during Kase's tenure. In order to achieve a diverse student body, successful efforts were made to recruit more women. The Mount Sinai graduating class of 1993 was the first in New York State to have a female majority. Women continue to represent half or more of each graduating class.

The recruitment of minority students to Mount Sinai had also been an early goal of the School, enunciated at the planning meetings in the 1960s. Unfortunately, the entire field of medical education has never been particularly successful in this area.[63] Still, Mount Sinai has developed a number of programs to bring these groups into healthcare careers, and it continues to try to achieve cultural diversity. The Class of 1992 had 8 percent of students from underrepresented minorities; the current figure is around 13 percent, which is above the national average. Since 1974, the Secondary Education Through Health (SETH) program has worked with New York City high schools to identify, recruit, and motivate students, bolster their academic performance, and facilitate entry into pathways leading to careers in medicine and health. There are also the Mount Sinai Scholars and Mentoring Program, the Biosciences Studies Institute, a Collegiate Summer Research Fellowship Program, and a Collegiate MCAT review program. In a related area, the Cultural Diversity in Medicine (CDM) program was founded in 1992. CDM is dedicated to educating medical students and residents about the impact of culture and ethnicity both on physician-patient relationships and on the delivery of quality patient care. This is now a separate Center in the School, building on the Office for Multi-cultural and Community Affairs, which was established in 1998.

The School has also worked on recruiting students from varying educational backgrounds to try to provide a diverse student body, establishing an Engineering and Medicine program during the tenure of Dr. Kase. But perhaps the largest effort to recruit a different kind of candidate was the development of the Carl H. Pforzheimer Jr. Humanities and Medicine Program in 1987. This program was originally established with five leading undergraduate universities to encourage students majoring in the liberal arts to pursue careers in medicine. Students are required to major in the humanities or social sciences and to complete one year of biology and chemistry with grades of B or better. Students are accepted into the program and receive an acceptance into the Medical School before the end of their second undergraduate year. The MCAT test is not required. Students then spend eight weeks at Mount Sinai during the summer after their sophomore year in college to receive exposure to clinical activities and to take a brief course in organic chemistry and the physics topics relevant to medicine. These students are not obligated to attend Mount Sinai, but, should they choose to do so, they are admitted after successful completion of their undergraduate education. The first nineteen students accepted into the program attended the summer program in 1989 and two years later became members of the Class of 1995. In 1996, Mary Rifkin, Ph.D., a member of the faculty of the Brookdale Center, became the Director of the program. In a ten-year review of this program, Rifkin and her colleagues found that, "although students in this program have more academic difficulties in the pre-clinical years, they excel in the clinical/community setting and have greatly enriched the medical school environment."[64]

The mid-1990s saw an increased national emphasis on primary care education for undergraduate and graduate medical students. This was of particular concern at Mount Sinai, which had previously trained large numbers of specialist physicians.[65] Mount Sinai addressed this

City Hospital Center at Elmhurst, one of Mount Sinai School of Medicine's
long-time affiliates. This image was taken after a renovation in 1995.

issue by having specialists lecture to generalists on the basics in their
field that generalists needed to know. It also reintroduced a first-year
interdisciplinary course and formed a combined Medicine/Pediatrics
residency program. There was an increasing emphasis on the use of the
School's affiliate hospitals, in particular the City Hospital Center at
Elmhurst, one of the leading institutions in New York's municipal
Health and Hospitals Corporation. Also, at this time, a Graduate Medical Education Consortium was established at the School to better integrate the residents' educational experience among all of the School's
affiliated teaching hospitals and to provide a broader base of clinical
experiences, including more time in primary care.[66] Training sites for
all groups were expanded to include practitioners' offices, more outpatient settings, and group practices to provide increased exposure to a
variety of primary care venues.

Technology also played an increasing role in the life of the School
during Dean Kase's tenure. Initially, computing support for the School
was provided by The City University, which housed a mainframe for
data storage and provided e-mail to the Mount Sinai community.
CUNY worked with Mount Sinai's Research Computing Unit (RCU)
of the Department of Biomathematical Sciences to accomplish this.
As technology advanced and its use became ubiquitous, it became

clear that Mount Sinai had to control its own computing environment. In 1990–91, the Levy Library, under the leadership of Lynn Kasner Morgan, took over the schoolwide hardware and software services that had been provided by the RCU. The Library created a division of Academic Computing to establish and manage the School of Medicine's hardware structure and to bring network technology to the Medical School components around campus. Moving quickly, by 1992 Academic Computing had locally mounted a Mount Sinai e-mail system and also provided servers to work as data centers for scientists and faculty. The division also established an upgraded dial-in capability so that faculty, students, and house staff could access School and library resources while in their offices, at home, or on rotation at the affiliate institutions.

At the same time, other divisions of the Library worked to acquire and license electronic information resources to support the educational and research efforts of the School. The Library's digital information resources initially included large databases such as MEDLINE, CINAHL, and CANCERLIT, as well as smaller, CD-rom–based products. Starting in the mid-1990s, the Library has increasingly moved its resources to

The Associated Alumni provided funds to establish the Electronic Information Center in the Gustave L. and Janet W. Levy Library. This 1995 image shows some of the computers that allow students, faculty, and staff to access the digital information products that the Library provides, as well as the wealth of resources available on the Internet.

Web-based products. The Medical School and the Mount Sinai Alumni Association contributed in a significant manner to help the Levy Library build and support the network, increase the number of computers available for student use, provide a help desk function, and construct a computer classroom for instruction.

A new instructional and assessment method was implemented at the Medical School in 1989 with the inauguration of the use of simulated patients in teaching. The Morchand Center for Clinical Competence, led by Mark H. Swartz, M.D., a graduate of the Mount Sinai School of Medicine, opened in 1991 to house this program. The Center uses actors to simulate patient situations. Students are required to do a history of the patient and to diagnose his current illness. The students are then evaluated not only on their ability to correctly diagnose the patient but also on their technique in dealing with the patient. This process helps to hone the students' communication skills with patients, as well as their clinical skills. The Center today hosts medical students from Columbia University's College of Physicians and Surgeons, Cornell University Medical College, Albert Einstein College of Medicine, the Health Science Center of the State University of New York (SUNY) at Stony Brook, New York Medical College, and the SUNY Health Science Center at Brooklyn. These institutions, along with Mount Sinai, form the New York City Consortium for Clinical Competence, the first major collaboration among medical schools in the New York metropolitan area. Each school has incorporated one or more of the Morchand Center programs into its own undergraduate curriculum. As part of the Consortium, faculty members from the schools meet regularly to confer, assess student progress, and make recommendations.

Another resource formally incorporated into the teaching of medical students under Kase was the use of professional nurses from the Department of Nursing of the Hospital as mentors, instructors, and role models. There had been a long history of interdisciplinary team training at Mount Sinai, especially as third-and fourth-year students went onto the patient floors and needed to learn the intricacies of inpatient management. During the 1980s, the Department of Nursing initiated a formal course for first-year medical students that focused on the theme of nurse-physician collaboration. The Department of Nursing continues to have ongoing responsibility for the teaching of Basic Cardiac Life Support (BCLS), Advanced Cardiac Life Support (ACLS), Pediatric Advanced Life Support, and techniques such as venipuncture, not only to medical students but also to house officers, fellows, and attendings who require certification in BCLS and ACLS. With the initiation of the "new" curriculum in 2000, the Department of Nursing plays an active teaching role in the course "Art and Science in Medicine."[67] Clinical nurse specialists in all inpatient and outpatient areas are also involved in medical school instruction.

In the early 1990s, a Center for Nursing Research and Education within the School of Medicine was formed. The Center grew out of an awareness of the important role nurses played with the medical students, coupled with the desire of Gail Kuhn Weissman, Ed.D., then Vice President for Nursing, to have an academic home for nursing to support its own educational and research efforts. The Center continues today under the leadership of Thomas Smith, Vice President for Nursing, Maria Vezina, Ed.D., Director of Nursing Education, and Mary Dee McEvoy, Ph.D., Director of Nursing Research. The numerous nursing research projects are reviewed by committees of the Department before submission to the Institutional Review Board. The Center offers many courses in nursing practice and is also responsible for inservice nursing education.

By the mid-1990s, the search for research funding was becoming more competitive. Mount Sinai remained aggressive, receiving increasing amounts of federal money each year. A breakthrough came in 1993 when ground was broken on Madison Avenue for another tower building, which became known as the East Building. In the words of Dr. Kase, "During a period of NIH uncertainty we did not circle the wagons but convinced the Trustees that a major new increment to research space was necessary. It was built and we filled it promptly."[68] The impact of this move can be seen in the growth of research programs and dollars after the building's completion in 1996: in 1995, 169 projects received NIH funding; in 1997, when the building was fully operational, 215 projects were funded.[69] Having the new space allowed Mount Sinai to enhance the research already ongoing at the School, as well as to recruit new scientists to bring their programs here. Over these years, Mount Sinai moved steadily upward in the NIH ranking of medical schools, based on the grants that were attracted there from thirty-fourth in 1994, to thirtieth in 1996, and twenty-fifth in 1998.

In 1997, Kase retired as Dean, resuming his role as Professor of Obstetrics, Gynecology, and Reproductive Sciences. He had indelibly changed the landscape at the Mount Sinai School of Medicine, making

The East Building, designed by Davis Brody & Associates, was completed in 1996. The eighteen-story, 736,000-square-foot building cost $200 million to build.

it a more academic, research-oriented institution, developing its own reputation and traditions separate from those of the Hospital. But, over the next few years, the close ties between the Hospital and the School would enmesh the School in tumult.

THE RUBENSTEIN YEARS, 1997–2001

Arthur H. Rubenstein, M.B.B.Ch., formerly Chairman of the Department of Medicine at the University of Chicago, was appointed to succeed Kase. The Mount Sinai that he joined was very different from the institution that had existed merely ten years before. As health care and its financing had evolved over the 1980s and 1990s, the leaders of Mount Sinai had tried not only to change appropriately with the times but also to anticipate the direction of the future in order to position the institution to thrive in the new environment. As a result, the Mount Sinai Medical Center had created the Mount Sinai Health System in 1995, a loosely tied group of clinical alliances with medical institutions from around the tristate area. Barry Freedman, then President of the Hospital, was appointed to lead this new entity. The purpose of the Health System was to have Mount Sinai, as the academic partner, provide faculty and expertise to the smaller hospitals, which would, in turn, provide the Medical School with sites for educational rotations and also refer more difficult cases to Mount Sinai for treatment.

Dean Arthur Rubenstein around the piano with medical students in the Levinson Student Center, 1999.

As time went on, Mount Sinai's leaders saw additional opportunity in establishing a closer partnership with another academic medical center. It was hoped that the combined institution would be able to merge business operations to save money, avoid some large capital purchases, and increase the market share and revenues for the hospitals. It was also thought that a combined center could favorably leverage its large size when negotiating with insurance companies, HMOs, and suppliers. In the mid-1990s, Mount Sinai began discussions with the trustees of other New York City institutions. New York University (NYU) Medical Center was most amenable to these talks, and plans began to take shape.

When the Medical Centers began talks in 1996, it was hoped that the two institutions might combine both their hospitals and the two schools of medicine. But it quickly became clear that there was a real disconnect between what the business-oriented lay leaders of the institutions were thinking and what the medical faculty considered to be in the best interest of each medical center. The NYU faculty was the most vociferous in its opposition to the merger, although the faculty at Mount Sinai were not unanimously in favor of these changes, either. The NYU faculty did not want to work on combining the two programs to create a new entity, nor did it feel it appropriate for NYU to sponsor two separate medical schools. There was the suggestion that NYU's academic reputation would suffer with the addition of Mount Sinai School of Medicine to the University. The NYU faculty was also concerned that, since it was clear that Dr. John Rowe, Mount Sinai's President, would be the CEO of the merged organization, Mount Sinai would exercise majority power over the NYU hospitals, while its own Dean would be weakened.[70]

In the face of this resistance, it was announced in February 1997 that the merger plans were off. However, the Trustees of both institutions were still convinced that the changing marketplace of New York health care demanded that some consolidation among medical centers occur, and so talks were resumed. In January 1998, the Faculty Council of the New York University School of Medicine actually began the process of suing NYU to stop the merger, but the case did not go forward.

On July 17, 1998, after extensive discussions, the creation of Mount Sinai-NYU Health was announced. As part of this reorganization, the two hospitals merged to form a clinical entity with a northern and southern "hemisphere," or campus, while the schools of medicine remained separate entities. The most important feature of the merger for the School was that on July 1, 1999, the Mount Sinai School of Medicine's affiliation changed from The City University of New York to New York University.

The impact on the School from the merger was extreme. The most wrenching change was that, after thirty years of trying to blend the School and the Hospital, Mount Sinai now had to try to define and separate what was "the Hospital" and what was "the School." Clinical departments, where the faculty had both School and Hospital appointments and where space was given to clinical, administrative, and research needs, now had to define functions and people as "Hospital" or "School." Administrative services such as human resources, purchasing, and finance had to be separated by Hospital and School functions. The School administration grew to handle the new demands, but the transition was not always smooth.

Still, as the Hospital began to suffer from increasing financial woes, the School, now more insulated from these issues, could point to positive developments and a focus on academic issues. One gratifying aspect of these years was the School's continuing climb up the NIH rankings in terms of grant money received (see figure 1.1). The other

Fig. 1.1 NIH Support

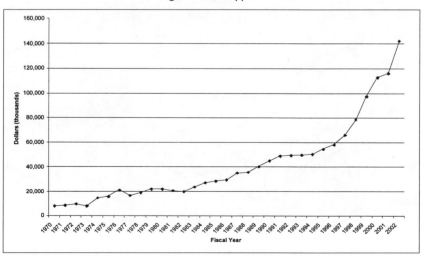

was a reorganization of the Faculty Practice Plan, which was returned from the jurisdiction of the President's Office to that of the Dean's Office. There was also the announcement, in 1999, of a staggering gift of $75 million from Mount Sinai Trustee Frederick Klingenstein and his family to support scholarships and translational research in the School. There was even talk of building a new educational space in the Annenberg Building and a translational research building to house expanded programs, but these plans ultimately did not progress due to financial concerns.

There were some changes in the traditional business of the School during Dean Rubenstein's tenure. A White Coat Ceremony was introduced for first-year medical students in September 1998 to formally induct them into the study of their chosen profession. Also, that year, Dean Rubenstein restructured the Appointments and Promotions (A&P) Committee with the goal of raising the academic bar, streamlining the A&P process, and providing better guidelines for the appointment and promotion of faculty. The School undertook the

The first White Coat Ceremony, held September 1998 for the Class of 2002. The official "coaters" shown are, from left to right, Drs. Barry Stimmel, Mark Swartz, and Richard Kane.

updating of its publications, introducing student bulletins and handbooks with a new, colorful look. During these years, the School
also enhanced its presence on the Internet with the development of a
new format for the School of Medicine's Web site and the creation of a
standard look for each page. Responsibility for building and maintaining the School's Web site was assigned to a new division of the Levy
Library, Web Development. The School's new publications were
mounted online, and there was an increased interest in using the Web
for recruitment.

At the end of the 1990s, on the basis of recommendation from the
accreditation team during its 1996 visit, a new curriculum plan was
developed that would begin in the fall of 2000. In many ways, this new
curriculum represented another step closer to a return to the intentions
of the 1960s plan: integrated courses, more small group lectures, self-
study, student work with patients from the first year, the use of the
most modern educational technology, and an overall emphasis on lifelong learning.[71]

ANOTHER BEGINNING, 2001–

In September 2001, Rubenstein resigned as Dean. This was a difficult
period for the School; the Hospital's merger with NYU began to
unravel, the President of Mount Sinai-NYU Health, John Rowe, had
left in 2000, and the financial situation of Mount Sinai, particularly
the Hospital, was deteriorating. At Rubenstein's departure, Nathan
Kase assumed the position of Interim Dean and Interim President of
Mount Sinai. His choice of the title Interim, as opposed to Acting,
showed his intention of being proactive during his limited tenure and
not merely responsive. It had been decided in early 2001 that Mount
Sinai and NYU would return to campus-based governance, and so
Mount Sinai's leadership began the process of reintegrating the School
and the Hospital. In February 2002, Kenneth I. Berns, M.D., Ph.D., was
recruited from the University of Florida as the new President and CEO
of the Medical Center, while the search for a new Dean continued.
Berns and Kase oversaw a series of evaluations by outside management consultants, including the Hunter Group and Cap Gemini. The
implementation of these sometimes painful recommendations resulted
in a streamlining of the School's organizational structure, with many

centers and institutes being folded back into the traditional departments. Programs were evaluated for their effectiveness and potential, and some projects were canceled, while others were changed to become more self-supporting. The areas of cancer, gastroenterology, the neurosciences, and cardiology were identified as fields where Mount Sinai would concentrate its efforts over the near future. The budget process was changed to reflect an emphasis on supporting areas essential to the mission of the institution.

In the fall of 2002, the appointment of a new Dean, Kenneth Davis, M.D., Chairman of the Department of Psychiatry at Mount Sinai since 1987, was announced. Davis is a graduate of the Mount Sinai School of Medicine, Class of 1973, and brings many years of insight to bear on Mount Sinai's problems. He assumed the deanship in January 2003. Two months later, it was announced that Berns would be stepping down as President of the Medical Center. Davis assumed that position as well, becoming both President and Dean. Merging the two positions returned

Convocation is the first academic event of each school year. This photo was taken in 2002, with Interim Dean Kase at the podium. The two men behind the table with the plaques are Dr. Kenneth Berns, President of the Mount Sinai Medical Center, and Peter May, Chairman of the Boards of Trustees. Photo credit: Russell Dian.

The Mount Sinai campus in 2003. Towering in the middle is the Annenberg Building, surrounded by eight other Mount Sinai facilities, including the Guggenheim Pavilion in the rear. To the right, across Madison Avenue, is the East Building.

Mount Sinai's leadership structure to what had existed when the School opened. At that time, the combined role assured that the larger, more entrenched hospital would not overshadow the young, emerging medical school. Forty years later, this organizational template acknowledges the parity of the Medical School with the Hospital and the critical role of the School in the vitality of this academic medical center.

2

The Curriculum

OF ALL THE tasks facing the founders of the new School, none was more daunting than the development of the curriculum. Yet, as Hans Popper, M.D., the founding Dean, pointed out, it was "an optimal time [to develop curriculum] because the vested interests of the faculty are not yet established and the opportunity for change diminishes fast as the school is organized."[1]

In 1965, Popper enunciated "The Mount Sinai Concept," which would serve as the foundation for the new school.[2] Referring to the traditional tripod of academic health centers—patient care, research, and teaching—Popper proposed creating another tripod out of the "teaching leg." This leg would balance three objectives in the training of the future physician: it would introduce quantitative biology into medicine (which Popper noted would lead to area specialization); it would broaden the influence of social sciences and human studies to counteract the presumed depersonalization of the organ-specialized physician; and it would strengthen the role of community medicine, which strives to provide good care to every patient beginning in the presymptomatic stage.

Also in 1965, Popper stated the principles that would underlie the new curriculum.[3] These included subject integration in the clinical as well as the basic science areas, relatively early career choice, the provision of adequate free and elective time, early exposure of the students to patients and clinical medicine through the provision of the course "Introduction to Clinical Medicine" to be given throughout the first two years, provision for basic science recall during the clinical years, and recognition of the significance of community medicine as an integral part of the overall curriculum. George James, M.D., the School's first President and Dean, noted that "the content of the current stan-

74

dard medical curriculum will be reduced . . . to allow the presentation of courses in behavioral science, social anthropology, philosophy, ethics and other fields of human science."[4] James went on to note that, because of the large clinical and laboratory staff of the Hospital, much of the teaching could be done on a tutorial basis. Popper's tenets had been presented to and accepted by the Liaison Committee on Medical Education (LCME), the national body responsible for the accrediting of medical schools.

Four curriculum coordinators were chosen in 1966 to lead the effort: Tibor Barka, M.D., Professor of Pathology and the Chair-designate of the Department of Anatomy; David Koffler, M.D., Associate Professor of Pathology and Assistant Dean for Academic Affairs; Howard Gadboys, M.D., Professor of Surgery (Cardiothoracic); and Fenton Schaffner, M.D., Professor of Medicine and Pathology. The group was chaired by George Christakis, M.D., Professor of Medicine and Associate Dean. The educational philosophy and the curricula of several existing and developing medical schools were studied, and, within a year, the group was able to present a four-year curriculum for discussion.[5] Teaching time was distributed by semesters for each of the four years. In the summary, the coordinators noted, "the proposed curriculum will necessitate a larger teaching staff than that required by the classical curriculum and continuous re-organization and supervision of teaching."[6]

In March 1967, the original group of curriculum coordinators was expanded into a curriculum committee. The committee included two nominees from each department represented on the Executive Faculty Committee and one nominee from all other departments of instruction. George James and Hans Popper cochaired the expanded committee. When it was fully constituted, the curriculum committee consisted of forty individuals. In April 1967, at the first meeting of the entire committee, it was "charged with the responsibility of proposing a curriculum for the Mount Sinai School of Medicine which takes into consideration the past excellence of The Mount Sinai Hospital's inpatient care, clinical teaching and research, the recent trends in medical education and the future needs of society."[7] At the time, the Departments of Biochemistry, Microbiology, and Laboratory Education were without chairs. As chairs were recruited to these disciplines, they were added to the committee. The coordinators' proposal was presented, and discussion over the contents was begun.

Chairmanship of the Curriculum Committee was taken over by Horace Hodes, M.D., Chair of the Department of Pediatrics, in June 1967, and three subcommittees were formed to define the contents of the "Introduction to Clinical Medicine" course, to decide the degree of integration of the study of disease processes and organ systems, and to determine the extent of free time in the curriculum. In the spring of 1967, approval was received to allow admittance of third-year students as well as a first-year class in September 1968. With the need to now refine the curriculum for both first- and third-year students, identify faculty, complete construction of the multidisciplinary laboratories in the 101st Street Basic Sciences building, and prepare instructional materials, the ensuing months were a time of hectic activity. The subcommittees met biweekly, and the full committee met monthly. For a period of time, some members of the committee harbored doubts that the task could be accomplished on time. The initial bulletin sent to prospective students contained little specific data related to course content, hours, or timing. The leadership, however, was resolute, and on Monday, September 9, 1968, the thirty-six students entering the first first-year class heard their first lecture, "Structure and Function in Medicine," given by Hans Popper, and they then engaged in a full day of classes and laboratory work (see figure 2.1).

Although every attempt had been made to adhere to Popper's tenets, the first curriculum did not differ that much in structure from the traditional first-year program that most medical schools had been conducting for decades. That being said, there was considerable involvement of the clinical faculty in basic science teaching. Clinicians gave many of the lectures and participated in many of the laboratory exercises. In the first thirteen-week trimester, there were seven hours of class or laboratory time five days a week, with only three hours per week available as free or elective time. The course now entitled "Introduction to Medicine" met for two hours weekly. Anatomy and biochemistry lectures and laboratory exercises took up all of the remaining time. In addition to gross anatomy, the Department of Anatomy also took responsibility for the teaching of embryology, histology, and cell biology. In the second trimester, however, some of the proposed innovations began to appear. Physiology, pharmacology, microbiology, and pathology replaced anatomy and biochemistry; there was more free and elective time, and "Introduction to Medicine" now met twice a week. In the third trimester, organ system instruction began with the

CLASS SCHEDULE
FOR
FIRST YEAR STUDENTS

Monday, September 9, 1968

9:00 - 9:50	Structure and Function in Medicine	Dr. H. Popper
10:00 - 10:50	Chemistry of Amino Acids, Peptides, and Proteins	Dr. P. Katsoyannis
11:00 - 11:50	Structural and Functional Organization of the Animal Cell	Dr. T. Barka
12:00 - 12:50	Cytological Methods	Dr. J. Kelly
2:00 - 2:50	Cell Membranes, I	Dr. T. Barka
3:00 - 4:30	Lab. A - Light, Lenses, and Cells	Staff
4:30 - 5:00	Introduction to Gross Anatomy	Dr. M. Levitan

Tuesday, September 10, 1968

9-00 - 9:50	Cell Membranes, II	Dr. T. Barka
10:00 - 10:50	Chemistry of Amino Acids, Peptides, and Proteins	Dr. P. Katsoyannis
11:00 - 11:50	pH and Buffers	Dr. Ginos
1:00 - 5:00	Lab: Instrumentation	Staff

Wednesday, September 11, 1968

9:00 - 9:50	Cytoplasmic Organelles and Inclusions,I	Dr. L. Ornstein
10:00 - 10:50	Chemistry of Amino Acids, Peptides, and Proteins	Dr. P. Katsoyannis
11:00 - 11:50	Amino Acids as Electrolytes	Dr. Ginos
1:00 - 5:00	Lab: Instrumentation, concluded	Staff

Thursday, September 12, 1968

9:00 - 9:50	Cytoplasmic Organelles and Inclusions,II	Dr. T. Barka
10:00 - 10:50	Chemistry of Amino Acids, Peptides, and Proteins	Dr. P. Katsoyannis
11:00 - 12:50	Introduction to Course: Diabetes in Young Adult; 1) Symptomatic, recently diagnosed; 2) Under prolonged treatment	Dr. H. Dolger
2:00 - 2:50	How to Use Indexes and Abstracts	Library Staff

Friday, September 13, 1968

9:00 - 9:50	Use of Special Bibliographies	Library Staff
10:00 - 10:50	Chemistry of Amino Acids, Peptides, and Proteins	Dr. P. Katsoyannis
11:00 - 12:50	Lab. B - Electron Microscopy	Dr. L. Ornstein
2:00 - 5:00	Biostatistics - What is Normal	Dr. P. Denson

teaching of the cardiovascular-respiratory system and the neurosciences. The following year, a biostatistics course was added.

The development of the "Introduction to Medicine" course was a story unto itself.[8] In the 1960s, efforts were under way in many medical schools to bring students into contact with patients earlier than the customary third-year clerkships. "Interdisciplinary teaching" was also widely discussed and became a buzzword, but few institutions were able to implement the concept, probably because of the difficulty in dealing with logistical and personnel (perhaps "personalities" is a more descriptive word) issues. That MSSM was able to produce a successful course was in large part related to the newness of the school with its enthusiastic faculty, none of whom had fixed ideas about their roles as teachers but all of whom wanted to contribute in any and every way possible.

One of the earliest courses of its kind in American medical education, the primary goals of "Introduction to Medicine" were to introduce the student to patients early in the first year and to demonstrate the relevance of the basic sciences. The course committee that developed the curriculum was composed of members representing many of the basic science and clinical departments and was chaired by Cecil Sheps, M.D., at the time the Director of the Beth Israel Medical Center, an affiliate of the young school. Bessie (Bess) Dana, M.S.S.A., the Curriculum Development Coordinator for the School, was appointed as the staff person for the committee.[9] Dana, a social worker by training, played a vital role in implementing the overall educational goals of the new Medical School and demonstrated a creative way of working with people of differing viewpoints and knowledge toward the achievement of a unified approach to medical education. It was this unique talent that encouraged collaboration among disparate elements of the faculty and led to the success of the course. It was not unusual to have a two-hour session for the students in which the faculty participants might, for example, include an anatomist and embryologist, a surgeon, a nutritionist, a member of the Department of Community Medicine, an ethicist, and a group of patients. Sheps relocated to North Carolina in 1968, prior to the opening of the school year, and Arthur Aufses, Jr., M.D., took over the chairmanship of the committee. Aufses, a surgeon in private practice at the time, was one of the two Department of Surgery representatives to the Curriculum Committee from the outset. He chaired the course committee until shortly before he left Mount Sinai in 1971 to become the Chief of Surgery at the Long Island Jewish Medical Center.[10]

The integrated systems teaching, beginning in the third trimester of the first year, was a prodigious effort to link basic and clinical science and to demonstrate the relevance of the former to the latter. The course was designed to present normal and abnormal structure and function of the various organ systems and pathophysiologic principles of disease processes, along with the concurrent introduction of clinical medicine in each area. This allowed an early introduction of the student to the complexity of medical science. A multidisciplinary committee unique to each system was responsible for the curricular content. As an example, eleven departments—Anesthesiology, Anatomy, Biochemistry, Neurology, Neurosurgery, Ophthalmology, Otolaryngology, Pathology, Pharmacology, Physiology, and Radiology—participated in the 162 classroom and laboratory hours of instruction in the neurosciences. Although there would be no second-year class until 1969, the curriculum for that year was created concomitant with that for the first year. The greater part of the second year was devoted to continued integrated study of the organ systems.

It was not only the first-year curriculum that had to be "ready to go" on opening day of the first school year. Twenty-three third-year students, either transfers from other four-year schools or recruits from two-year medical schools, also began their traditional clerkships on September 9, 1968. Despite the large and variegated patient population at The Mount Sinai Hospital, four additional hospitals affiliated with the Medical School were utilized to provide rich and varied teaching resources. These included the Beth Israel Medical Center, the Bronx Veterans Administration Hospital, the Hospital for Joint Diseases and Medical Center, and the City Hospital Center at Elmhurst, in total representing more than 5,000 patient beds. As with the first two years, the third and fourth years were planned together. Fifty-nine weeks were devoted to mandatory clerkships in Medicine, Pediatrics, Surgery, Obstetrics and Gynecology, Psychiatry, Neurology, Ophthalmology, and Community Medicine. An additional six weeks were taken up with two combined clerkships; one included Otolaryngology and Urology, while the other combined Orthopaedics, Rehabilitation Medicine, and Dermatology. Students were also expected to spend another four weeks in clinical areas of their own choice. Twenty-two additional weeks of electives completed the program leading to graduation.

Vital to the creation, development, and progress of the new school and its new curriculum was the Department of Laboratory Education. In 1967, Edra Spilman, Ph.D., was appointed Professor and Chairman. Recruited from Western Reserve University where he was Director of the multidiscipline laboratories, Spilman played a major role in the development of Mount Sinai's Basic Science and Annenberg Buildings and the multidiscipline laboratories. These laboratories were considered the most modern and efficient facilities of their kind in the nation. The students were assigned to a laboratory space for two years, and the faculty from the various basic sciences came to them. The teaching faculty and the students found the design of the laboratories to be almost ideal for a variety of teaching and study activities.

The Department of Laboratory Education also encompassed the Division of Biomedical Communications, which immediately began producing teaching videotapes and movies for use on the School's closed circuit television system; the Medical Arts Studio, which rapidly became operational; and a Syllabus Office, which produced more than 250,000 pages in the first three months of the School's operation. Within two years, more than 1.5 million pages would be produced annually for the students. In 1969, the Department was renamed the Department of Medical Education, with Dean James as the Acting Director. The Department's responsibilities were expanded to embrace not only medical education, with emphasis on research and development, but also administrative matters such as class scheduling and the logistics of the teaching program. Spilman was appointed Associate Dean for Special Services and administered the Department. The Department has never received full departmental status, and when Thomas Chalmers, M.D., became President and Dean in 1973, he assumed the position of acting Chair. Barry Stimmel, M.D., who had joined the faculty in the Department of Medicine in 1967, had been named Assistant Dean for Admissions and Student Affairs, in 1970, and then Associate Dean for Academic Affairs, in 1975, was named the Acting Chair of the Department, in 1979, a post he held for fifteen years. Throughout his tenure in the role of Dean for Academic Affairs, Stimmel was at the forefront of research in medical education. Constantly evaluating the results of the School's educational efforts, Stimmel produced many publications on various aspects of the educational process and spearheaded the constant evaluation and reevaluation of the curriculum.

As a new school in 1968, nothing at Mount Sinai could be cast in stone. As noted earlier, the original curriculum coordinators warned that "continuous re-organization" would be required. The full committee also recognized that everything was preliminary:

Finally, a word must be said about the finality of the scheme proposed. One of the duties of the subject committee, particularly the chairman is to evaluate the material they present in terms of worthiness of the time spent. This will have to also be done independently by the curriculum coordinators and by the departmental chiefs. From such observations changes will come, some in the form of minor adjustments, and some as major upheavals. Only by keeping flexibility as a keynote can our curriculum continue to serve its purpose, to always provide for the training of the physicians for the years ahead.[11]

Student overload was also recognized as a potential problem in both the preclinical years and in the clerkships. The curriculum committee noted: "Since it has become obvious that all of medical knowledge cannot be taught, but only principles, an attempt will be necessary on the part of the subject committees to see that students are exposed to as many [principles] as can comfortably fit into the crowded schedule."[12] It became evident early on, for example, that there were far too many lectures throughout the curriculum; members of the original committee remember well the admonitions given at the end of every year to "reduce the number of lectures." Unfortunately, too few course directors heeded the call, and many students complained of the heavy burden.

Having developed the program of instruction, the Curriculum Committee was also charged with developing methods of evaluation and grading. Examinations were "viewed as instruments for the promotion of learning, as an encouragement to each student to achieve his maximum potential, and as an opportunity to develop the student's capacity for self-knowledge and self-evaluation."[13] The number, types, and frequency of examinations were left to the discretion of the course directors.

Considerable discussion took place regarding the grading of students. From the outset, it was agreed by all that, in order to provide appropriate letters of recommendation for internship and residency and to allow selection of students for Alpha Omega Alpha (the honor medical fraternity), clerkships would be graded as honors, pass, or fail. In the preclinical years, there was consensus that examinations would be anonymous and that only those who failed would be identified, but there was much debate as to the grading system. In the end, the school opened with letter grades for the first year. But not for long! The thirty-six first-year students met weekly during lunch hour at a "town meeting," and they also met weekly with Dean James.[14] Before the year was over, grading for the first two years was changed to pass-fail, a situation that has continued to the present. Students were also expected to take the examinations given by the National Board of Medical Examiners, Part 1 during the second year and Part 2 during the fourth year, but in the beginning there was no requirement that students had to receive a passing grade either to be promoted or to graduate. The National Board Examinations were later supplanted by the United States Medical Licensing Examinations (USMLE).

The curriculum was evaluated annually and revised when necessary on the basis of student performance, student evaluations of courses, and faculty recommendations. New courses were added and others removed. Unfortunately, the number of lecture hours never decreased in any meaningful way and, in fact, increased during the early 1980s. Total curricular hours remained reasonably stable. A first-year course in First Aid was instituted in the late 1970s, leading to certification in Basic Life Support (BLS) for every medical student. "Introduction to Medicine" underwent many iterations over the years, changing its format and reducing the number of allocated hours.

In 1978, Stimmel organized a major curriculum retreat. In his introduction to the participants he stated:

The School of Medicine began its first academic year in September 1968, with the admission of both a first and a third year class. Hardly a year went by when a curriculum retreat was organized to review the educational process on the basis of what must have been one of the shortest "follow-ups" in medical history. Although the overall feelings of both faculty and students at this retreat [1969] were quite positive with respect to the future of the School, it is of interest that a full ten years have elapsed without a feeling of a need for a similar retrospective analysis of our curricular activities.[15]

The retreat reviewed student data collected assiduously over the years, discussed recommendations for improvements in media education and computer science, and discussed the evaluation process of student performance. As a result of the 1978 retreat, Computer Sciences and Biostatistics merged to form the Department of Biomathematical

Student Mark Dannenbaum, Class of 1986, is shown at a computer terminal in the Levy Library utilizing a student self-testing program, 1983.

Sciences. There was considerable expansion of the audiovisual section of the Levy Library, including the installation of terminals for computer-assisted instruction (CAI), stimulating a significant increase in the production of CAI programs for student use.

Stimmel conducted another retreat in 1980, at which time he stated, "the facility with which a major curricular change can be effected has been compared to the difficulties faced in attempting to relocate a cemetery."[16] He noted that, over the twelve years of teaching, there had been an almost 50 percent increase in block-time curricular hours during the basic science years, including several new courses. In the clinical years, required subinternships in medicine and pediatrics in the final year had been created. Current teaching in the surgical specialties appeared unproductive and required revision. The students perceived that the preclinical curriculum hours were excessive and prevented "sufficient time for thinking and independent learning."[17] The two major questions raised were diametrically opposed to each other. Was the curriculum too rigid? Should the existing elective time be decreased to allow the inclusion of subjects considered important but not currently represented?

Changes in the senior leadership of the School over the next several years precluded sweeping changes. Dean Chalmers, however, established the Department of Geriatrics and Adult Development, in 1982. Robert Butler, M.D., the Chairman of the new Department demanded and was allowed to create a mandatory clerkship. The new clerkship necessitated changing the number of weeks of some of the other clerkships, a situation that caused more than minor friction.[18]

Shortly after Nathan Kase, M.D., was appointed Dean in 1985, another retreat was held not only to address specific concerns in the curriculum but also to review and respond to a report of the Association of American Medical Colleges that was critical of all medical school curricula and that called for a drastic change in the way medicine should be taught. In addition, the School was preparing for an accreditation site visit in 1986 and was required to conduct an intensive self-assessment that included the curriculum. An Ad Hoc Committee to Review the Curriculum was created and issued its initial recommendations in 1986.[19]

Changes recommended by the Committee included a reduction in the number of lecture hours, the development of interactive computer-based programs for self-assessment and independent learning, and the development of a comprehensive, practical clinical examination to assess student skills. Other recommendations also acted upon included a 20 percent decrease in required curricular time in the basic sciences; a resequencing of courses so that all of the basic sciences except pharmacology and organ system pathology were taught in the first year; a shift to starting the school year in August rather than after Labor Day; and the creation of examination weeks for "block" examinations, with study time beforehand. New courses ("Student Well-Being," "Brain and Behavior," "Life Cycle," "Clinical Examination and Critical Assessment," and an integrated ethics curriculum) were either instituted or rapidly replaced existing courses. In some courses, small-group sessions and seminars replaced up to 50 percent of structured course time. Stimmel would shortly note that, "despite the consistent decrease in structured hours," the student's scores on Part 1 of the National Board Examinations "remained unchanged or have slightly, but not significantly, increased."[20]

The "Student Well-Being" course, first offered in 1987 and directed by Joyce Shriver, Ph.D., was taught to incoming first-year students during their first weeks of school. The course, which incorporated First

Aid and certification in Basic Life Support, was based on the assumption that the primary prevention of impairment and the promotion of student well-being would be a most effective way of training competent and compassionate physicians. The program's objectives were:

> to sensitize, alert, and educate students about their vulnerability to stress, substance abuse, disruptive interpersonal relationships, burn out and impairment; to teach students skills for personal stress management, relaxation, coping, time management and more effective communication, studying and test-taking; to teach students basic nutrition and the value of and approaches for improving the quality of their lives through the personal application of this information together with regular exercise; to emphasize well-being and the benefits accruing from them; and to create a collegial and open atmosphere—in which students can begin to feel part of a medical and educational community and to create a non-academic setting— in which students can meet and interact with fellow classmates and faculty.[21]

The course was eminently successful, and, indeed, in 1993, the School received an award from the American Medical Student Association for fostering student well-being.

The development of the integrated ethics curriculum for the School has been notable. Ethical issues had been discussed on occasion with students in the "Introduction to Medicine" course and other venues from the School's earliest days, with input from the members of the Hospital's Ethics Committee, the latter chaired continuously by Kurt Hirschhorn since its creation in 1975. In 1981, a seminar series for faculty was started, created by the psychiatrist James Strain, the neurologist Daniel Moros, the neurosurgeon Russ Zapulla, and the philosopher Stefan Baumrin, Ph.D., a member of The City University Graduate Center (CUNY) faculty. Moros and Baumrin then began combined CUNY/MSSM seminars in medical ethics. They were joined by Rosamond Rhodes, who, in 1984, became Mount Sinai's first ethics fellow. As an ethics fellow, Rhodes introduced ethics sessions for the clinical clerkships in medicine and neurology, joined service rounds in a number of the other departments, and organized an annual ethics conference. Rhodes completed her Doctorate in Philosophy and joined the faculty as a lecturer in 1988; she rose rapidly through the ranks and was promoted to Professor in the Department of Medical Education in 2003.

Starting from scratch, Rhodes developed a multidisciplinary curriculum that spans all years of student education. Recognizing that medicine, as a profession, rests on a set of underlying moral commitments, the ethics curriculum, which starts in the first weeks of school and continues throughout the four years, is designed to provide the student with the conceptual tools needed to navigate ethical issues commonly encountered in practice, to develop skills in critical reasoning, and to foster the habits of critical reflection and discussion about ethical issues. The ethics faculty makes use of every teaching modality including lectures, small-group discussions, case presentations, computer-assisted instruction (CAI), and writing assignments to nurture and reinforce the basic skills.

Under Rhodes's leadership, the ethics faculty now offers three graduate degree programs, faculty seminars in philosophy and medicine, centerwide ethics lunches, and a moral theory group. Local, regional, and international conferences are today a regular part of the ethics activities of the School, as are the many projects involving resident education, ethics committee participation, and consultation and research in the Hospital.

Implementation of the 1986 recommendations of the Ad Hoc Committee led to a sweeping restructuring of the preclinical years and a major effort, led by John Thornborough, Ph.D., and associates, in developing computer-assisted instructional materials in the basic sciences. Critical to all of the School's efforts, especially in the area of student technical support, has been the staff of the Levy Library, its Director, Lynn Kasner Morgan, and Merril Schindler, Director of the Media Resource Center. The library has stayed abreast of educational technologies, beginning with the provision of terminals dedicated to student self-testing and then the installation of personal computers in the Library's Media Center. In 1994, the Levy Library took over the task of implementing computer-assisted instruction programming for the School, hiring programmers to work with faculty to bring their subjects into the digital age.

In 1987, the School created an Honors Research Track for students who apply for and are able to complete an intensive, full-time, original-research effort of at least four months' duration, and usually longer, conducted during elective periods and separate from the requirements of any other courses. The students must submit a scientific paper suitable for publication or present the work at a national meeting. Students who fulfill all the requirements of the program earn the added honor of "Distinction in Research" on their diploma. In another bold step to stimulate and increase scholarship, Dean Kase, in 1988, instituted a requirement that students pass Part 1 of the licensing examination to be promoted to the third year and that fourth-year students pass Part 2 in order to graduate.

In 1994, the Office of Student Research was established under the direction of Karen Zier, Ph.D., to stimulate the interest of more students in research. The success of the program is measured by the more than sixty-five abstracts that are usually submitted and presented as either papers or posters during the annual Student Research Day. For example, members of the Class of 2003 were either the author or coauthor of almost one hundred papers published prior to their graduation.

The desire to create a comprehensive examination to determine clinical competency prior to graduation led to the creation of a program using standardized patients to assess clinical performance. Under the

Mark Swartz, M.D., is shown reviewing a clinical session in the Morchand Center for Clinical Competence, 1991.

leadership of Mark Swartz, M.D., and bolstered by a major gift from Marietta Morchand in honor of her family, the Morchand Center for Clinical Competence opened in 1991. A state-of-the-art facility, the Center, with its many patient stations and observation theater, has improved the history taking, physical examination, and communication skills of younger medical students. It has also enabled the development of a reproducible clinical assessment examination. With the knowledge that, in 2004, the USMLE added a clinical assessment examination to its requirements for licensure, the School is well positioned to have its students do well.

Swartz also went on to establish the New York City Consortium on Clinical Competence, a major collaboration among the medical schools of New York City to improve student education.[22] More than 1,200 fourth-year medical students per year from other schools are examined in the Center. The program has also been expanded to allow incoming house officers from all the hospitals in Mount Sinai's Graduate Education Consortium to be tested on their clinical assessment skills.[23] In 1998, Swartz became the first MSSM faculty member to receive the Robert J. Glaser Distinguished Teacher Award of the Association of American Medical Colleges.

In 1995, a self-study assessment, led by Alex Stagnaro-Green, M.D., then Dean of Medical Education, was begun in preparation for a 1996 site visit by the Liaison Committee on Medical Education (LCME). What emerged was the recognition that even top students, with an extensive fund of medical knowledge, were challenged, when asked to apply that knowledge at the bedside.[24] Lawrence Smith, M.D., Chair of the Executive Curriculum Committee, cochaired a New Curriculum Design Initiative and presented a plan for reform to the LCME. The School was awarded a seven-year accreditation, the maximum allowable at the time. The Accreditation Committee noted, however, that the School continued to have a very high number of formal teaching hours and depended heavily on lectures. Among the Committee's recommendations were a maximum of two lectures per day, an increase in small-group teaching, an increase in CAI materials, more free time, and discontinuation of printed syllabi wherever possible.

By the time that Arthur Rubenstein, M.D., arrived as Dean of the School, in 1997, curriculum redesign was under way. Over the ensuing years, administrative changes were numerous. Smith was named the new Dean for Medical Education, effective January 2001. Suzanne Rose, M.D., who had been responsible for third- and fourth-year students, was appointed Associate Dean for Medical Education, overseeing the clinical curriculum.

Implemented with the students who entered the School in 2000, "Curriculum 2000" incorporated many of the LCME recommendations. Courses were restructured to provide interdisciplinary teaching throughout all four years, and new, unique opportunities for research and in-depth scholarly activities were provided.[25] Interestingly, a new course, "The Art and Science of Medicine," incorporates facets of the former "Introduction to Medicine" course, including topics such as sociology, healthcare delivery, medical ethics, and medicolegal issues. The course takes students to multiple ambulatory settings, emergency rooms, nursing homes, and hospices and is presented throughout the preclinical years. Small-group sessions in the course are led by basic science and clinical faculty, members of the Division of Social Work of the Department of Community and Preventive Medicine, and members of the Department of Nursing. In a perfect example of the expression, "what goes around, comes around," the "Art and Science of Medicine" course incorporates all of the curricular principles espoused by Hans Popper almost forty years earlier. The new curriculum also includes a course called "Molecules and Cells," which encompasses Biochemistry, Molecular and Cell Biology, Genetics, and Cell Physiology. A novel six-week block entitled the "Integrative Core" at the end of the first year consists of an in-depth analysis of selected topics relating science to medicine, conducted in seminars. "Courses without Walls" were designed to implement integration of topics such as medical ethics, infectious diseases, medical informatics, human sexuality, communication skills, nutrition, and evidence-based medicine throughout the curriculum. In response to the greater reliance on the ambulatory setting for clinical teaching, students are assigned to Mount Sinai's new Visiting Doctor program, where they accompany the faculty and the nurse practitioners of the Department of Nursing on their home visits. In the fourth year, a block of time is set aside for USMLE preparation.

Technology continues to play an increasing role in the School's curriculum as computer-based programs and simulations are implemented. In 2001, the School began using a course management system

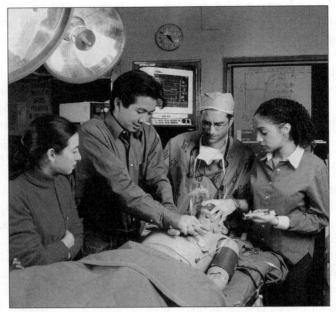

Adam Levine, M.D., is showing the medical students "Hal," a mannequin that simulates the body's response to anesthesia, 1999.

called WebCT to facilitate student access via the Web to course schedules, syllabi, practice questions, lecture notes, case simulations, slide reviews, journal articles, still images, and videos. Web-Ed, as Mount Sinai's program is known, supports online discussions and e-mail for each course, as well as calendars, online testing, evaluations, and grade distribution. The program is supported by the Levy Library, and faculty members are encouraged to work with Library staff to optimize their course material for the Web. Every course has a presence on Web-Ed, with different courses using it more intensely than others. For several courses, interactive programs in Flash or html have been created and embedded in Web-Ed. Because of its virtual nature, the students are able to access their course material from any site, at any time.

In August 2004, the Mount Sinai School of Medicine admitted its thirty-seventh class of first-year students. They will be the beneficiaries of a curriculum that not only utilizes the newest technology but also preserves the best of one-on-one and small-group teaching from a dedicated group of basic scientists and compassionate, caring physicians.

3

The Graduate School of Biological Sciences

Terry Ann Krulwich, Ph.D.*

THE GRADUATE SCHOOL of Biological Sciences currently encompasses the Ph.D., the M.D./Ph.D. (the Medical Scientist Training Program [MSTP]), the Summer Undergraduate Research Programs (SURP), and the Post-Baccalaureate Research Education Program (PREP). In the beginning, however, it was a far more modest undertaking, but one with lofty goals that made it essential to the new Mount Sinai School of Medicine.

THE PH.D. PROGRAM

One of the goals in the establishment of the Mount Sinai School of Medicine (MSSM) was to position the institution to maintain the tradition of important, innovative research that had been a hallmark of The Mount Sinai Hospital during the years when clinical, observational, and pathological studies had been prime. Following World War II, more and more medical research emanated from university and medical

---

*Terry Ann Krulwich, Ph.D., is the Sharon and Frederick A. Klingenstein-Nathan G. Kase Professor in the Department of Pharmacology and Biological Sciences. She was Executive Officer of the Ph.D. Program and the founding Director of the M.D./Ph.D. Program (the Medical Scientist Training Program, MSTP) from 1973 to 1999 and Dean of the Graduate School of Biological Sciences from 1981 to 2002. See also chapter 4, on the basic sciences.

school laboratories. Biology was beginning to move into the molecular age. By the late 1950s, it became apparent that a stand-alone hospital setting could no longer sustain the level of research activity required to nourish and to build upon this heritage. In this new era, even a hospital with a remarkable history of contributions in biomedicine now required a critical mass of basic scientists, along with their predoctoral graduate students and postdoctoral fellows, to maintain its cutting-edge capabilities. It was also evident that basic scientists were indispensable to the education of medical students. In turn, their attraction to Mount Sinai and the ultimate quality of their research activities in newly emerging areas of science would depend upon their opportunity to collaborate with research-oriented medical students, residents, fellows, and clinician-scientists. This symbiotic relationship would facilitate the identification of problems of biomedical importance and maximize the likelihood that novel findings would be translated into applications for either disease prevention or therapeutic strategies. It was also clear that the new generation of clinician-scientists wanted the benefit of having excellent predoctoral students in their laboratories, as well as postgraduate M.D. and Ph.D. fellows. Recognizing that it would benefit all segments of the faculty, a Graduate School program, offering basic science training toward the Ph.D. degree, was designed side by side with the planning of the new School of Medicine. It would be a key component of the new School.[1]

And so, with the granting of the charter by New York State in 1963, the Graduate School of Biological Sciences was established within the School of Medicine. In 1965, Irving Schwartz, M.D., a graduate of the New York University School of Medicine (NYU), was recruited from the University of Cincinnati, where he served as the Chairman of the Department of Physiology, to be the first Dean of the Graduate School of the Biological Sciences. He was also appointed the founding Chairman of MSSM Department of Physiology.[2] In an interview shortly after his arrival, Schwartz was enthusiastic about the challenge of developing a new school:

> I see the graduate school as a key element in creating a top flight medical center and educational institution. . . . The climate created by the presence of outstanding graduate students helps to attract a top-flight faculty of scholars and scientists who need this source of stimulation. In addition, a strong reputation as a research and graduate education

center brings visitors from all over the world, and it breeds collaborative projects with other excellent institutions. All this escalates the
vitality and the quality of the intellectual environment—it creates a
gestalt that adds to the research potential and the teaching capacity of
the entire institution at all levels.[3]

Schwartz would later go on to state, "My colleagues and I believe that
the best way to develop the atmosphere of a graduate school within
a medical school is through the juxtaposition of two such schools
physically, operationally and philosophically. . . . We hope to achieve
vital interdisciplinary interactions among clinicians, basic scientists,
medical students and graduate students within one institution."[4] As
Dean of the Graduate School and the Chairman of the Department of
Physiology, Schwartz played a vital role in the formulation of the

Irving L. Schwartz, M.D.,
the first Dean of the Mount
Sinai Graduate School of
Biological Sciences and
Professor and Chairman of
Physiology.

curriculum for the Graduate School and also of the basic science curriculum in the medical school.

Mount Sinai's affiliation agreement with The City University of New York (CUNY) affected primarily the Graduate School. Originally named the Biomedical Sciences Doctoral Program, the Ph.D. program was administered by Mount Sinai's own Graduate School leadership on the Mount Sinai campus in accordance with the rules governing all doctoral programs of the CUNY Graduate School. Regardless of their areas of specific concentration, students completing their studies received a CUNY Ph.D. degree in Biomedical Sciences at the CUNY graduation. The Executive Officer of Mount Sinai's Doctoral Program interacted with the CUNY Graduate School leadership, including President Harold Proshansky and, later, President Frances Degan Horowitz. The Executive Officer also worked closely with the admissions, registrar, bursar, and foreign students' offices, as well as with the curriculum committee of the CUNY Graduate Center, thereby ensuring coordination and approval of the developing program on the Mount Sinai campus. Mount Sinai's first Executive Officer was Professor Harold Burlington, Ph.D., of the Department of Physiology. The Doctoral Program faculty, now known as the Graduate School faculty, is a subset of the School of Medicine faculty. Nomination to the Doctoral Faculty was based on the leadership of an active, cutting-edge, funded research program, as well as on a commitment to the efforts required to support the education of graduate students.

The program established in 1968 was similar to other medical school-based Ph.D. programs of the time: a confederation organized around the five basic science departments of Anatomy, Biochemistry, Microbiology, Pharmacology, and Physiology. This predoctoral research training model mirrored the way in which the basic sciences were taught to medical students.

Most of the Ph.D. candidates in the initial group were students who had chosen a faculty mentor at another institution and then came to the new Graduate School at Mount Sinai after their mentor was recruited to be part of Mount Sinai's original basic science faculty. The first seven Ph.D. students were distributed among the five basic science departments. With the faculty fully engaged in designing the medical school curriculum, it was not possible to create a separate curriculum for the graduate students. As a consequence, the graduate students took part in the medical student courses taught by their parent depart-

Terry Krulwich, Ph.D., second Dean of the Mount Sinai Graduate School of Biological Sciences and Professor of Biochemistry.

ment. It was uncommon to participate in introductory or advanced courses offered by other departments. Fortunately, however, it was possible to develop a small but excellent selection of more advanced courses that were appropriate for Ph.D. training; medical students were encouraged to take these courses as electives. In addition, most Ph.D. students took at least one advanced course on the CUNY Graduate School campus.

During the early years of the Graduate School, the day-to-day paperwork and support of admissions and registrar functions on the Mount Sinai campus were handled by Senta Frank and James Fuss. Mrs. Frank remained associated with the Graduate School for many years and is still fondly remembered by early alumni.

In 1970, Terry Ann Krulwich, Ph.D., joined the Medical School faculty as an Assistant Professor in the Department of Biochemistry. A New Yorker, she received her Ph.D. in Bacteriology from the University of Wisconsin. This was followed by a postdoctoral fellowship in molecular biology at the Albert Einstein College of Medicine. Shortly after joining the Mount Sinai faculty, she was invited to join the Graduate

School faculty. A Ph.D. candidate joined Krulwich's laboratory as soon as it was up and running. As a result of her interactions with this student, Krulwich became involved in curriculum development for the Ph.D. program, and in 1974 she became the second Executive Officer of the Biomedical Sciences Doctoral program and also assumed much of the responsibility for leading the Ph.D. program.

Between 1968 and 1973, five to ten graduate students matriculated annually. Early on, the Graduate School was at a considerable disadvantage competitively because it did not have the resources to offer financial packages comparable to those provided at other institutions. It had long been customary for the better Ph.D. science programs to support their students by waiving tuition and by offering a stipend to enable students to defray living costs. It was believed that the absence of such support would discourage the best and the brightest from entering the biological sciences and hurt America's competitive edge in this arena. In the 1960s, many programs typically offered a stipend of $3,000 to $3,500 a year, the amount then in effect for support of fellows funded by the National Institutes of Health (NIH) and/or the National Science Foundation (NSF). By contrast, the stipend offered by Mount Sinai during its first two years of operation was variable and was negotiated by the department chairs and the individual mentors but did not exceed $900 a year. Fortunately, in 1970, the stipend rose to $3,400, enhancing the attractiveness of the program. By the early 1980s, this amount had increased to $5,000 a year, and within a decade it had reached $15,000 a year.

Before selecting a Ph.D. mentor, graduate students generally receive support from fellowships provided by either the School or generous foundations. After the choice of a preceptor has been made, the mentor's research grants also help to provide support. As the Graduate School programs became stronger, several Departments were able to obtain training grants in support of their students. Human Genetics (Robert Desnick, M.D., Ph.D., Program Director), Cell and Molecular Biology (Francesco Ramirez, Ph.D., Program Director) and Neurosciences (Bernard Cohen, M.D., Program Director) were among the early successful competitors for these prestigious awards.

Over the years, the training, curriculum, and course organization for the Graduate School, as well as the organization of the basic science curriculum of the Medical School, were transformed into more integrated models. These current models reflect how contemporary scientific research is actually conceived and conducted; they also identify the importance of interdisciplinary collaboration among the biomedical sciences in both the learning process and the conduct of research.

The median program length for students who entered the Graduate School during its first decade was between 4.2 and 4.9 years. Graduates from this early period gravitated to careers in diverse academic and research institute settings or positions in pharmaceutical (and later biotechnology) firms.

During the tenure (1973–1983) of Thomas Chalmers, M.D., as President and Dean of the Medical School, a Department of Biomathematics was instituted. As a result, the Graduate School's training of biomathematicians predated a trend that has now become a strong focus of the contemporary biological sciences. A biostatistics course conceived and offered by the Biomathematics faculty became a CUNY-wide requirement.

During the 1970s, the applicant pool for the Ph.D. program expanded steadily, in part fueled by the initiation of special scientific seminars and by the dissemination of information about the program to the large number of undergraduates who found their way into Mount Sinai's laboratories to participate in summer research projects. By the early 1980s, the size of a typical entering group of Ph.D. candidates had increased to more than a dozen. The curriculum was greatly enriched by new majors in Human Genetics and Pathobiology. In addition, improved options for the introductory course work began to appear, especially an advanced Biochemistry track that replaced the traditional Medical School biochemistry course for the Ph.D. cohort. In fact, the growth of the doctoral programs created an increased level of beneficial interplay between the Ph.D. students and the medical students. When working side by side with the basic researchers, the medical students appeared to become more productive and to have greater satisfaction in their work.

In 1981, Krulwich was appointed Dean of the Graduate School, succeeding Irving Schwartz. The latter remained productive in science as the Director of the Mount Sinai Center for Polypeptide Research until his retirement in 1989. Schwartz continued to actively support the functions of the Graduate School and, through his worldwide network of scientific colleagues, was responsible for inviting many special speakers and distinguished guest scientists to interact with faculty and students.

During the twelve-year tenure (1985–1997) of Nathan Kase, M.D., as Dean, the Mount Sinai School of Medicine enjoyed a substantial expansion of its basic research effort, focused on establishing multidisciplinary Centers of Excellence, with particular growth in Neuroscience and Molecular Biology, as well as new programs in Cell Biology. These disciplines created major areas of training for graduate students. Over the years, many applicants to the Ph.D. program expressed an initial interest in several potential majors and wanted to explore available options before entering the specific laboratory where they would conduct their thesis project. As a consequence, the option of flexible entry into the Graduate School without a commitment to a particular major was extended to entering Ph.D. students. This option had long been available in the M.D./Ph.D. program (see later discussion). Currently, students are required to rotate through at least two laboratories before choosing a faculty mentor for their dissertation project.

As expected, with the growth of the basic science and clinician-scientist faculty, there was a corresponding increase in the number of graduate students. As of 2002, there were 147 students enrolled in the Ph.D. program, and twenty-four to twenty-five new students were matriculating annually. Potential applicants' first introduction to the Graduate School is now most often through the School's Web pages.[5]

A major development for the Graduate School during the 1990s was the introduction of a mandatory core curriculum for all graduate students. Designed to ensure that every student mastered the fundamental concepts of the contemporary molecular-based sciences at the onset of their careers, major courses were created in Cell Biology and Molecular Biology. Either General Human Biochemistry or Advanced Human Biochemistry was also required, placing a heavy course burden on the incoming students. Francesco Ramirez, Ph.D., Mary Rifkin, Ph.D., Gianni Piperno, Ph.D., and Arthur Cederbaum, Ph.D., of the basic science faculty, played critical roles in the creation and success of these courses. The bar was set high for students; they were also required to successfully complete an advanced course in their area of particular interest. In addition, mean final course grades could not be below B. The core curriculum requirement remains today an initial academic hurdle, providing an early assessment of the academic readiness of the beginning student.

Throughout his or her graduate school career, each student was guided by a three-member advisory committee that provided counseling in arranging laboratory sequences and assistance with initial course work problems that might arise. After the choice of a research preceptor was made (the faculty member who would supervise the dissertation work), that mentor became the head of the student's Advisory Committee. Committee membership was assigned so as to maximize the student's research potential. All students were required to meet with their Committee semiannually and received written reports from the Committee following each meeting.

The explosion of new research fields and technical advances and revolutionary changes in the exchange and analysis of information in the past decade mandated a reconsideration of how the Ph.D. programs were configured. With the addition of an Immunobiology major in the early 1990s, there were eleven major tracks within the Graduate School program, many of which had student and faculty cohorts that were too small to sustain a full range of seminars, journal clubs, and student works-in-progress. The potential for new programs was also growing rapidly, as the School of Medicine added greater strength in the areas of cancer biology, gene therapy, structural biology, computational biology and bioinformatics, and behavioral and cognitive neurosciences. Ongoing programs in genetics/genomics, developmental biology, cell signaling, rational drug design, virology, and host-pathogen interactions also thrived.

It was also becoming apparent that students were moving into narrow major areas of study after completion of the core curriculum and missing out on the multidisciplinary components that were representative of the School as a whole. The old disciplinary boundaries were administratively useful but were no longer optimal for education in the new research environment.

As a consequence, a faculty committee was convened in the early 1990s to examine graduate school programs across the country that had adopted a cross-departmental, multidisciplinary training model for Ph.D. study. The intent was to design a model that would expose the student to both interdepartmental clusters of faculty that spoke common scientific languages and to smaller subgroups that would provide intensive exposure and rigorous mastery of specialized subject areas. This resulted in the Multidisciplinary Training Area (MTA)/Flexible Entry model that was developed during 1995–1998. It is the current organizational model for both Ph.D. and M.D./Ph.D. research education in the Graduate School. This plan established six overlapping MTAs, allowing for either early or later choice of a specific research area and mentor and encouraging early completion of the core curriculum. The six MTAs include Biophysics, Structural Biology and Biomathematics, Genetics and Genomic Sciences, Molecular, Cellular Biochemical, and Developmental Sciences, Mechanisms of Disease and Therapy, Microbiology, and Neurosciences. Each MTA, codirected by two faculty members, supervises the Advisory Committees within the area and organizes the examination committees that are responsible for the format of the two-part exercise necessary for advancement to candidacy toward the Ph.D. degree. The first part is a qualifying examination that examines the ability of the candidate to synthesize and to use the broad knowledge obtained in their general area of concentration. The second part of the exercise is the creation by the student of an NIH-style proposal with respect to the research planned for the dissertation. That proposal is submitted to and defended before a faculty committee. The MTA also organizes the final public research seminar and dissertation defense of each student.

The rapid accretion of knowledge in the biomedical sciences and the adoption of the MTA model have led to a continuous reassessment of the core curriculum. New specialized topics have been introduced, while overall class time has been trimmed. Paul Wassarman, Ph.D., Chair of the Department of Cell, Molecular, and Developmental Biology; Jeanne Hirsch, Ph.D., and Diomedes Logothetis, Ph.D., have become Directors of the curriculum cores. An Introduction to Journal Club has been initiated in which students are able to enhance both their presentation and their critical analysis skills. As part of this effort, students prepare manuscripts and research proposals and participate in journal clubs and seminars throughout their programs.

Many opportunities are provided for graduate students to cultivate their teaching skills and to enjoy the rewards of teaching. As a result, numerous senior graduate students mentor junior students in research and serve as teaching assistants in many of the Medical School and Graduate School courses. In 2002, a series of four teaching workshops, led by Kenneth Baine, Ph.D., Director of the Center for

Excellence in Teaching of New York University (NYU), was offered for credit. This well-received series provides the more than twenty-five participating students with provocative and helpful exposures to contemporary thinking about the nature of learning and the actual useful practices that emanate from those insights.

Over the years, the expanding Graduate School activities have required increasing administrative support. During the 1980s, this was admirably provided by Mark Shepard and Gloria Gibson. In 1990, Gita Bosch assumed the position of Graduate School Administrator, and as Associate Dean she helped to shape the administrative structure underpinning the complex array of programs and courses available.

During the 1980s, both the scientific community and the public became increasingly sensitive to issues of responsible conduct of research (RCR), namely research ethics. Mount Sinai's Graduate School introduced a program of research ethics training before it became a requirement of all NIH-funded training programs. With the goal of increasing the students' ability to recognize and resolve ethical issues that arise during research, a series of required RCR seminars was instituted, utilizing role-playing exercises as a vehicle to focus on the issues. Paul Friedman, M.D., of the University of California at San Diego School of Medicine, was an invaluable consultant as the program got under way.[6] The success of the program led to its adoption by many other schools and has been widely displayed on the Internet. The course includes the following topics: fraud, fabrication and plagiarism; problems of self-delusion; the need for highly honed experimental and critical skills to avoid missteps; record keeping and data retention; mentoring issues; authorship issues; conflicts of interest; grants, fellowships, and related ethical issues; use of animals in research; use of human subjects in research; technology transfer; and where and how to obtain help in sensitive situations as members of an academic community. Since its inception, participation in the RCR course has raised the consciousness of both faculty and students regarding these issues. The RCR curriculum also includes sessions specific to each major training area for that group's students.

For more than a decade, the Graduate School has conducted an annual retreat, including a mix of academic presentations, visits by alumni to talk about their careers and career paths, recreational activities, and break-out groups to discuss Graduate School curricular issues. These important community-building events provide a day-long informal break from the laboratory and classroom. Most of the retreats also include role-playing exercises relating to training and ethical issues and require student and faculty interaction.

In recent years, other activities have been created to enhance student life in the school. An Orientation Committee offers a comprehensive opportunity for incoming students to bond as a class, develop relationships with advanced student mentors, and adjust to their new setting. During the 1980s, the Graduate School initiated its first awards, namely two Dissertation Awards given annually to graduating Ph.D. or M.D./Ph.D. students. More recently, annual academic awards from each MTA and service awards from the Graduate School have been created. Major efforts have been made to provide information to students about career development. These efforts include a symposium presented as part of the initial student orientation, support of a student Career Development Association, workshops and alumni encounters at retreats, and cosponsorship of a large, multi-institutional Career Development workshop that is a day-long event at one of the New York City medical school campuses each spring. Since 1999, the Mount Sinai Alumni has funded a Graduate School Travel Fund that partially defrays expenses for numerous students to participate in important scientific meetings and special courses. Graduate students participate together with medical students in social activities and community service activities.

In 1999, the Mount Sinai School of Medicine changed its university affiliation from The City University of New York to New York University as part of a larger agreement in which some elements of the NYU and Mount Sinai healthcare efforts were consolidated. Mount Sinai's School of Medicine remains a distinct School from the NYU School of Medicine, and Mount Sinai's Graduate School of Biological Sciences continues to administer its graduate programs independently on its own campus. All graduate degrees are now awarded at a joint commencement exercise held with the Mount Sinai School of Medicine. The change in affiliation has encouraged the further augmentation of joint programs already in effect with NYU. These include a mathematics program in conjunction with the Courant Institute, genetics and neurosciences collaborations with the NYU Washington Square campus faculty, and a joint virus-host training program led by Peter Palese, Ph.D., Chair of the Mount Sinai Department of Microbiology, which includes the Mount Sinai School of Medicine, the NYU School of Medicine, and the Biology Department of the NYU campus. Despite the change in university affiliation, several productive and long-standing interactions with CUNY campuses and programs have been retained. Ph.D. students also continue to be part of a citywide consortium and, with appropriate approval, may take a graduate course of special interest on campuses throughout the area without extra tuition cost.

The 147 Ph.D. students enrolled in 2002–2003 will complete their Ph.D. programs in just over five years. They have a wider array of career choices than ever before. The vast majority will continue to pursue postdoctoral research training; some will gravitate to postdoctoral programs in industrial research venues or in settings focused on the newly emerging threat of bioterrorism.

In the last analysis, a school is measured by the success of its students. Alumni of the Mount Sinai Ph.D. program are highly successful in positions that reflect both the broad perspectives and the intensive expertise they attained through their graduate programs. The majority of the graduates have gone on to careers in academic and research institute settings or positions in pharmaceutical research. They have also taken advantage of a multitude of other settings in which their talents can be deployed. These include, for example, positions in science administration, science writing, and business/law applications of science, as well as in the new cutting edge efforts in biotechnology. Many occupy prestigious academic positions in educational and research institutions. A sizable number of the graduates have been chosen to join Mount Sinai's faculty, bringing added distinction to the Graduate School.

Many international students have remained in the United States; others have returned to their home countries and bring honor from afar. The Graduate School benefits from the return visits of many alumni who meet with current students in seminars and in career development and retreat sessions.

THE MEDICAL SCIENTIST TRAINING PROGRAM (MSTP)

From the opening of the Medical School, one of the goals was to eventually train highly qualified students as both clinicians and research scientists, awarding them both the M.D. and the Ph.D. degrees. At about the time that Krulwich began her efforts in the Graduate School in 1971, Mark Sobel, then a talented medical student, was discovering the attraction of biomedical research. He was exactly the type of student for whom an M.D./Ph.D. program would be a natural choice, but Mount Sinai did not yet have one. The needs of Sobel, and then of other medical students, led Krulwich to obtain approval to initiate M.D./Ph.D. training. Sobel became the first of a pioneering group of students to receive both degrees.

During the 1960s, a significant number of undergraduates who were talented in "hard science" and who exhibited early proficiency in research settings chose to pursue careers in medicine, rather than enter Ph.D. programs. This nationwide trend led the National Institute of General Medical Sciences of the NIH to initiate the Medical Scientist Training Program (MSTP) grants to support the tuition costs and stipends of M.D./Ph.D. candidates through both degree programs. It was reasoned that providing financial assistance during M.D./Ph.D. training would attract exceptional students to careers in either basic biomedically oriented science or in the vital clinical-basic science research interface. However, the M.D./Ph.D. student willing to devote the additional years required to become skilled in both scientific research and clinical medicine would most likely fall behind his or her peers in potential future earnings and/or accumulate heavy loan burdens that would discourage career paths in basic research. The MSTP program would help alleviate this problem.

There is no doubt that the prestigious Medical Scientist Training Program enriches those schools fortunate enough to obtain these highly competitive entities. The program attracts outstanding students who form an academically strong and influential research-oriented component in their medical school classes. They are also a uniquely gifted component of the Ph.D. program cohorts, forming bridges and collaborations between clinical and basic scientists. Nationwide, MTSP graduates have a significantly higher success rate in obtaining and sustaining nationally funded research programs than typical medical school graduates.[7] Mount Sinai's own MSTP graduates are no exception, and they have assumed diverse roles at the forefront of academic medicine as basic biomedical scientists, clinical researchers, and leaders of multifaceted research programs. For all these reasons, the expansion of this highly successful program was one of several approaches proposed at the national level to address a crisis-level shortage of well-trained physician-scientists.[8,9]

In 1977, the Mount Sinai M.D./Ph.D. program, then one of about two dozen extant, received an MSTP training grant. That pivotal milestone not only provided crucial financial support for an elite and expensive program but also was a compelling endorsement of the Graduate School's programs that are the foundation of the scientific training of the MSTP student. Moreover, it was an affirmation of the success of the early efforts at Mount Sinai to create an integrated dual degree track for outstanding students.

Mount Sinai's MSTP grant support has grown over the years and has been uninterrupted. Krulwich served as the Program Director of Mount Sinai's MSTP from the beginning of the program until 1999. Once the Training Grant support was secured, applications to the MSTP began to increase rapidly. Like most schools with MST programs, Mount Sinai includes all of its M.D./Ph.D. students in the Program, even though the grant supports the stipend and tuition costs of fewer than one-third of the students at any given time. Most MSTP students traditionally enter the first year of the combined Program. Additional qualified students may enter following matriculation in either the M.D. or the Ph.D. program, after they discover that they want to add an additional dimension to their career goals.

At Mount Sinai, students enter the MSTP without a formal commitment to any Graduate School training area. Like their Ph.D. counterparts, they conduct laboratory rotation sequences that either cross department lines or are highly focused. During their first two years, they complete the basic science requirements and the introductory work in physical diagnosis in the Medical School and also complete most of the core requirements of the Ph.D. curriculum. They also choose a research mentor.

Although there is some variability and flexibility of the timing and ordering of program elements, most MSTP students complete their advanced Ph.D. course work and conduct their dissertation project before entering the third and fourth years of the Medical School curriculum. They receive intensive guidance from both their Ph.D. advisory committees and liaison advisers of the MSTP Steering Committee.

The MSTP has unique program elements to constantly encourage students and to reinforce the diversity of the academic enterprise. These include special seminars and dinner meetings with physician-scientist role models, many of whom are MSTP graduates themselves. There are clinical continuity and refresher programs for students who are conducting Ph.D. work and who need to improve or augment their clinical skills in preparation for entry into clerkships. These elements were greatly enhanced after Mark Taubman, M.D., assumed the MSTP Directorship in 1999. Taubman, a cardiologist and Director of Vascular Biology in the Cardiovascular Institute, initiated an annual MSTP Retreat. Guidance and enrichment in the clinically related aspects of the Program were strengthened, and growth of the program was accelerated. Critical to the success of the M.D./Ph.D. program has been the retention of the historically close relationship with the larger Ph.D. program. Following the appointment of Lawrence Smith, M.D., as Dean for Medical Education in the Medical School, in 2001, improved options for scheduling of the Medical School requirements for MSTP students were put in place.

In addition to their involvement with the Director of the MSTP, students work with Associate and Assistant MSTP Directors who, along with Krulwich, have most recently included Robert Desnick, M.D., Ph.D.; Margaret Baron, M.D., Ph.D.; Edward Fisher, M.D., Ph.D.; Jonathan Licht, M.D.; and Anne Moscona, M.D.

Like its Ph.D. counterpart in the Graduate School, the MSTP has flourished. In the academic year 2002–2003, sixty-five students were enrolled. The ninety-two MSTP alumni have gone on to careers in basic biomedical research and varied medical specialties in outstanding academic settings. They have brought added luster to this most important component of the Mount Sinai School of Medicine.

THE SUMMER UNDERGRADUATE RESEARCH PROGRAM (SURP)

Undergraduate students from many universities have been welcomed into Mount Sinai's research laboratories since the opening of the School. They have also played vital roles in many clinical research programs. Originally a catch-as-catch-can or first-come, first-served arrangement, the program was formalized, expanded, and reorganized during the 1980s with the intent of heightening the Graduate School's profile among research-oriented undergraduates. A formal application and admission process was introduced, and the program was advertised among undergraduate schools.

Named the Summer Undergraduate Research Program (SURP), its core is an intense ten-week period of research. Each SURP fellow chooses a research mentor from a carefully selected group of faculty who volunteer to participate, and then the student carries out a short-term project in the mentor's laboratory. The fellows are provided with a summer stipend and may choose to live in Mount Sinai housing during the fellowship period.

Program Co-Directors are named annually, and the activities of the program have been expanded to also include a broader range of social/recreational pursuits, facilitating a greater interaction of the SURP fellows with the graduate students. The SURP program culminates in a poster session in which each SURP fellow presents a poster on his or her work. This successful construct attracts more than three hundred applicants for eighteen to twenty-five places each summer.

The SURP has been and continues to be a valuable source of excellent applicants to the Ph.D. and MST Programs. It is also an important instrument that allows the Graduate School to reach out to underrepresented minority undergraduate students. It has provided a meaningful number of these students with the opportunity to work in a cutting-edge research environment focusing on biomedical problems. Many minority students have been recruited either to Mount Sinai's graduate programs or to other graduate programs. All of these students credit their summer experience as SURP fellows with having played a crucial role in their future success.

THE POST-BACCALAUREATE RESEARCH AND EDUCATION PROGRAM (PREP)

The Graduate School has long been committed to increasing the number of underrepresented minority scientists in biomedicine. In 2001, the School competed for and was awarded one of eight NIH training awards for a period of postbaccalaureate research education. The PREP scholars must be recent college graduates who exhibit motivation for basic research and who must plan to enter a Ph.D. program. They are matched with a research laboratory in which they spend 75 percent or more of their time on an intensive research experience over a one- to two-year period. That experience is usually supplemented with initial graduate course work, skills development activities, and exposure to role models and areas of science/health policy that are of special relevance. PREP provides Mount Sinai with another opportunity to ameliorate a shortage of minority biomedical scientists. The initial experience with the program has proved exciting. In 2002, there were six PREP scholars in the program. Two alumnae of the first entering group were matriculated in the Mount Sinai Ph.D. program, one of whom had already been successful in securing an independent NIH fellowship in support of her work.

At the end of 2002, Terry Krulwich announced her intention to step down from her administrative position as Dean to concentrate on her laboratory research and teaching. For the first thirty-seven years, leadership of the Graduate School had been in the hands of only two individuals—Irving Schwartz and Terry Krulwich. During their combined tenure through 2002, 378 students received Ph.D. degrees and 92 students graduated with M.D./Ph.D. degrees, majoring in sixteen different scientific disciplines (see table 3.1).

In January 2003, Diomedes Logothetis, Ph.D., Professor of Biophysics and Physiology, assumed the Deanship of the Graduate School of Biological Sciences and was also named the Director of the M.D./Ph.D. program. As the leader of more than 200 graduate school faculty and more than 200 graduate students, Logothetis, an outstanding researcher, teacher, and mentor, has been given the unparalleled opportunity and challenge to further broaden, strengthen, and elevate the programs of Mount Sinai's Graduate School of Biological Sciences.

Table 3.1.
Graduate School Specialization

Degree Area of Specialization	Ph.D.	MSTP
Biochemistry and Molecular Biology	50	13
Biomathematical Sciences	3	5
Biophysics, Structural Biology, Biomathematics	5	2
Cell Biology/Anatomy	16	3
Cell and Molecular Pathology	8	1
Genetics and Genomic Sciences	8	0
Human Genetics	31	3
Immunobiology	1	0
Molecular Biology	18	5
Molecular, Cellular, Biochemical Development	23	5
Mechanisms of Disease and Therapy	20	4
Microbiology	57	14
Mount Sinai Microbiology	6	1
Neurosciences	40	19
Physiology/Biophysics	41	7
Pharmacology	51	10
TOTAL	378	92

PART II

4

The Basic Sciences

INTRODUCTION

LONG BEFORE THE existence of the Medical School, Mount Sinai had an overriding interest in basic science. The Hospital leadership demanded that house officers be well versed in the basic sciences. When admission to the House Staff was based on a lengthy oral examination, basic science questions were routinely asked of the applicants, and examiners duly noted when the answers were not up to the required level. In a 1934 report to the Medical Board, the Chairman of the Committee on Examinations commented: "It has been the custom, originally decided by the Medical Board, to examine candidates in gross and topographical anatomy. Instruction in anatomy at the various schools has become so superficial that it hardly pays to continue to examine candidates in this subject. It is unfortunate that this should be the case because anatomy will always be one of the basic subjects in medicine."[1] The founders of the Mount Sinai School of Medicine (MSSM) were not about to allow basic science instruction to be short-changed.

When the School of Medicine opened its doors to admit the first students in 1968, both the medical students and the graduate students were greeted with five conventional basic science departments: Anatomy, Physiology, Biochemistry, Microbiology, and Pharmacology. Because the faculty had not as yet had time to develop separate courses for the graduate students, both student groups attended the same classes during their first two years. In 1968, the combined faculty of these departments numbered thirty-six. By the end of 2002, the basic science faculty numbered close to three hundred. Major organizational modifications occurred as a result of changes in leadership and personnel, the evolution of molecular biology, and the changing needs of individual faculty members and, to some degree, because of financial considerations. Consequently, departmental names were changed, the Departments of Pharmacology and Biochemistry merged, and additional centers of research excellence were added. These include the Brookdale Department of Molecular, Cell and Developmental Biology, the Department of Biomathematics, the Carl C. Icahn Institute for Gene Therapy and Molecular Medicine, the Dr. Arthur M. Fishberg Institute for Neurobiology and the Kastor Neurobiology of Aging Laboratories, the Institute for Immunobiology, and the Derald C. Ruttenberg Cancer Center. The Department of Microbiology was the only one of the original departments that entered the twenty-first century with exactly the same name that it had thirty-five years earlier. The Department of Human Genetics spans both School and Hospital and is discussed elsewhere,[2] as is the Department of Biomathematics.[3]

This chapter traces the evolution of the original basic science departments in the School and focuses on the teaching programs, the research themes of the departments, and the projects of many of the individual faculty members. Because of their interests and their collaborations, many of the basic science faculty have joint appointments, either in two basic science departments, a department and a center, or a secondary appointment in a clinical department.

Teaching has always been a part of a basic science faculty member's obligation to the School, but scientific productivity is also expected.[4] Research is a critical component of every basic science department's activities, and, in fact, in terms of academic promotion for basic scientists, scientific productivity has unquestionably been of greater importance than teaching ability.

One of the marks of excellence of a school is the degree to which faculty are able to secure extramural funding for their projects. MSSM has had consistent success in this regard. In the fiscal year ending September 2002, the School, competing with 125 medical schools, had obtained $142 million in competitive funding from the National Institutes of Health (NIH) and was ranked number twenty-two in the nation. But even before the Medical School was created, the staff of The Mount Sinai Hospital did well in securing research funding. As noted in *This House of Noble Deeds: The Mount Sinai Hospital, 1852–2002*: "In the 1960s, The Mount Sinai Hospital was ranked twenty-seventh in the country for receiving research funds from the federal government, a tremendous feat for a stand-alone hospital."[5] At that time, the Hospital was competing with more than eighty medical schools.

THE DEPARTMENT OF ANATOMY

THE DEPARTMENT OF CELL BIOLOGY AND ANATOMY

THE CENTER FOR ANATOMY AND FUNCTIONAL MORPHOLOGY

Tibor Barka, M.D., a "founding father" in the field of histochemistry, created the School's Department of Anatomy. Born and trained in Hungary, Barka was the first in that country to use radioactive isotopes in biological research. Forced to flee his native land in 1956, he emigrated to Sweden and joined the staff of the Karolinska Institute. In 1958, Barka was invited by Hans Popper, M.D., Ph.D., to join the Hospital's Department of Pathology, and, in 1966, he received one of the School's first faculty appointments as Professor of Pathology. The following year he was named Chairman of the Department of Anatomy and was given the tasks of developing the curriculum and of recruiting the faculty for the new department.

On Monday, September 9, 1968, the first day of classes for the thirty-six first year students in the new School, Barka and his five faculty colleagues began instruction in anatomy. The course, which extended through all three semesters of Year One, consisted of traditional lectures and laboratory dissection. Max Levitan, Ph.D., assumed responsibility for the anatomical dissection laboratory. The faculty also took on the responsibility for teaching embryology and histology. The three courses continue today as requirements for first-year students.

Barka published the first texts on histochemistry in Hungary, and later his book on the subject, written with Paul Anderson, M.D., became the classic reference work in the field.[6] Also in collaboration with Anderson, Barka did work in the development of histochemical techniques that led to one of the most frequently cited publications of the period from 1960 to 1980.[7] He was also the first to report on the effect of

Tibor Barka, M.D., the first
Professor and Chairman of
the Department of
Anatomy, 1967.

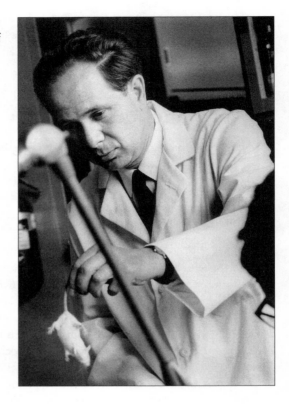

isoproterenol on induced cell proliferation.[8] Barka has served on the editorial boards of several journals in his field, including eight years as Editor-in-Chief of the *Journal of Histochemistry and Cytochemistry*.

In 1971, Levitan, a geneticist, published the *Textbook of Human Genetics* in association with Ashley Montagu.[9] The volume became a standard text and went through three editions. Levitan, one of the School's most popular teachers, continues on the faculty to this day. Levitan has studied the genetics of the *Drosophila* for more than four decades.[10] The primary focus of the research concerns the factors that determine the arrangement of genetic materials on the chromosome and the effect of chromosomal aberrations on development. Particular emphasis is on following the evidence that certain combinations of linked gene arrangements confer selective advantage on their carriers. He and his colleagues are also studying the molecular basis of a chromosome breakage factor and are investigating the suppressors of wing mutations in natural populations of *Drosophila*.[11-15]

The first faculty in Anatomy also included Joyce Shriver, Ph.D., who went on to become the Associate Dean for Student Affairs for the Pre-Clinical Years and who received the first Excellence in Teaching Award given by the students in 1973. She spent innumerable hours on various facets of the neurosciences curriculum over the years and in the 1980s designed the "Student Well-Being" course for entering first-year students.[16]

In 1969, Barka noted the diverse functions of the Department: "Anatomy ranges from the molecular to the macroscopic level and includes cell biology, developmental biology, and embryology, microscopic anatomy, neuroanatomy and neurobiology, and gross anatomy with its ramifications. The underlying, unifying concept of the discipline is the existence of a developmental hierarchy of structural organization."[17] In the same year, the Department also offered a course in anatomy to graduate students from The City University of New York.

Over the ensuing years, many of the Department's research activities concerned analyses of the mechanisms of cellular differentiation and organ development and growth. The organs studied included the salivary glands, testes, colon, brain, and muscle.

In the 1980s, working with the salivary glands, Barka and his colleagues were the first to demonstrate the effects of B-adrenergic stimulation on c-fos gene expression in the mouse submandibular gland. They were also the first to document retroviral mediated gene transfer into the salivary glands.[18, 19] In collaboration with Phyllis Shaw, Ph.D., and colleagues, Barka carried out studies on the role of cysteine proteinase inhibitors (cystatins) in physiologic and pathophysiologic processes, and in the control of cell replication.[20-22] Shaw's laboratory continues to study the regulation of cystatin gene expression during fetal development.[23] In 2003, Shaw and her associates reported the first demonstration of the expression of the homeobox protein, Hmx3, in the postnatally developing rat submandibular gland.[24] A prime example of translational research, the work has relevance because two human diseases, an autosomal dominant form of cerebral hemorrhagia seen in Icelandic patients and an autosomal recessive form of progressive myoclonus epilepsy, are known to result directly from genetic alterations of cystatins.

Shaw, who joined the faculty of the Department in 1984, has directed the Medical School's Histology/Cell Biology course since 1991. The first-year course teaches the basic concepts of cell biology and the structure and function of specialized cells and their organization into tissues and organ systems. Recent revisions and curricular innovations have included the initiation of a Teaching Assistant Program utilizing second-year students, coordination of the teaching time with the Physiology course, and the introduction of clinical correlates and computer-based learning. A dedicated educator and researcher, Shaw has had significant extramural funding and has been the recipient of numerous teaching awards.

Additional research in the Department of Anatomy begun during the first half of the 1980s included studies of the hormone serotonin and its action in the brain; the transplantation of pancreatic islet cells into the brain cavities of diabetic rats, as well as efforts to understand diabetes (Type I) as an autoimmune disease on a cellular level; and the mysosin isoforms in cardiac muscle as the heart moves from the embryonic to the adult stage.

In 1985, Barka stepped down as Chairman, and Robert McEvoy, M.D., became the Acting Chair, a position he held until Paul Lazarow, Ph.D., assumed the chair in 1990. At that time, the name of the Department was changed to the Department of Cell Biology and Anatomy in recognition of Lazarow's position as a preeminent cell biologist. As noted in the *Annual Report* for 1989: "The renaming of this department . . . reflects the fact that the department is about to begin a period of rapid growth in cell biology. A major new focus of research will concern intracellular protein translocation, vesicular transport and membrane and organelle assembly."[25] As the biology of the cell has been elucidated, a number of subcellular structures called "organelles" have been identified, each with specific properties and functions. Pathology and genetic defects of individual organelles have been found to be the etiology of many diseases.

Recruited from Rockefeller University, Lazarow had made seminal contributions to the understanding of how peroxisomes form within cells and what functions they perform. Peroxisomes are cytoplasmic cell organelles that contain enzymes for the production and decomposition of hydrogen peroxide and that are found in virtually every cell. They play a major role in cell respiration and in the metabolism of fatty acids and other lipids. Lazarow and his colleagues discovered that the organelle forms by the importation of newly made proteins into preexisting peroxisomes, followed by organelle division. Over subsequent years, this discovery allowed Lazarow and others to obtain a wealth of information on the molecular machinery, involving some twenty-three gene products, responsible for this complex phenomenon.

Inherited defects in peroxisome assembly cause multiple enzyme deficiencies that result in serious human diseases, for example, Zellweger syndrome, neonatal adrenoleukodystrophy, and infantile Refsum disease.[26] These neurological problems frequently manifest themselves at birth, and some patients have a life span of only weeks or months. Many of the genes responsible for these conditions have now been identified, and Lazarow and his group discovered the gene for rhizomelic chondrodysplasia punctata.[27] Lazarow also discovered that peroxisomes are important for the oxidation of fatty acids in the rat liver, and they are now known to catabolize diverse compounds including prostaglandins and xenobiotics. Peroxisomes also contribute to the synthesis of bile acids, cholesterol and ether lipids.[28]

Gianni Piperno, D.Sc., recruited by Lazarow from Rockefeller University in 1991, has had continuous NIH support for his research on intraflagellar transport (IFT), a major factor in the assembly of proteins in cilia and flagella. Cilia and flagella are cellular organelles involved in motility and sensory transduction. Piperno and his colleagues have established a method for the analysis of motion of IFT particles *in vivo* and have identified temperature-sensitive mutants that are defective in IFT.[29–32] This work has significant clinical relevance, since dysfunction of cilia and flagella may cause disease of the respiratory tract or kidney, abnormalities of vascular endothelium, and sterility. In the School, Piperno directed the Cell and Developmental Biology Course from 1992 to 1997.

Lazarow also played a vital role in the advancement of research at Mount Sinai. Within the Department, he oversaw the redesign and rebuilding of the research space, recruited a dozen new faculty members in cell biology, and initiated a Ph.D. training program. In 1991–1992, he also chaired a Dean's Advisory Committee on New Scientific Directions for Research in the new research building (the East Building) and served on the committee charged with the construction of the building. In 2000, Lazarow stepped down as Chair of the Department, but his laboratory remains active and productive, studying the biogenesis of peroxisomes.

By the year 2000, cell biology had become a major theme in all of the basic science departments. In 2001, in an effort to streamline the structure of the basic science departments, faculty members were given primary appointments in reshaped departments. Anatomy continued to encompass those scientists truly focused on anatomy. The name of the Department was changed to the Center for Anatomy and Functional Morphology in order to more accurately reflect the major goals and function of the Department's activities. Jeffrey Laitman, Ph.D., was named the Director of the Center. In the interregnum between Lazarow and Laitman, the Department was cochaired by Laitman and Liliana Ossowski, Ph.D.

Ossowski, who received her Ph.D. in Genetics from the Weizmann Institute in Rehovoth, Israel, had been a long-time member of the Department. With the overall goal of developing targeted cancer therapy, her research focused on the properties of cancer cells that distinguish them from normal cells. Ossowski and her colleagues have demonstrated the vital need for the urokinase receptor in tumor growth and metastasis.[33-36] With the reshaping of the Department, Ossowski transferred her primary faculty appointment to the Department of Medicine.

Laitman, a native New Yorker, was recruited, in 1977, from Yale University, where he had obtained his doctorate. Trained in anatomy and physical anthropology, Laitman focused his research on the comparative anatomy, development, and evolution of the mammalian upper respiratory tract and on the contiguous areas of the base of the skull. He has studied the functional anatomy of the region in an array of mammals from rodents to whales to humans. In collaboration with Joy Reidenberg, Ph.D., the anatomist, and members of the Department of Otolaryngology where Laitman has a joint professorial appointment, he has made great strides in the understanding of the breathing, swallowing, and vocalizing patterns of human infants. The work has had significant implications in helping to understand basic human anatomy and the events that occur in clinical conditions such as Sudden Infant Death Syndrome, also known as crib death.[37]

Reidenberg, a former student of Laitman's and a graduate of Mount Sinai's Graduate School of Biological Sciences, has centered her studies on the comparative anatomy of the upper respiratory tract in a wide range of mammals, with a particular emphasis on cetaceans (whales, dolphins, and porpoises). Cetaceans were chosen as a "natural experiment" to understand the evolutionary forces leading to a highly modified upper respiratory tract adapted for an aquatic existence.[38-40] Investigations are also in progress to examine the role of the cetacean larynx in sound production for communication or echolocation. This research, carried out in the field, often finds Reidenberg studying a beached whale at oceanside.

Laitman and Reidenberg's work on the evolution of the aerodigestive tract and the study of fossil remains has led to the development of new techniques to reconstruct the human vocal tract and increased the understanding of the origins of speech and language.[41] Laitman, for example, has shed light on how early human ancestors, such as the enigmatic group known as Neanderthals, may have differed from present-day humans.[42,43] Such findings have often put his studies in the public eye, and Laitman and members of his laboratory have frequently been the subject of newspaper and television coverage.

Laitman has been widely honored for his research, but, in the eyes of many, his greatest contributions to Mount Sinai have been in the realm of teaching. A charismatic teacher of anatomy and always an advocate of interdisciplinary teaching, Laitman, along with his colleagues, has dramatically changed the way gross anatomy is taught. Designed to meld classic dissection with today's technology used to view the body, the current course introduces first-year students to minimally invasive instruments, ultrasonography, multiaxial computerized image reconstruction, magnetic resonance imaging, and plastinated prosections. The gross anatomy course is taught by faculty from fifteen clinical and basic science departments, including chairs and voluntary and full-time clinical faculty. A highly successful teaching assistant program composed of upper classmen and advanced graduate students brings an added dimension to the course. The anatomy teaching programs have received national and international recognition.[44]

Many of Laitman's graduate students, as well as junior faculty that he has mentored, have gone on to leadership positions as anatomy teachers and course directors in the United States and abroad. In recognition of his seminal research and teaching accomplishments, Laitman was appointed a Distinguished Professor of the School in 2002, the first anatomist in the School's history to be so honored.

The members of the Center for Anatomy and Functional Morphology have been well recognized for their expertise in teaching and have received numerous Excellence in Teaching awards from the student body. In 1987, the American Association of Anatomists created the Basmajian/Williams and Wilkins Award to recognize members of the Association teaching gross anatomy "who have made outstanding accomplishments in anatomical sciences research and have demonstrated excellence and commitment to the teaching of gross anatomy."[45] Laitman was the recipient of the first Basmajian Award, and three other members of the faculty—Marilyn McGinnis, Ph.D., David Colman, Ph.D., and Reidenberg—have won the award in subsequent years.

McGinnis, who joined the faculty in 1981, focused her laboratory's research on the identification of gene products that are regulated by androgens, on how those androgens act in the brain to alter genomic expression, and on discovering specific neural loci of androgen receptor activity involved in male reproductive behavior.[46, 47] The goal is to understand the mechanisms of hormone action in the brain and to provide a basis for a direct link between the biochemical response of the neuron and the behavioral response of the organism. A second research focus involves the effects of anabolic steroids in the brain. These efforts have already had significant impact by establishing that both testosterone and estrogen are necessary for the expression of male sexual behavior; that the ventromedial hypothalamus is critically involved in the expression of male reproductive behaviors; and that long-term exposure to anabolic steroids increases aggressive behavior, particularly in response to provocation, and alters the normal patterns of circadian rhythm.[48-50]

Colman, who was recruited to Mount Sinai in 1993 and who also had joint appointments in the Department of Neurology and in the Fishberg Research Center for Neurobiology, has made major discoveries regarding nerve cell development and regeneration, how these cells communicate with each other, and the role of cell adhesion molecules not only in brain development but also in cancer.[51-54] His studies on myelination and the protective effect of the myelin sheath in the brain and the peripheral nervous system have profound significance for spinal cord injury. In September 2002, Colman left Mount Sinai to become the Director of the Montreal Neurological Institute.

Another member of the Department who has won numerous teaching awards during his almost thirty year tenure on the faculty is Kai Mark Mak, Ph.D. Based part-time at the Bronx Veterans Administration Medical Center, he has collaborated with Charles Lieber, M.D., in numerous studies of alcoholic liver disease and of the metabolism of alcohol that have been most productive.[55-57] They have recently shown that a soy extract may attentuate the damage to liver cells induced by alcohol.[58]

The Center for Anatomy and Functional Morphology has continued its innovative teaching efforts. In conjunction with the Department of Radiology, and codirected by William Simpson, M.D., Assistant Professor of Radiology, and Laitman, it introduced a mandatory fourth-year clerkship, Anatomic Radiology, in 2003. This novel return to the anatomy laboratory while learning practical, clinical radiology reminds the student about to graduate of the importance and relevance of basic science in his or her future career in medicine, and the course has been well received by the student body.

The Center also has major teaching commitments for teaching in the Graduate School of Biological Sciences, with responsibility for required courses in the core curriculum, elective courses, and opportunities for training in the faculty's research laboratories.

The Center for Anatomy and Functional Morphology, with its blend of innovative and multidisciplinary undergraduate and graduate teaching programs and its cutting-edge research, continues today as a vital element in the fabric of the basic sciences.

THE DEPARTMENT OF PHYSIOLOGY

THE DEPARTMENT OF PHYSIOLOGY AND BIOPHYSICS

In early 1965, very shortly after his recruitment to Mount Sinai as Dean of the Medical School, Irving Schwartz, M.D., left that position and was appointed as both Chair of the Department of Physiology and Dean of the Graduate School of Biological Sciences.[59] The medical and lay leadership of the new School and Schwartz himself recognized that his talents as a brilliant physiologist and educator could be better utilized in these new dual roles. And so Schwartz and his physiology faculty began the task of developing the curriculum for the course to be given to the first-year medical students and graduate students in September 1968. The goals for the course were clearly spelled out:

> Excellence in any aspect of the practice of medicine requires an understanding of the fundamental principles, concepts and observations

that pertain to how living cells, tissues and organs function. This body of basic knowledge constitutes the subject matter of physiology which, therefore, establishes for the student the foundation of his study of medicine and provides for the physician the tools for his assimilation, evaluation and participation in the new discoveries that are continually changing the content of medical science and the nature of medical practice.[60]

The course curriculum included:

a survey of the general attributes of living tissues (such as the phenomena of excitability, membrane transport, energy transformation, biosynthesis, replication, adaptation) as well as the more specialized functions and interrelationships among the cardiovascular, respiratory, renal, gastrointestinal, exocrine, endocrine and nervous systems. The treatment of the subject matter is analytic in large part using examples of function drawn from diverse sources throughout the animal kingdom. However, the overview is synthetic and the most general goal of our teaching program is to motivate students to establish an enduring interest in and quest for broader and deeper understanding of the nature and interactions of the processes which comprise the complete, integrated living system which is man.[61]

In his appointments prior to coming to Mount Sinai, Schwartz conducted research on renal function, sweat metabolism, and the chemical and biological activity of neurohypophyseal peptides. In 1961, he and his colleagues at Brookhaven National Laboratories began investigations that localized the satiety center of the brain to the ventromedial hypothalamus, work that was completed at Mount Sinai.[62]

At Mount Sinai, Schwartz assembled a most productive group of scientists who expanded the work on neurohypophyseal hormones. This group became internationally recognized in the field and published some three hundred papers, about one-third of which dealt with the structure-function relationships of the neurohypophyseal hormones. Methods were developed for the analysis of the hydro-osmotic and contractile actions of these hormones and also for the characterization of the processes that terminate hormone action. Carboxamidopeptidases, a class of enzymes that degrade oxytocin and other peptide hormones, were defined by the group.[63]

Irving L. Schwartz, M.D., the first Professor and Chairman of the Department of Physiology and the first Dean of the Mount Sinai Graduate School of Biological Sciences, 1968.

Utilizing high-resolution nuclear magnetic resonance spectroscopy, Schwartz and his colleagues were among the first to determine the conformation (three-dimensional molecular structure) of peptides in solution.[64] Up to that time, X-ray crystallography was the only technique available to determine molecular structure, but this was two-dimensional only. The new technique paved the way for extensive study of molecular function and structure in hundreds of laboratories throughout the world. Other major accomplishments of Schwartz and his colleagues in their studies of neurohypophyseal hormones included the creation of highly biologically active isologs of these hormones and the development of photo-affinity analogs of neurohypophyseal hormones used to study the function of the corresponding receptors. Another major contribution was the separation of the plasma membrane envelope of epithelial cells of the mammalian kidney into luminal and contraluminal components, which allowed the investigation of the sequence of events occurring during the action of antidiuretic and parathyroid hormones.[65]

In 1973, the Department was renamed the Department of Physiology and Biophysics in recognition of the changing face of its work. Six years later, in 1979, Schwartz stepped down as Chairman and was invested as the first Harold and Golden Lamport Distinguished Professor and Director of the Mount Sinai Center for Polypeptide and Membrane Research, where he continued his research for another decade. Schwartz received many awards throughout his distinguished career and is honored at Mount Sinai by the Irving L. Schwartz lectureship, endowed by one of his former students.

Other early members of the Department also played vital roles in its development. Harold Burlington, Ph.D., was instrumental in the creation and the teaching of the integrated cardiovascular-respiratory curriculum for the medical students. With a focus on the regulation of blood cell production, Burlington collaborated with scientists at the Brookhaven National Laboratory, on Long Island, New York, for many years.[66-68] Harold Lamport, Ph.D., who was intimately involved in systems analysis and in the application of mathematics and physics to physiology and medicine, also served as Chairman of the Committee on Biomedical Engineering of the School of Medicine. William Brodsky, M.D., one of the original and long-serving members of the Department's faculty, studied physiological and biochemical water and hydrogen ion transport processes in secretory tissues of humans, lower mammals, amphibia, and reptiles.[69-72] Joseph Eisenman, Ph.D., a neurophysiologist, joined the faculty in 1970. A mainstay in the Department's teaching who directed the neurophysiology portion of the curriculum, Eisenman also investigated the role of temperature in neuronal function.[73-75] He also played a major role in the work of Mount Sinai's Institutional Review Board.

Patrick Eggena, M.D., and Roderick Walter, Ph.D., who both came to Mount Sinai with Schwartz, played critical roles in the elucidation of the mechanism of action of peptide hormones, especially vasopressin, at both the cellular and the molecular levels and in the demonstration of the three-dimensional structure of oxytocin and vasopressin in solution.[76-78] The toad bladder was the experimental model for much of this work, and, in 1986, Eggena described a precise and sensitive method for measuring the effects of the neurohypophyseal hormones on urea permeability of this organ.[79]

A superb teacher who has received numerous teaching awards, Eggena was named Acting Chair of the Department in 1979, a position he held until 1985. Director of the Physiology course for more than twenty-five years, and convinced that the best teaching of physiological principles occurs in the setting of patient care, Eggena has devoted more and more of his efforts in recent years to the development of case-based, Web application teaching materials that utilize patient scenarios. His text *The Physiological Basis of Primary Care*, now in its second edition, is similarly case-based.[80] He remains a most valued member of the Department. Basil Hanss, Ph.D., a member of the faculty since 1994, who is responsible for the teaching of renal physiology, became the Codirector of the Physiology course in 2002.

In January 1985, Harel Weinstein, D.Sc., a Professor in the Department of Pharmacology, was appointed Chairman of the Department of Physiology and Biophysics. Born in Europe and educated in Israel, where he received his doctorate in Theoretical Physics and Chemistry and served in the military, Weinstein joined the faculty of the Johns Hopkins University in 1973. As a theoretical physicist and theoretical chemist interested in biology as a frontier, Weinstein was studying the relation between the structure of neurotransmitters and drugs and their effects on biological systems, work that was similar to the research being conducted by Jack Peter Green, Ph.D., M.D., Chairman of Mount Sinai's Department of Pharmacology, and his group. Green arranged for Weinstein to spend the summer of 1974 at Mount Sinai; the collaboration was most successful, and Weinstein remained on the faculty, becoming Professor of Pharmacology in 1979.[81]

Weinstein's appointment as Chairman of Physiology came at a propitious time in science. The tools that he used for his research were becoming more powerful, and biology was becoming more and more molecular. As Weinstein has noted, "I became more and more interested not in the molecules themselves, like acetylcholine and histamine, but in the systems that they were activating. And I understood that I needed to create a view of integrated systems of physiology that would sustain that understanding and would create new knowledge. So for that reason, physiology and biophysics became an area in which I wanted to work."[82]

Utilizing supercomputers and computational graphics machines, Weinstein's laboratory seeks, at the molecular level, a mechanistic understanding of the structures and properties of cellular components and physiological functions. Weinstein's research has been continuously focused on the use of computational biology to discover the structure, dynamic, and electronic determinants of biological processes that underlie physiological functions.

The approach includes theoretical determinations of molecular structure and properties and computational simulations of molecular mechanisms and processes that can be studied with great accuracy. These theoretical studies are designed to complement experimentation in providing mechanistic insights about systems of ever-increasing size and complexity and to guide pointed experimental exploration of cellular processes and functions in numerous collaborative studies. Based on methods of quantum and statistical mechanics, the theoretical methods used are continuously being refined and tested in the study of biomolecular systems and implemented in novel algorithms. A unifying theme is the understanding of mechanisms triggered by molecular recognition and leading to signal transduction.

In 1993, Weinstein and one of his graduate students, Daqun Zhang, provided the first publication on computational simulations of the activation of a G protein coupled receptor, work that was based on a three-dimensional molecular model constructed by the student.[83] Three years later, Weinstein and his group presented the first demonstration of the complexity of ligand binding to G protein coupled receptors and of the mode in which the orientation in the binding site determines the extent and nature of the receptor activation.[84] More recently, similar computations and simulations were used to determine the molecular basis of partial agonism of the human serotonin receptor.[85] The methods employed by Weinstein and his colleagues have also provided the first description of the computational identification of the structural details for the pairing of G protein coupled receptors in the cell membrane and the mode in which proteins that regulate gene expression are able to find their specific (and small) sites of action on the long stretches of DNA on which they reside.[86, 87] Computational analysis has also allowed for the definitive description of the manner in which G protein coupled receptors can be modeled and their function analyzed.[88]

The Department flourished during Weinstein's seventeen-year tenure. The number of faculty members increased from seven to twenty-seven; in fiscal year 2002, the Department had in excess of $11.5 million in extramural funding from the NIH, and publications were appearing in leading science journals on a regular basis. At that time, it was pointed out that:

> The focus of research is on a variety of mammalian and invertebrate neuronal systems; cellular mechanisms in cardiovascular and renal systems; molecular endocrinology and regulation of gene expression; mechanisms of signal transduction, ion channel structure and function; cellular receptor structure and function; cell dynamics and motility; membrane dynamics; structural biology and structural genomics; and structure-function relations in protein-ligand interaction and protein-DNA complexes. Major experimental approaches include biochemistry and biophysics of proteins/nucleic acids/membranes, electrophysiology, molecular biology, imaging microscopy, NMR spectroscopy, X-ray crystallography, and extensive computational modeling and simulation.[89]

The Department is at the cutting edge of research. Prior to his recruitment to Mount Sinai in 1996, Robert Margolskee, M.D., Ph.D., discovered gustducin, a taste-cell-specific protein that is vital to the perception of taste.[90] The ability to taste the sweetness of carbohydrate-rich foodstuffs is critical to human nutritional status. Since coming to Mount Sinai, Margolskee and his colleagues have employed a multifactorial approach to gain new and important insights into the peripheral and central mechanisms responsible for taste transduction and coding and how they regulate gustatory behavior *in vivo*, and into the identification of gene networks that coordinately transduce taste sensory information to the central nervous system. Results from the laboratory also suggest that gustducin is a principal mediator of both bitter and sweet taste.[91–94] In 2001, in a prime example of the modern efforts in biomedical gene discovery utilizing bioinformatics, molecular biology, computational modeling, and whole animal physiology, the group identified the gene responsible for tasting sugar.[95] In 2001, Margolskee provided the first major review of the taste field for the general scientific community in forty years.[96]

Interested in the identification and functional characterization of the gene cascade activated during memory formation, Cristina Alberini, Ph.D., employs DNA microarrays and other techniques to identify where in the brain and when following learning this cascade is activated and has made significant contributions to this area of research.[97] Catherine Clelland, Ph.D., a member of F. Carter Bancroft's group (discussed later), is engaged in developing novel approaches employing DNA microarray technology in the diagnosis of disease. She and her colleagues have identified a gene in mouse embryos whose expression patterns suggest a role in the specification of multiple cell types, in particular, the fetal skeleton.[98]

Madeleine Kirchberger, Ph.D., a member of the Department for more than thirty years, has devoted her research efforts to the factors regulating membrane ion transport in cardiac muscle. Her discovery of the regulation of the cardiac sarcoplasmic/endoplasmic reticulum Ca2+-ATPase (SERCA) by a small protein named phospholamban has great clinical significance, since it translates into the critical role played in the positive inotropic effect of certain catecholamines on the heart.[99–101]

The Department also has an active group of investigators in the area of neurobiology, where Klaudiusz Weiss, Ph.D., the Lillian and Henry Stratton Professor of Physiology and Biophysics, has devoted more than a decade at Mount Sinai investigating the neural basis of behavior. Recruited from Columbia University in 1990, Weiss, his colleagues Elizabeth Cropper, Ph.D., Ferdinand Vilim, Ph.D., and Vladimir Brezina, Ph.D., and their associates employ multidisciplinary approaches (behavioral, morphological, electrophysiological, and molecular biological) to study the marine mollusc Aplysia, whose simple nervous system facilitates studies of the plasticity of feeding behavior due to changes in animal motivational state. They have identified and purified novel neuropeptides that modify the relationship between muscle contraction amplitude and relaxation rate so as to maintain optimal motor output when the intensity and frequency of feeding behavior changes. More recently, the group has used mathematical modeling and complementary experiments to study the dynamics of modulation in an accessory muscle of the mollusc.[102–109]

Weiss serves on the editorial boards of the *Journal of Neurophysiology* and *Invertebrate Neuroscience*, is an Assistant Editor of *The Mount Sinai Journal of Medicine*, and also serves on several committees of the NIH. He has been a recipient of a Research Scientist Development Award and of a MERIT Grant and a Senior Scientist Award from the National Institute of Mental Health and has received a prestigious McKnight Investigator Award. At Mount Sinai, in addition to many committee assignments, including service on the Appointments and Promotions Committee, which he now chairs, Weiss has codirected the Signal Transduction Program (1993–1995) and also codirected the Graduate Program in Neuroscience (1990–2001).

In Biophysics, a group of structural biologists that includes Aneel Aggarwal, Ph.D., Katherine Borden, Ph.D., Roberto Sanchez, Ph.D., and Ming-Ming Zhou, Ph.D., utilize experimental and theoretical biophysical techniques such as X-ray crystallography, nuclear magnetic resonance imaging, and theoretical modeling, combined with other biochemical and molecular biological techniques, to carry out investigations of macromolecular structure and function. Physiological systems/structures studied by individual members of this group of investigators include gene transcriptional and translational regulation, oncogene products and the organelles they form, protein/nucleic acid interactions, cell-surface receptor activation, and signal transduction.[110–116] Roman Osman, Ph.D., originally a member of the Department of Pharmacology, applies computational approaches for his research on the structure and function of macromolecules involved in physiological events, including cooperativity of DNA repair enzymes, transmembrane receptor signal transduction, and the molecular bases of blood coagulation and autoimmune diseases.[117–120]

Recruited from Harvard Medical School to Mount Sinai in 1993, Diomedes Logothetis, Ph.D., and his colleagues have devoted their efforts to researching the G protein-mediated transduction mechanisms that regulate cell excitability. This signaling activates a class of potassium channels that, in response to external signals, are responsible for slowing the heart rate and inhibiting neurotransmitter release. Experimental models have included isolated cardiac and neuronal cells and expression of recombinant proteins in mammalian cell lines and Xenopus oocytes. The group has made a number of fundamental observations. It provided the first example of the effector function of the beta-gamma subunits of GTP-binding proteins and has also discovered that a large family of potassium channel proteins function by direct interactions with the phospholipid PIP2 in the plasma membrane. Signaling pathways that regulate the level of this phospholipid critically affect the function of several potassium channels.[121–125]

An outstanding teacher, Logothetis has been the recipient of numerous teaching awards from the graduate students, and, in 2003, he was named Dean of the Mount Sinai Graduate School of Biological Sciences and Director of the Mount Sinai School of Medicine M.D./Ph.D. Program.[126]

In 2001, Weinstein established the Institute for Computational Biomedicine (ICB) within the Mount Sinai School of Medicine. The ICB was designed to foster interdisciplinary activities in all areas of modern biomedicine, to which it will bring approaches from applied mathematics, computer science, and computer-based technologies. The goal of the ICB is to create and promote the development of novel computational research tools in bioinformatics, computational biology, modeling, and the simulation of integrative biological systems. In collaboration with the Department of Physiology and the Department of Pharmacology and Biological Chemistry, the Institute has developed an information system designed to support quantitative studies on the signaling pathways and networks of the cell.

In October 2002, Weinstein stepped down from the Chair to become Chair of the Department of Physiology at Weill-Cornell University Medical College. F. Carter Bancroft, Ph.D., was appointed as the Interim Chair.

Recruited to Mount Sinai by Weinstein from Sloan-Kettering Memorial Cancer Center in 1985 as a full Professor, Bancroft's previous faculty appointments were at Columbia and Cornell Universities. His early interest in the study of growth hormones led him to the first use of molecular biological techniques to study pituitary function, yielding identification of the messenger RNA for growth hormone.[127] Other major areas of investigation had focused on the hormonal regulation of prolactin gene expression. In recent years, Bancroft has been heavily involved in what he calls "DNA-based technologies," seeking uses for DNA that are not inherently biological but instead employ the biochemical properties of DNA and its enzymes to develop novel technologies.[128] Among the technologies that Bancroft and his collaborators have developed and patented are one for the use of DNA as the basis for a computer (DNA-based computation)[129] and another, termed "DNA-Based Steganography," for the use of "secret-message DNA" to hide messages within the great complexity of the DNA genome of humans (and/or other organisms).[130]

Four of Mount Sinai's Shared Core Facilities are housed in the Department of Physiology and Biophysics: the Cell Imaging Laboratory (a satellite of the Microscopy Shared Research Facility), the Molecular Modeling Core, the Computational Core, and the Xenopus Oocyte Core.

In addition to its pivotal function in teaching cellular/molecular and system physiology in the Medical School, the Department of Physiology and Biophysics offers graduate education and training in research in many areas of cell and molecular physiology, molecular and membrane biophysics, and neurophysiology and integrative neurobiology, as well as various aspects of bioinformatics and computational biology.

It is evident that the Department is well positioned to continue to carry out its dual mission of performing cutting-edge research in a number of important areas of modern-day physiology and biophysics and teaching the discipline to both medical and graduate students.

THE DEPARTMENT OF MICROBIOLOGY

In 1967, as preparations for the Medical School were proceeding, a Department of Microbiology was established with Hans Popper, M.D., adding the title of Acting Chairman to the many other roles that he was filling. One year later, when the School opened its doors to the first students, Stanley Schneierson, M.D., took over as Acting Chairman and was named the School's first Professor of Microbiology. Schneierson, a member of the Hospital's Department of Microbiology since 1934 and Director of that Department since 1960, was an outstanding clinician/microbiologist. He had developed antibiotic sensitivity testing to a fine degree, established one of the first full-service virology laboratories in a general hospital, and published more than eighty papers.[131] His *Atlas of Diagnostic Microbiology*, first published in 1971, became a standard text in medical schools and laboratories.[132] During his short tenure as Acting Chairman, Schneierson worked with planning and curriculum committees and assisted Edra Spilman, Ph.D., in preparing the Department's space in the Basic Science Building. Schneierson remained an active teacher in the Department's courses and continued to direct the Hospital's Department of Microbiology until his retirement in 1973.

In January 1969, Edwin Kilbourne, M.D., was appointed Chairman of the Department of Microbiology, beginning a twenty-three-year

Edwin D. Kilbourne, M.D., the first Professor and Chairman of the Department of Microbiology, 1984.

association with Mount Sinai. Following receipt of his undergraduate and M.D. degrees from Cornell University, Kilbourne trained in internal medicine and then, following service in the United States Army Medical Corps, embarked upon a career in science that would bring him international recognition and add luster to the new Medical School.

Kilbourne was recruited from Cornell University Medical College, where he had served as Professor of Public Health and Director of Virus Research. Kilbourne's top priorities upon arrival at Mount Sinai included the recruitment of full-time faculty members and development of the curriculum. Aided by Schneierson, other members of the Hospital's Department of Microbiology, and new faculty recruits, Kilbourne created courses in microbiology and infectious diseases for the medical students, as well as a number of well-received courses in the Graduate School of Biological Sciences. The Department was also given the responsibility for teaching immunology. Moreover, Kilbourne was immediately able to restart his research on the influence of steroids on laboratory-induced viral infection and on the influenza virus and other respiratory viruses using the Department's laboratories, which he characterized as "probably unsurpassed in terms of major equipment and facilities necessary for modern biomedical research."[133] The equipment included a newly developed protein sequencer that "can do the work in one day that it took the full attention of a skilled researcher to do in two weeks."[134]

Within a year, Kilbourne reported a landmark breakthrough in influenza virus research.[135] As Kilbourne subsequently noted:

> In 1960 while carrying out studies of influenza virus recombination, I made the observation that the growth potential of a virus could be enhanced through genetic recombination with a standard laboratory virus with good growth potential. In 1969 I deliberately recombined a strain of the new Hong Kong subtype endemic virus with a standard laboratory strain and isolated a reassortant virus called X-31. . . . Because of the resistance of the FDA to the use of a man-made recombinant virus as a vaccine, I found it necessary to carry out an elaborate experimental trial of X-31. The vaccine, manufactured in England by Evans Limited, was tested in mice, volunteers and in military populations with notable success.[136]

In 1969, recombinant technology had not yet come into its own, leading Kilbourne to add: "X-31 was engineered in every sense of the word in representing a hybrid containing genes of two viruses for the purpose of facilitating virus production."[137] A series of papers then documented the feasibility of using genetic recombination for the design of vaccine viruses with optimal characteristics, and, as the expression goes, "the rest is history."[138–141] Vaccines against influenza had been made with difficulty and in limited quantity since the early part of the twentieth century. Put simply, Kilbourne's contribution was to mate portions of the virus that cause the illness with portions of a nonpathogenic virus that grows well in culture. The result was a vaccine that grew well, had the ability to confer immunity, and did not cause disease.

This work, which predated genetic engineering, led to the worldwide acknowledgment of Kilbourne as a leader in influenza vaccine development. He received many awards and honors, including the R. E. Dyer Lectureship Award of the National Institutes of Health for outstanding achievement in research important to medical science, in 1973; the Borden Award of the Association of American Medical Colleges for outstanding research in medical science the following year; and election to the National Academy of Sciences in 1977.

Many of the early faculty members recruited by Kilbourne would play vital roles in the endeavors of the Microbiology Department and the School for decades. Jerome Schulman, M.D., Kilbourne's long-time colleague at Cornell, joined the faculty in July 1969. Working independently and also collaboratively with Kilbourne, Schulman had already made significant contributions in several areas of viral research, areas that were all continued at Mount Sinai. Major publications dealt with the factors that influence the experimental transmission of influenza virus infection; the role of antineuraminidase antibody in the development of immunity to influenza virus infection; and the selection of viruses with the unique properties that allow for rapid growth and induction of immunity.[142–145] In one of the first publications from Mount Sinai by the collaborators, Schulman and Kilbourne pointed out the independent variations in the nature of hemagglutinin and neuraminidase in the various influenza viruses.[146] The finding was another milestone in the evolution of vaccine development.

Schulman not only continued his own research on the influenza virus but collaborated with many members of the Department, providing medical and biological expertise to the basic scientists, as they approached virus research from the chemical, biochemical and molecular point of view.[147] In addition to his many teaching and committee assignments in the School and his outside organizational activities and service as an associate editor for the journal *Viral Immunology*, Schulman served for many years as a member of the Board of Directors of Mount Sinai's Lambda Chapter of Alpha Omega Alpha.

In 1971, Kilbourne recruited Peter Palese, Ph.D., to the faculty. Born in Central Europe, Palese received a classical education, including many years of Greek and Latin study, and then obtained his Ph.D. in Chemistry from the University of Vienna. Emigrating to the United States in 1970 as a postdoctoral fellow in molecular biology, Palese became interested in viral structure and virus replication. Stimulated by Kilbourne, he turned his attention to the influenza virus. In collaboration with Schulman, Palese identified the role of the neuraminidase of the influenza virus, and he and his colleagues were the first to show that neuraminidase inhibitors were able to inhibit viral replication in tissue culture.[148] The influenza virus is a negative-strand RNA virus, one of a broad group of pathogenic animal viruses that include measles, mumps, rabies, Ebola, and hantaviruses. Palese was the first to map the influenza virus genome, identifying the eight genes of the virus and what they coded for.[149]

The Department continued its growth under Kilbourne's direction until 1986, when, in one of a series of administrative leadership changes under Dean Kase, he stepped down as Chairman. Kilbourne remained at Mount Sinai until 1992, when he moved to New York Medical College, where he continued his research on the influenza virus for another ten years. In 1991, Kilbourne received Mount Sinai's highest honor, the Jacobi Medallion of the Mount Sinai Alumni Association, the first basic science chairman of the School to be so honored.

Palese took over the Chairmanship of the Department in 1987 and set about recruiting additional faculty members, many of whom work in collaboration with him. Palese's unique contributions to microbiology have led to his receiving many awards and honors, culminating in his election to the National Academy of Sciences in 2002.

Continuing research on the influenza virus has been extraordinarily productive. Working with mutations of the influenza virus, Palese moved on to develop techniques to reconstruct viruses from DNA. He and his colleagues described a system that allowed the use of recombinant DNA technology to modify the genome of the negative-strand RNA influenza viruses and that also provided the ability to engineer vectors for the expression of foreign genes.[150, 151]

Adolfo Garcia-Sastre, Ph.D., joined the Department in 1991 as a postdoctoral fellow. In 1994, he was the first author of a publication describing a new methodology to increase the coding capacity of influenza viruses without altering the structure of the viral proteins.[152] Five years later, the group recreated the influenza virus from recombinant DNA without the need for a helper virus, essentially building the virus from molecules found in the native viral structure.[153] This work paves the way for the insertion of foreign genes into viruses, creating any desired mutation, and the creation of viruses that can function as vaccines.[154] These techniques have also enabled the study of the role of individual viral proteins in pathogenicity.[155]

Utilizing lung sections taken from victims of the 1918 influenza pandemic and stored by U.S. Army pathologists, Palese's laboratory has now focused on finding the genomic sequence of that particular influenza virus and trying to recreate it. The ultimate goal of Palese's work is to understand the underlying mechanisms responsible for the variations in pathogenicity of viruses and to devise measures to counteract these viruses. The host's immunologic defenses play a major role in combating viral infections, and Christopher Basler, Ph.D., and his colleagues have created an assay that they have already applied to the Ebola virus, permitting the identification of viral molecules that inhibit specific host defenses.[156]

John Blaho, Ph.D., a member of the Department since 1994, has focused his research on the herpes simplex virus (HSV), with the goal of understanding the molecular events involved in the regulation of the replication of this human virus and also the mechanisms by which the virus causes cell death. Herpes simplex viruses are neurotropic viruses that cause a variety of infections, remain latent in the neurons of their host for life, and can be reactivated at any time to cause lesions at or near the initial site of infection. In 1999, Aubert and Blaho discovered that the major regulatory protein of HSV is required for the prevention of programmed cell death (apoptosis) in infected human cells.[157] Blaho and his associates have also defined the biology of the major protein component of the HSV virion[158–160] and have discovered that HSV first induces and then prevents apoptosis during its replication cycle, a process the group has termed "Apoptosis Modulation by HSV."[161–163] An understanding of the processes of replication and reactivation is of great clinical import because of the possibility of designing effective therapy that would interfere with one or another molecular pathway as a way to cure HSV infection. Blaho and his colleagues also hope to translate their findings into novel strategies for limiting uncontrolled cell growth such as is found in tumor cells by using a process they recently discovered called "viral oncoapoptosis."[164]

Since coming to Mount Sinai, Blaho has played a significant role in the teaching activities of the Department. He has served on numerous student advisory and examination committees and has been responsible for the training of more than twenty Ph.D. and M.D./Ph.D. students. He has directed the Graduate Microbiology Training Program and, since 1998, has been both the director of the "Introduction to Microbiology" core Graduate School course and codirector of the Graduate School's Microbiology Multidisciplinary Training Area. In 1999, Blaho led the preparation of the training grant proposal, funded by the NIH, that established a program in Mechanisms of Virus-Host Interactions. An active participant in science at the national level, Blaho serves on the editorial boards of the major virology journals and is also a permanent member of the NIH virology study section.

In 1974, James Wetmur, Ph.D., joined the Department as an Associate Professor of Microbiology. Wetmur received his Ph.D. in Chemistry at the California Institute of Technology; his thesis advisor was Norman Davidson, Ph.D. Their paper on the kinetics of renaturation of DNA,[165] with Wetmur as the first author, became a Citation Classic.[166] In 1996, Davidson received the National Medal of Science for "breakthroughs . . . which have led to the earliest understanding of the overall structure of genomes."[167] The work of Wetmur and Davidson "established the principle of nucleic acid renaturation, a key element in molecular biology with important uses in determining the structure and function of genes."[168] Davidson credited his medal to "the bright young people who have chosen to work with me over the course of my career."[169]

At Mount Sinai, Wetmur continued his research on the kinetics and renaturation of DNA and also applied this work and other DNA research techniques that he developed to the investigation of bacteria, viruses, and the DNA profiles of specific diseases.[170–175] He also analyzed DNA-protein interactions involved in the basic biological processes of replication and recombination, later using proteins from thermophiles as tools.[176–181] With his background in bacterial genetics, Wetmur established collaborations with members of the Departments of Genetics and Community and Preventive Medicine that led to the characterization and cloning of specific proteins that have implications for molecular screening of individuals for genetic susceptibility for lead poisoning.[182–184] He and his colleagues have also developed a murine model of genetic susceptibility to lead bioaccumulation.[185] The team's more recent work in molecular epidemiology has contributed to the development of a technique that uses pooled DNA for genetic epidemiologic studies and to the recognition of the unique susceptibility of certain infants to the toxicity of organophosphate pesticides.[186, 187]

Active in numerous organizations, including the New York Academy of Sciences, where he has held leadership positions, Wetmur has also carried major responsibility for teaching microbiology and physical chemistry in both the Medical School and the Graduate School.

One of the first immunologists to be recruited to the basic science faculty of the new Medical School in 1969 was J. Donald Capra, M.D., a graduate of the University of Vermont School of Medicine. Capra, his colleague J. Michael Kehoe, Ph.D., and their students and associates made fundamental breakthroughs in the identification of the structure of the immunoglobulins.[188–192] In 1977, Capra moved to the University of Texas Southwestern Medical School, and in 1997 he was appointed the President of the Oklahoma Medical Research Foundation.

The immunologist Constantin Bona, M.D., Ph.D., joined the Department of Microbiology in 1979. Recruited from the Centre National de la Recherche Scientifique in Paris, Bona had already made several seminal discoveries. In 1972, he and his colleagues demonstrated the transfer of antigens from macrophages to lymphocytes; two years later, he was the first to demonstrate that a water-soluble mitogen extracted from bacteria could cause blastic transformation of mouse splenic lymphocytes.[193, 194] Bona's early studies at Mount Sinai focused on the gene markers (idiotypes) that regulate the immune response.[195] His research gradually shifted to the investigation of autoimmune systems, the characterization of the genes that code for autoantibodies, and the establishment of the criteria necessary to define autoantibodies.[196, 197] Bona and his associates also constructed chimeric molecules, the features of which suggested that these molecules could be used to produce vaccines.[198, 199] More recently, he cloned and sequenced the mutant fibrillin-1 gene that is responsible for the mouse "tight-skin disease," an animal model of scleroderma or systemic sclerosis. These mice also produce autoantibodies to the fibrillin-1 gene.[200] Similar autoantibodies have been found in a significant number of patients with systemic sclerosis.[201] Further studies have shown that the abnormal gene can be disrupted, thus diminishing fibroblastic growth.[202] This finding has led to the development of a drug that has antifibrotic properties and that shows promise as a therapeutic agent in diseases marked by excessive collagen synthesis.[203]

During his tenure at Mount Sinai, Bona spent two productive sabbatical years, as a visiting professor at the Massachusetts Institute of Technology in 1984, and at the University of Kyoto, Japan, in 1996.

In addition to almost four hundred publications in peer-reviewed journals, Bona has published thirteen books, contributed almost sixty book chapters and monographs, served on numerous editorial boards, and lectured extensively throughout the United States and abroad. Along with his continuous responsibility for the teaching of immunology and microbiology in both the Medical School and the Graduate School, Bona has served as thesis director for a dozen Ph.D. candidates and mentored more than forty postdoctoral students. Two of these former students, Thomas Moran, Ph.D., and Sofia Casares, Ph.D., are currently on the faculty, having been appointed by Palese. Others have gone on to major academic positions throughout the world.

Moran has concentrated his research on the responses of the immune system to viral infection; namely, what molecular events take place after infection with a virus, and what is required for the body to generate an appropriate immune response. A striking finding was that the clearance of influenza virus from mouse lungs was severely inhibited by the presence of the cytokine interleukin-4 (IL-4), because IL-4 suppresses the generation of cytotoxic T cell precursors.[204] More recent studies have investigated the interaction of dendritic cells (the antigen presenting cells) with respiratory viruses.[205] These data, showing that viruses trigger dendritic cells by a pathway known to stimulate the synthesis of type I interferon, may be useful in the design of more effective and less dangerous vaccines. Additional studies of the maturation of dendritic cells may have an important impact on the association of respiratory virus infection and allergy.[206]

Moran serves on the editorial board of the *Journal of Virology* and on several NIH study sections. For more than a decade, he has directed the Hybridoma Core Facility, generating monoclonal antibodies for use by all members of the faculty involved in the study of autoimmune disease. Moran has also been the course director of the Medical School's Medical Immunology course since 1994.

Casares, Bona, and Teodor-Doru Brumeanu, M.D., have collaborated on many projects that have scrutinized the biology of autoreactive T cells. Their goal is to develop therapeutic strategies that will selectively inactivate pathogenic T cells and thereby allow the induction of tolerance in immune-mediated autoimmune diseases, such as Type 1 diabetes and multiple sclerosis, and also protect against graft rejection. In 1997, the group reported the ability to create chimeric molecules utilizing genetic engineering techniques.[207] Four years later, they formulated a molecule containing the chemotherapeutic agent doxorubicin that targeted and destroyed a specific T cell population.[208] In 2002, they were able to report the selective destruction of the T cells responsible for attacking pancreatic islet cells in autoimmune diabetes, thereby protecting the animal model from developing diabetes.[209]

A member of the Department faculty since 1988, Lu-Hai Wang, Ph.D., who also has a joint appointment in the Ruttenberg Cancer

Center, has spent more than a quarter of a century in the study of oncogenes. Prior to coming to Mount Sinai, Wang and his colleagues mapped the location of the first known oncogene (src) on the Rous sarcoma virus genome and subsequently completed that virus's gene map. Later research at Mount Sinai identified several additional oncogenes and also demonstrated the oncogenic potential of the human insulin receptor.

Wang and his associates have continued to make fundamental observations on the molecular events involved in the transformation and growth of malignant cells. Utilizing human cancer cell lines in culture and animal models, the group has reported on how the human insulin receptor and other proteins can be activated to become oncogenes. They have also identified and characterized the function of unique signaling proteins and have demonstrated differential sensitivity of breast cancer cells toward the inhibitors of signaling molecules that are activated in the cancer cells. In addition, the team has described the important signaling functions for anchorage independent growth of cells, one of the most important properties of cancer cells.[210–217]

Over the years, the Department of Microbiology has created an ever more sophisticated teaching program. The Microbiology Training Program, codirected by Blaho and Garcia-Sastre, is highly research oriented. The Department also has major teaching responsibilities for instruction at all levels, accepting high school and college students for research experience and attracting outstanding graduate and M.D./Ph.D. students. This program provides training in antivirals, autoimmune disease, bacterial genetics, environmental microbiology, immunology, molecular virology, cellular and viral oncogenesis, vaccine development, and other related topics. The faculty includes not only members of the Department but faculty from many other departments, centers, and institutes of the Medical Center. The Training Program in Mechanisms of Virus-Host Interactions is a joint venture with the New York University School of Medicine, designed to educate pre- and postdoctoral trainees in virology. This program also sponsors two symposia each year at the New York Academy of Sciences.

The lethal nature of the materials that the faculty of the Department deals with on a daily basis constitutes an ever-present biohazard.

Extreme caution is the watchword. In recognition of the outstanding contributions of Palese and his faculty to the knowledge of viral and bacterial infectious disease, Mount Sinai received a major grant from the NIH in the summer of 2003 to construct a level 3 bio-safety laboratory within the Department of Microbiology. There are only two higher level bio-containment facilities in the United States; both are government operated. As Palese noted, "This laboratory will enable us to make a difference in even more fields within microbiology and infectious diseases—e.g., West Nile virus, SARS, and other emerging diseases."[218]

The Department of Microbiology has been at the cutting edge of scientific research and innovative teaching for thirty-five years. The current leadership and faculty have amply demonstrated their commitment to continuing the tradition established by the founding members of the Department.

THE DOROTHY H. AND LEWIS ROSENSTIEL DEPARTMENT OF BIOCHEMISTRY

THE DOROTHY H. AND LEWIS ROSENSTIEL DEPARTMENT OF PHARMACOLOGY AND BIOLOGICAL CHEMISTRY

Everlasting credit is due to Dean Irving Schwartz for appointing Panayotis Katsoyannis, Ph.D., as the founding Chairman of the Department of Biochemistry in 1967. Recruited from the Brookhaven National Laboratory where he served as Head of the Division of Biochemistry, Katsoyannis had already achieved a measure of fame for his studies on the synthesis of oxytocin and vasopressin and as the first to chemically synthesize human insulin.[219, 220]

Katsoyannis accepted the position in May 1967 but did not move to Mount Sinai until January 1968. This left him little time to hire faculty, design the curriculum, write the course syllabus, and oversee the construction of the Department laboratories in the Basic Sciences Building and equip them, all the while continuing his research at Brookhaven, until he moved to Mount Sinai in January 1968. Nevertheless, on Day 1, (September 9, 1968), Katsoyannis began the teaching of biochemistry with a lecture on the chemistry of amino acids, peptides,

Panayotis G. Katsoyannis, Ph.D., the first Dorothy H. and Lewis S. Rosenstiel Professor and Chairman of the Department of Biochemistry, 1977.

and proteins. Like many of his colleagues on the faculty, Katsoyannis was more than a little chagrined at the appearance of the students: "If you recall, '68 was the students' revolt. . . . I remember some students came with head bands and without shoes."[221] Recognizing that "a proper understanding of biochemistry may be considered one of the fundamental requirements for a proper understanding of modern medicine and to be, perhaps, the fundamental requirement for significant progress in modern medicine,"[222] Katsoyannis and his fellow department members presented a course that, in their view, "would best serve the future interests of the students both as medical scientists and practitioners."[223]

Since the original class had only thirty-six students, it was possible to conduct many of the laboratory exercises in the laboratories of the faculty, thereby providing an in-depth research experience for many students. The Department also offered courses in enzymology and biomacromolecules at the Graduate School level; these courses were attended by house staff and clinical faculty as well. Within a year of the opening of the School, the Department instituted small-group teaching conferences, each group consisting of ten students with one faculty member. An offering unique to Mount Sinai was a basic biochemistry course for the practicing clinicians at the Medical Center. In addition, numerous elective courses were developed for undergraduate and graduate students over the years. In 1970, a laboratory tutorial was instituted in the basic biochemistry course. One year later, a series of "Clinical Correlates," presented conjointly by the basic science faculty and the clinical faculty, was introduced to cover in depth the biochemistry relevant to the particular disorder being discussed. Individual tutorials were arranged for students who had inadequate backgrounds in biochemistry or who required remediation. Concomitant with undergraduate teaching, the Department developed courses for the students in the Graduate School of Biological Sciences. Over the years, these courses were increased substantially in number as well as content.

Teaching was always given a high priority in the Department, and the effect was salutary. Mount Sinai students achieved high grades on the National Board and on the U.S. Licensing Examinations in biochemistry and on at least one occasion (in the 1980s) ranked number one among all medical schools. As class size increased, it became necessary to move the laboratory exercises into the student laboratories, but small-group teaching continued.

Despite the difficulties inherent in the renovation of the Cummings Basic Sciences Building, the faculty was able to engage in research from the outset. Katsoyannis and his collaborators, Monty Montjar, Ph.D., James Ginos, Ph.D., Gerald Schwartz, Ph.D., and Anthony Trakatellis, Ph.D., some of whom had moved with him from Brookhaven, continued their studies of insulin and how modifications of its structure affected its chemical and biological properties. Over the next thirty-five years, Katsoyannis and his group, with continuous extramural funding, extended their work on the actions of insulin and insulin growth factors and also developed many insulin analogues. A 1975 study of the sites of insulin binding in subcellular fractions of the liver suggested that there may be several types of intracellular membranes to which insulin can bind besides the plasma membranes.[224] Extensive studies were carried out seeking the requirements necessary for high biological activity of insulin, the synthesis of highly potent and superactive insulins, and the effects of changes in protein structure on insulin activity. A single substitution of an amino acid residue of human insulin can result in a superactive hormone. It is therefore possible to prepare "tailor-made" insulins with high potency.[225–230] A spectrally enhanced insulin was synthesized to function as a probe in the study of hormone-receptor interaction.[231] Trakatellis also studied the effects of antibiotics on the process of protein synthesis.[232, 233]

Katsoyannis holds several patents and his peer-reviewed publications number more than one hundred. He is the associate editor of the *Journal of Protein Chemistry* and has been widely honored for his ground-breaking work. In 1972, on the fiftieth anniversary of the discovery of insulin, he received the Commemorative Medallion of the American Diabetes Association, and in 1997, he received an honorary degree from the University of Patras, Greece. The Mount Sinai Alumni honored Katsoyannis in 1995, bestowing the Jacobi Medallion.

Seymour Koritz, Ph.D., recruited from the University of Pittsburgh, also remained on the faculty for his entire career. He conducted pioneering research on the mode of action of adrenocorticotrophic hormone and later also investigated the role of cyclic AMP in related metabolic processes.[234–239] Jacob Chanley, Ph.D., a long-time member of the Hospital Department of Chemistry and a collaborator of the psychiatrist Seymour Rosenblatt, M.D., was appointed an Associate Professor in the new Medical School Department of Biochemistry and continued his research on the chemistry of the nervous system, especially in relation to psychiatric disorders.[240–243]

Diana Beattie, Ph.D., another member of the original faculty, and also from the University of Pittsburgh, where she had collaborated with Koritz, remained on the faculty until 1985. At that time, she was appointed Chair of the Department of Biochemistry at the University of West Virginia. Beattie, whose investigations centered on mitochondrial biogenesis and metabolism, was extraordinarily productive, publishing more than seventy-five papers during her tenure at Mount Sinai.[244–249] A most effective teacher, Beattie lectured extensively in both medical school and graduate school curricula and mentored numerous Ph.D. candidates.

Prior to his recruitment to Mount Sinai in 1974, George Acs, Ph.D., had already made seminal contributions to understanding the linkage between transfer RNA and amino acids, hormone-induced messenger RNA synthesis, the purification of the plasminogen activator, and the conservative replication of the Reo virus DNA.[250–252] In 1987, while conducting research on the hepatitis virus, Acs and his colleagues established the first stable cell system in which Hepatitis B virus replicated and then demonstrated that the virus produced by those cells could cause hepatitis in chimpanzees.[253, 254] Over the years, more than two hundred schools and pharmaceutical and biochemical companies have requested samples of that cell line. Acs has remained active and productive to the present.

Recruited to Mount Sinai immediately after completion of her postdoctoral fellowship in molecular biology, Terry Krulwich, Ph.D., has played a vital role in the Department, the School of Medicine, and the Graduate School of the Biological Sciences since 1970.[255] With a background in bacteriology, she did her early research on the regulation of sugar and amino acid transport in bacterial cells and isolated membranes. Major work has been accomplished in her laboratory on the bioenergetics of alkaliphilic bacteria (bacteria that grow in a highly alkaline environment) and on the proteins responsible for the development of bacterial antibiotic resistance.[256–262] A charismatic teacher at both the medical school and the graduate school levels, she has inspired her trainees to go on to highly successful careers throughout the world. Krulwich was awarded an honorary Doctor of Science degree from her alma mater, Goucher College, in 1987, and the following year received Mount Sinai's Outstanding Faculty Achievement Award in the Basic Sciences. In 1997, Krulwich was invested as the Sharon and Frederick A. Klingenstein-Nathan G. Kase, M.D., Professor of Biochemistry and Molecular Biology. Two years later, she was the recipient of the Jacobi Medallion, awarded by The Mount Sinai Alumni.

Arthur Cederbaum, Ph.D., currently Professor of Pharmacology and Biological Chemistry, joined the faculty as a research associate in 1971. The author or co-author of more than 250 publications, Cederbaum has had continuous extramural funding for more than three decades. His research, often in collaboration with pathologists and clinical hepatologists, has concentrated on the Cytochrome P450 (CYP) system of enzymes of the liver and the mechanisms by which ethanol and other drugs interfere with CYP and cause hepatotoxicity. Most recent efforts in Cederbaum's laboratory have developed molecular models to evaluate the production of an increased state of oxidative stress by ethanol and the deleterious effects thereof. There are many forms of CYP, most of them found in the liver, but they are also involved in vascular autoregulation, particularly in the brain, and are also vital to the formation of cholesterol and steroids. CYP enzymes are also intimately involved in the degradation and metabolism of many drugs by a process of oxidative metabolism, and the interaction of drugs with CYP can produce serious drug reactions.[263–271]

Cederbaum has been the recipient of two NIH Research Career Development Awards and since 1989 has held an NIH/NIAAA (National Institute on Alcohol Abuse and Alcoholism) Merit Award. In 2003, he held no fewer than four NIH R-01 grants. An outstanding teacher, Cederbaum has received numerous awards from both graduate and undergraduate students. He has also been the recipient of the Outstanding Faculty Achievement Award in the Basic Sciences.

Joining the Department immediately following the completion of his postdoctoral fellowship in 1986, David Bechhofer, Ph.D., has played a prominent role in the teaching activities of the Department as both a lecturer and as a small-group leader in the Biochemistry course. Bechhofer served on the planning committee to create the "Molecules and Cells" course in the preclinical years of the new curriculum and has served as the Course Director since 2001. Bechhofer's research focuses on the mechanism of RNA processing and messenger RNA decay in the bacterium *Bacillus subtilis*. Utilizing a combination of genetic and biochemical approaches, he and his associates are pursuing the identification of new bacterial genes that are involved in RNA processing.[272–275]

In 1998, Katsoyannis's retirement as Chairman led to the first of a series of major organizational changes in the structure of basic science at Mount Sinai. At the same time, Robert Lazzarini, Ph.D., Chairman of the Brookdale Center for Molecular Biology, also stepped down, and the Department of Biochemistry merged with the Brookdale Center under the combined leadership of Francisco Ramirez, Ph.D., and Mary Rifkin, Ph.D., both members of the Center's faculty.[276]

Because of the commonality of research interests in biochemical mechanisms, the Dorothy H. and Lewis Rosenstiel Department of Pharmacology and Biological Chemistry was created in 2001, bringing together many of the former members of the Department of Biochemistry and the Department of Pharmacology under the Chairmanship of Ravi Iyengar, Ph.D.

THE DEPARTMENT OF PHARMACOLOGY

THE DOROTHY H. AND LEWIS ROSENSTIEL DEPARTMENT OF PHARMACOLOGY AND BIOLOGICAL CHEMISTRY

The recruitment of Jack Peter Green, Ph.D., M.D., in 1968, just prior to the School's opening, as the founding Chairman of the Department of Pharmacology paved the way for the development of a world-renowned research department. Because the teaching of pharmacology to the first-year students did not begin until the second trimester, Green and his colleagues had several months in which to prepare a curriculum consisting of lectures and discussion groups. Uniquely, social and economic issues related to drugs and drug use were included in the course, and it was noted that "quantum mechanics and its application to biological systems is also being taught to the first year medical students; no other medical school does this."[277]

Throughout his distinguished career, Green focused his research on the role of neurotransmitters in the brain and on the mechanisms

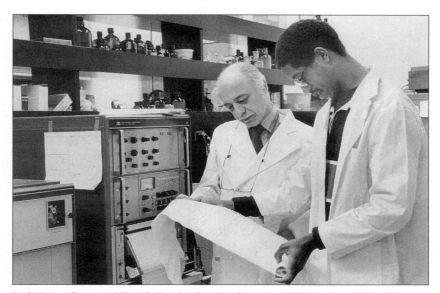

Jack Peter Green, M.D., Ph.D., the first Professor and Chairman of the Department of Pharmacology, with a student.

by which hallucinogens affected the brain. He was one of the first to identify the role of histamine and its metabolites as neurotransmitters and neuromodulators in the brain. Key investigations by his group included studies on the identification and distribution of brain histamine and led to the demonstration of the rapid turnover of brain histamine, consistent with its function as a putative transmitter.[278] The work on histamine receptors and their action identified the coupling of the histamine H2 receptor to adenyl-cyclase in the brain.[279, 280] Green's laboratory also developed chemical methods for the detection of minute amounts of acetylcholine, allowing the separation of that enzyme from other substances with cholinergic activity. Green was also the first to understand that there were multiple subtypes of serotonin receptors, and extensive research was carried out on the synthesis of inhibitors that could then become drugs and therapeutic agents. Green and his colleagues also made fundamental contributions to the understanding of the actions of the hallucinogens on the brain.[281–288]

Green was among the first to apply quantum mechanical techniques to the study of pharmacological problems, principally to the actions of indoles and hallucinogens, research that continued throughout his career.[289] Within one year of the creation of the Department, Green stated:

> Quantum mechanics revealed the molecular characteristics that determine the biological effects of hallucinogenic drugs like LSD, mescaline, and THC, which is the active material in marijuana. Defining the electronic factors that cause hallucinations is a necessary prelude to the design of drugs that can antagonize the effects of the hallucinogens and thereby provide a means of treating acute intoxication by these agents.[290]

Later, Green, along with Harel Weinstein, Ph.D., who joined the Department in 1974, and their colleagues, instituted the use of computational techniques to study drug-receptor interactions to predict biological activity, thereby providing a model for the development of rational drug use.

Green and Weinstein became strong collaborators who were at the forefront of utilizing chemical physics in the description of the biological processes related to medicine. In 1980, they organized one of the first conferences of its kind to consider the conceptual basis and practical role of quantum chemistry in biomedical sciences.[291] The conference included three presentations by members of the Department.[292–294] In 1976, Weinstein and his colleagues described, for the first time and long before the first receptor molecule was cloned, a discrete molecular mechanism for the action of a neurotransmitter on its receptor, a mechanism that is still considered valid today.[295] As noted earlier, Weinstein became Chairman of the Department of Physiology and Biophysics in 1985.

Green recruited a superb faculty. Marian Orlowski, M.D., and Sherwin Wilk, Ph.D., began a collaboration in the early 1970s that culminated in their discovery of the multicatalytic protease complex, commonly known as the proteasome.[296–298] The proteasome, an intracellular enzyme complex responsible for protein degradation, is critical to cell survival and cell death and is found in all eukaryotic cells. The name "proteasome" was originally proposed by Alfred Goldberg, Ph.D., of the Harvard Medical School.[299] Orlowski and Wilk continued their investigations into the proteasome throughout their careers, both separately and in collaboration. In 1990, a major review by Orlowski provided an up-to-the-minute synthesis of the work to date.[300]

Orlowski was born in Poland, where he received his M.D. degree. Prior to joining the Mount Sinai faculty in 1972, he had already made major contributions in clinical enzymology, including the introduction of the gamma-glutamyl transpeptidase test in the clinical diagnosis of liver disease.[301–303] He also isolated and defined the biochemistry of the enzymes involved in the synthesis and degradation of glutathione.[304] At Mount Sinai, Orlowski identified the inhibitors of L-glutamate decarboxylase as convulsive agents and also identified additional enzymes and designed potent specific inhibitors to them.[305–307] Proteasome inhibitors have been developed as anticancer drugs, subjected to animal studies and clinical trials, and at least one has now been approved for use in the management of advanced myeloma.[308–310] These studies of the anticancer effects of proteasome inhibitors were conducted by Robert Orlowski, M.D., Marian Orlowski's son.

Wilk had spent five years as a research chemist in the circulatory physiology laboratory of Mount Sinai's Department of Medicine in the 1960s while pursuing his Ph.D. At that time, he played a key role in the development of assays for the measurement of amines, a method that became the gold standard for the diagnosis of pheochromocytoma.[311] Following a postdoctoral fellowship, he returned to Mount

Sinai and has remained on the faculty, continuing his research on purifying, characterizing, and cloning the endopeptidases that make up the multicatalytic enzyme complex. In addition, he has developed a number of inhibitors of endopepidase function.[312] Wilk and his associates have also characterized and cloned aspartyl aminopeptidase, a widely distributed enzyme that is an important factor in the metabolism of angiotensin.[313]

In 1971, Joseph Goldfarb, Ph.D., joined the faculty and, for more than three decades, has been a mainstay of the Pharmacology teaching program. Goldfarb has also been active in pharmacology teaching at the national level, serving on educational committees of the Association for Medical School Pharmacology for more than a decade. He has also been a long-time member of committees of the National Board of Medical Examiners, including the Pharmacology and Subject Test Committee, which he has chaired since 2001. Goldfarb has also taught in the MSSM ethics curriculum since its inception. Goldfarb's pharmacology research provided one of the earliest indications that opioid antagonists alone could have physiological effects, presaging the discovery of endogenous opioids.[314] He and his associates were also the first to identify a functional role for 5-HT3 receptors,[315] and more recently they demonstrated that drugs can differentially activate multiple effector pathways at a single receptor site.[316]

Ravi Iyengar, Ph.D., was recruited to the Department by Green in 1986 because of their common interest in cell signaling. With a background in membrane biochemistry, Iyengar emigrated to the United States in 1974 from India and, following receipt of his Ph.D., joined the faculty of Baylor College of Medicine, where he remained until joining Mount Sinai as an Associate Professor.

A world-renowned expert in the field of cell signaling, Iyengar has made seminal contributions to the understanding of how cells respond to extracellular signals, and how those signals are routed through complex networks to regulate important cellular functions. Critical to the signaling process is a group of "G" proteins, so-called because they bind guanine nucleotides. Iyengar and his associates have described the G protein pathways:

> The heterotrimeric (made of three different subunits) guanine
> nucleotide-binding proteins (G proteins) are signal transducers that
> communicate signals from many hormones, neurotransmitters, chemo-

kines, and autocrine and paracrine factors. The extracellular signals are received by members of a large superfamily of receptors with seven membrane-spanning regions that activate the G proteins, which route the signals to several distinct intracellular signaling pathways. These pathways interact with one another to form a network that regulates metabolic enzymes, ion channels, transporters, and other components of the cellular machinery controlling a broad range of cellular processes, including transcription, motility, contractility, and secretion. These cellular processes in turn regulate systemic functions such as embryonic development, gonadal development, learning and memory, and organismal homeostasis.[317]

Iyengar's group has also described how signaling pathways intercommunicate with each other, and they have been at the forefront in the development of computational models and in the integration of these models experimentally to provide valuable insight into the behavior of these complex systems.[318–320] This work is of immense clinical relevance, as cell signaling molecules constitute about 80 percent of the targets of currently known drugs, and components of these cellular networks are potential targets for new drug therapies in a variety of disease states.

Iyengar is the principal investigator of an Advanced Research Center funded by the New York State Office of Science and Technology and also serves on the editorial boards of the *Journal of Biological Chemistry* and the Signal Transduction Knowledge Environment section of the journal *Science*. He has carried a heavy teaching and administrative load since coming to Mount Sinai and has served on numerous peer review committees of the NIH, the American Cancer Society, and other national organizations. In 2003, Iyengar was appointed Dean for Research in the School. Since 1990, Iyengar has edited six volumes of the *Methods in Enzymology* series on the G proteins, their effectors, and their pathways. Three were published in 2002.[321–326]

After an illustrious thirty-year career as Chairman of the Department of Pharmacology, Jack Peter Green retired in 1997. As a physician who spent his career in basic science research, a common trait of the School's first chairmen, Green had an intense interest in physicians practicing other occupations, especially literature, and has published on the subject.[327] Green has been the recipient of numerous awards. He is one of only seven individuals elected to honorary membership in the European Histamine Research Society. At Mount Sinai, he is honored by the annual Jack Peter Green Lecture.

Iyengar was named the Acting Chair and was appointed the Chairman in 1999. New faculty members were recruited, a postdoctoral training grant from the National Institute of Drug Abuse was renewed for years 20–25, and a predoctoral training grant was awarded by the National Institute of General Medical Sciences. In 2003, the predoctoral training grant was renewed for an additional five years (until 2009). With a provision for ten trainees, this program represents one of Mount Sinai's largest training grants. Iyengar was invested as the Dorothy H. and Lewis Rosenstiel Professor of Pharmacology and Biological Chemistry in 2001, the same year that Biochemistry and Pharmacology merged, and he has taken the Department of Pharmacology and Biological Chemistry to a new plateau.

The "new" department has fifteen independent research laboratories under the leadership of new faculty as well as individuals from the original Biochemistry and Pharmacology Departments. Although overlapping in scope, three main areas of pharmacologic investigation have evolved: cell signaling, the biology of proteases, and systems biology.

Joining Iyengar in the investigation of cell signaling are Maria Diverse-Pierluissi, Ph.D., Emmanuel Landau, M.D., Ph.D., and Daniel Weinstein, Ph.D. Relevant to the understanding of transmitter release in addiction and pleasure behaviors, Diverse-Pierluissi and her associates are studying the integration of the multiple signaling pathways that converge to modulate calcium channel activity and how they could ultimately alter the timing of signaling in the nervous system.[328, 329] The research of Landau, who is a Professor of Psychiatry in addition to his appointment in Pharmacology and Biological Chemistry, is aimed at discovering the biochemical machinery of neuronal plasticity with the goal of finding pharmacological tools that influence learning and memory.[330] A second area of research in Landau's laboratory is to define a receptor for amyloid peptide, a major component of the plaques associated with Alzheimer's disease that has been implicated in the neurotoxicity of the disease.[331] The broad goal of Daniel Weinstein's research is to explore the molecular and cellular basis of early patterning events in the vertebrate embryo, especially the induction of the mesodermal germ layer and the division of the ectoderm into neural and nonneural domains.[332, 333]

Since the characterization of the proteasome more than twenty years ago, the biology of the proteases and the study of the proteasome have been a major focus of investigation in the Department. In addition to the ongoing studies of Orlowski and Wilk, Dennis Healy, Ph.D., a member of the faculty for more than fifteen years, and his colleagues concentrated their studies on the role of aminopeptidases in the regulation of the renin-angiotensin pathway and its significance in the homeostatic control of blood pressure.[334–336]

Members of the Department have diverse interests, but, like all of the faculty, their ultimate goal is to understand molecular interactions that may be employed to produce new or better therapeutic drugs. Christopher Cardozo, M.D., a former associate of Orlowski and Wilk, is investigating the mechanism of androgen receptor degradation with a view to developing therapeutic strategies against resistant prostate cancers.[337] He and his laboratory colleagues are also exploring the mechanisms of action of anabolic steroids, with the goal of suggesting novel uses for these drugs in clinical practice.

Dan Felsenfeld, Ph.D., researches the functions of integrated systems at the subcellular as well as the cellular level in various tissue and organ systems.[338] Lakshmi Devi, Ph.D., in her research on heterodimers, is seeking new combinations of existing drugs or developing new drugs to manage chronic pain.[339] Benjamin Chen, M.D., Ph.D., is investigating the molecular mechanisms involved in viral assembly with an emphasis on HIV. The insights gained will be used to develop antiviral strategies.

As a result of the success of the Graduate School of Biological Sciences, the "new" curriculum of the Medical School, and the merger of the Departments of Biochemistry and Pharmacology, the Department of Pharmacology and Biological Chemistry has assumed an enormous teaching commitment, both in time and in personnel.

When the Medical School opened in 1968, pharmacology was taught in the first year, but within two years it was switched to the second year. Other than changes in the number of teaching hours, pharmacology has been and continues to be one of the most rigorous courses of the preclinical years. In the 1998–1999 academic year, ninety-five hours of pharmacology were taught in the second half of the second year. In August 2000, the new curriculum was introduced to the entering first-year students. The new first-year course "Molecules and Cells" and the second-year course in pharmacology are directed by David Bechhofer, Ph.D., and Joseph Goldfarb, Ph.D., respectively, both members of the Department of Pharmacology and Biological Chemistry.[340]

The goal of "Molecules and Cells" is for students to understand the mechanisms by which cells receive and process extracellular signals, regulate gene expression, control organellar biogenesis, and divide or differentiate. The study of these mechanisms constitutes the first two-thirds of the course. The fundamentals of carbohydrate, fatty acid, and nitrogen metabolism constitute the final third of the course. The relationship of these processes to human disease is emphasized throughout. The course continues the tradition of small-group teaching established in both biochemistry and pharamacology. Each group consists of ten or eleven students and is led by two faculty members, one a basic scientist and the other a clinician. The second-year pharmacology course emphasizes the mechanisms and principles of pharmacokinetic (how the body handles drugs) and pharmacodynamic (how drugs affect function) drug action.

The Department plays a major role in teaching in the Graduate School. Obligations include the teaching of advanced pharmacology courses, participation in the core curricula and in the advanced courses of other disciplines, and research training in the laboratories of all of the faculty. A project titled "Training Program in Pharmacology of Drugs of Abuse: Interdisciplinary Training in Drug Abuse Research," begun in 1978 and funded by the National Institute of Drug Abuse, has been expanded over the years to include faculty members from clinical and basic science departments, both on campus and from other educational sites.

In response to the rapid evolution of biomedical science and in keeping with the mandate of the School, the Department has developed an extensive array of shared research facilities, departmental cores, institutional centers, and other resources, including uniting the facilities of the former Departments of Biochemistry and Pharmacology.

The Department of Pharmacology and Biological Chemistry represents the amalgamation of two venerable original departments of the Medical School. Under Iyengar's strong leadership, the Department's multiple academic functions are carried out by an outstanding faculty committed to research and teaching.

5

The Centers and Institutes

INTRODUCTION

THE LATE 1980S witnessed the creation, emergence, and growth of new and productive basic science entities within the Mount Sinai School of Medicine. In addition, as noted earlier,[1] some of the existing basic science departments underwent major and often dramatic reorganization in both structure and leadership, a process that continued into the new millennium. Almost all of this occurred, or at the least had its origin, during the thirteen-year tenure, from 1985 to 1997, of Nathan Kase, M.D., as Dean of the School.

Kase believed that Mount Sinai's overall basic science research had not changed sufficiently from the phenomenology and pathology of the "old science" to be competitive in the new era of genetics and molecular biology.[2] Moreover, he felt that the existing departments were too "discipline focused" in that they were not exploiting the multi-disciplinary approaches of the "new science" to basic biologic problems and were not maximizing the applied scientific opportunities available through the use of the rich clinical resources of the Hospital. In order to achieve the envisioned multidisciplinary expansion, Kase felt that it would be academically and politically easier to add the new elements side by side with the existing basic science and clinical departments rather than to totally replace the existing traditional vertical and intra-disciplinary framework. Also, wherever possible, the new basic science entities were juxtaposed geographically to their clinical counterparts— for example, molecular biology and medicine or neurobiology and neurology. Molecular biology, neurobiology, and immunobiology were the focus of the new centers that were established. Genetics, which existed as a division within the Department of Pediatrics, was given departmental status and was strengthened in both its research and

clinical capabilities. Mount Sinai's cancer initiative was enhanced by the creation of a multidisciplinary cancer center. In order to attract outstanding candidates for the leadership positions, the new centers and institutes were designed administratively to be equivalent to the traditional departments. Each had its own budget and the authority to proffer both primary and secondary faculty appointments.

As with all new academic enterprises, money and space were critical. The Board of Trustees, then chaired by Frederick Klingenstein, committed significant capital annually for a period of five years to begin the process. It was envisioned that a successful project would generate enough funds to allow the funding of a subsequent project. As Kase put it: "It was pump priming and staying just one step ahead."[3] In addition, outstanding philanthropic support provided greatly needed startup funds. Major gifts from the Ramapo Trust, Dr. Lucy Moses, and Derald Ruttenberg and his family established the Brookdale Center for Molecular Biology (now the Brookdale Department of Molecular, Cell, and Developmental Biology), the Dr. Arthur M. Fishberg Research Center in Neurobiology, and the Derald H. Ruttenberg Cancer Center, respectively. A substantial gift from the estate of Lillian and Henry Stratton allowed the creation of several endowed professorships that were indispensable for high-level recruitment.

Research space had been at a premium for years. Although it would have been possible to find bits and pieces of space in the Annenberg Building, an additional facility was needed. Once again, Klingenstein and the Board gave their approval, funds were raised, and the East Building was planned and built on the premise that it would soon be filled with outstanding scientists whose grants, through their indirect overhead support, would help the building pay for itself.

CENTER FOR MOLECULAR BIOLOGY

THE BROOKDALE CENTER FOR MOLECULAR BIOLOGY

THE BROOKDALE DEPARTMENT OF MOLECULAR, CELL, AND DEVELOPMENTAL BIOLOGY

The first of the new centers, the Center for Molecular Biology, opened in 1986. Terry Krulwich, Ph.D., Professor of Biochemistry and Dean of the Graduate School, was named the Acting Director. Two laboratories were established, and research projects were instituted to investigate the molecular events involved in early embryonic development.

In 1988, Kase completed the recruitment of Robert Lazzarini, Ph.D., then Chief of the Laboratory of Molecular Genetics in the National Institute of Neurological Diseases and Stroke (NINDS) of the National Institutes of Health (NIH), to head the Center for Molecular Biology. In addition to developing its own research agenda, the Center's mandate also included establishing a multidisciplinary environment within the institution and creating core facilities that could be shared by all members of the research community irrespective of their primary departmental affiliation.[4]

Lazzarini, a native New Yorker, received his Ph.D. from the University of California, Los Angeles, and his postdoctoral training at Johns Hopkins University. From there, Lazzarini moved to the NIH, where he remained for twenty-five years, until his relocation to Mount Sinai. His early research involved the field of bacterial nitrogen metabolism, but he then moved on to studies of the vesicular stomatitis virus (VSV), a close relative of the rabies virus, specifically investigating the defective interfering particles of the VSV. These particles are natural inhibitors of viral infections. Lazzarini and his colleagues cataloged a number of these particles and characterized them at the molecular level, ultimately leading to a highly acclaimed comprehensive theory of their origin and of their inhibitory activity.[5–8] In his last years at the NINDS, Lazzarini's research interests shifted to the investigation of myelin, the biological material that is focally lost in multiple sclerosis plaques. He had cloned a number of the structural genes of myelin and had also produced transgenic mice carrying the human genes for these myelin components or altered forms of them, in effect creating models for the investigation of the human disease.[9, 10] Lazzarini continued these studies at Mount Sinai, and the work expanded to encompass the development of the oligodendrocyte, the cell that lays down the myelin and then wraps it around the nerve cell.[11, 12]

In 1989, Francesco Ramirez, Ph.D., joined the Center as the Dr. Amy and James Elster Professor of Molecular Biology (Connective Tissue Disorders), Professor of Pediatrics, and Deputy Director of the Brookdale Center. In 1991, Ramirez and his colleagues were the first to demonstrate and report a conclusive genetic linkage between the candidate gene for fibrillin 1 and the Marfan syndrome locus.[13] That same publication was also the first to report the unsuspected existence of a second fibrillin (fibrillin 2) gene, which was also genetically linked to congenital contractural arachnodactyly (CCA or Beals syndrome). In 1997 and 1999, two subsequent publications reported the creation of different strains of mutant mice that faithfully replicated the full clinical spectrum of human Marfan syndrome.[14, 15] These studies provide novel insights into the cellular events that exacerbate the collapse of the fibrillin-deficient aortic wall during aneurysm progression, insights that may provide new targets for therapeutic intervention. These cellular events spatially coincide and temporally include medial calcification, intimal hyperplasia, and adventitial inflammation. In addition to his scientific contributions, Ramirez also held senior administrative positions in the Medical School that included the post of Dean for Research, which he held from 1999 to 2002. In 2002, he left Mount Sinai to become Director of the Laboratory of Genetics and Organogenesis at the Hospital for Special Surgery in New York.

Having established developmental biology as an overarching theme, the Center leadership set about recruiting molecular geneticists. Within a short period of time, the Center assembled an outstanding faculty, three of whom investigate the molecular biology of the early development of the *Drosophila* fruit fly. Leslie Pick, Ph.D., who joined the faculty in 1990, and her laboratory colleagues have attempted to answer the question of how a complex organism, composed of numerous differentiated cell types and integrated organ systems, develops from a fertilized egg. Using the *Drosophila melanogaster* as a model system, Pick and her coworkers have studied how the master regulatory genes (the homeobox genes) establish the basic body plan of the fly during early embryogenesis.[16-18] This group has also investigated how neuronal connections are established in the central nervous system during development, and it has recently discovered that the *Drosophila* insulin receptor functions as an axon guidance receptor in the developing visual system.[19]

Manfred Frasch, Ph.D., recruited from the Max-Planck Institute for Developmental Biology, and his co-investigators have studied the processes of mesoderm patterning and mesodermal tissue development in *Drosophila*, with the goal of defining the identities and functions of regulatory molecules that spatially subdivide the mesodermal cell layer and determine the primordial cells of the heart, skeletal and visceral musculature at defined locations in the early embryo.[20-22] Most of these molecules and regulatory pathways are evolutionarily conserved, and it appears that mesoderm patterning and muscle/heart development in insect and vertebrate embryos involve many closely related mechanisms. Therefore, *Drosophila* can serve as an excellent model system for developmental processes of the heart, skeletal, and gut muscles, as well as for organogenesis of other internal organs in vertebrate embryos.[23]

Marek Mlodzik, Ph.D., and the members of his laboratory also study the *Drosophila*, focusing on the molecular understanding of the signaling pathways that regulate the generation of epithelial planar cell polarity and the signaling pathways that lead to the differentiation of the eye and the antenna in early development.[24,25]

In 1991, James Bieker, Ph.D., one of the Center's earliest faculty members, whose laboratory studies the regulation of gene expression during erythropoiesis, identified a novel, erythroid-specific gene, erythroid Krüppel-like factor (EKLF), that plays a critical and necessary role in establishing adult beta-globin gene expression.[26] The beta-like globin gene cluster is developmentally regulated, with genetic changes in expression from embryonic to fetal to adult life. EKLF consolidates the switch to the adult variant and as such merits designation as the "switching factor" that was a long-sought protein in the field. Genetic ablation leads to a profound beta-thalassemia because of the absence of adult beta-globin expression. The identification of EKLF stimulated investigators in many laboratories around the world to search for EKLF-like homologues, and the KLF-like family now comprises fifteen members.[27]

Two other early faculty members of the Center were Mitchell Goldfarb, Ph.D., and Thomas Lufkin, Ph.D. Goldfarb's laboratory has focused on the control mechanisms signaled by growth factors and growth factor–like proteins in the development and function of the vertebrate nervous system. Seeking to uncover the signaling pathways that promote neuronal differentiation, the laboratory has identified a family of related receptor-associated signaling adaptors, termed SNTs, which may play a central role in triggering differentiation.[28, 29] The group also studies the functions of neuronally expressed proteins that are similar in sequence to fibroblast growth factors (FGF-homologous factors [FHFs]), and it has discovered that FHFs signal inside cells as components of a neural-specific protein kinase signaling molecule.[30, 31] The major interest of Lufkin and his group has been the understanding of the molecular mechanisms that govern normal embryonic development in mammals. Utilizing transgenic mice, these investigators have concentrated on how the homeobox gene affects the development of the central nervous system and the skeleton.[32–34]

Lazzarini and Richard Gorlin, M.D., then Chairman of the Department of Medicine,[35] in a combined venture that was the epitome of Kase's thinking, created a Division of Molecular Medicine within the Department of Medicine. Housed in Center space immediately adjacent to Department of Medicine laboratories and funded primarily by the Department of Medicine, the Division attracted outstanding young medical scientists. An early focus of the Molecular Medicine initiative was molecular cardiology, a study at the molecular level of those events that are important to cardiac function. Mark Taubman, M.D., and Andrew Marks, M.D., from the Harvard University School of Medicine, were the first to be recruited to this Division.[36] The third member of the Division, but with different interests, was Jonathan Licht, M.D., who was recruited from the Dana Farber Cancer Institute in 1991. A hematologist and medical oncologist who also holds an appointment in the Derald H. Ruttenberg Cancer Center, Licht performs research that seeks to understand how abnormal gene regulation may cause a variety of tumors, including acute leukemia, Wilms tumor, and multiple myeloma.[37-42] In 2002, Licht was appointed Vice Chairman for Research in the Department of Medicine and in June 2003 was named the Director of the newly combined Division of Hematology/Oncology.

On the basis of his experiences at the NIH, Lazzarini brought other innovations to Mount Sinai. In association with the leadership of the Center for Neurobiology and under the direction of Kevin Kelley, Ph.D., the Mouse Genetics Shared Research Facility (SRF) was opened in 1999 to provide Mount Sinai faculty with novel, genetically manipulated mouse lines. Kelley's own collaborative research interests span several areas of developmental biology that use the mouse as a model system. The first DNA Sequencing Core Facility for the entire School was also established within the Center. Lazzarini additionally encouraged faculty to use common equipment housed in Center space and created the "Brookdale store," where restriction enzymes were available for purchase by other researchers at a significant discount from usual vendors.

In 1998, after ten years as Director of the Brookdale Center for Molecular Biology, Lazzarini informed Kase of his desire to step down from the position. As Kase himself was retiring, Lazzarini waited until Arthur Rubenstein, M.B.B.Ch., was installed as Dean and then retired from the position. Lazzarini moved further into administration, first taking on the responsibility of directing the renovation of the Bronx Veterans Administration Medical Center research laboratories and then replacing Francesco Ramirez as head of a committee charged with allocating research space on the Mount Sinai campus more equitably.

At the same time (1998) that Lazzarini was stepping down from his position, Panayotis Katsoyannis, Ph.D., Chairman of the Department of Biochemistry since the opening of the School, also retired. This led to a number of major changes in the structure of the various departments. First, Biochemistry and the Brookdale Center merged into a single entity with Ramirez and Mary Rifkin, Ph.D., as codirectors for a short period of time. Rifkin had joined the Center as Associate Director in 1992 following a number of years on the faculty of The Rockefeller University. Her research focused on the molecular biology of parasites. Rifkin played a significant role in Graduate School teaching, and in 1996 she became the Director of the Medical School's Humanities and Medicine Program.[43] In 1999, she was named Dean for Academic Affairs.

Over the ensuing years, other changes took place, and in 2001 Biochemistry separated and merged with the Department of Pharmacology. A number of cell biologists in the Department of Anatomy and in other basic science areas joined the Center, which was renamed the Brookdale Department of Molecular, Cell, and Developmental Biology.

Jeanne Hirsch, Ph.D., a member of the Department of Anatomy and Cell Biology, was one of those who transferred to the Brookdale Department, where she has continued her studies on signal transduction pathways mediated by heterotrimeric G proteins utilizing the yeast cell as a model. Because the mechanism of G protein activation is conserved in all eukaryotes, information obtained from the yeast system is likely to have broad implications for signaling pathways in a wide variety of biological systems. Hirsch's work involves the pathways responsible for nutrient regulation of growth through extracellular signals and the pathway that mediates the response of haploid yeast cells to mating pheromones. The pheromone signal results in changes that prepare the cell for mating.[44, 45]

In February 2002, Paul Wassarman, Ph.D., was appointed Chairman of the Brookdale Department after having served more than two years as the Interim Chair. Recruited to Mount Sinai's Department of Anatomy and Cell Biology in 1996, Wassarman originally trained as a biochemist with postdoctoral training in molecular biology and crystallography. He joined the faculty of the Department of Biological Chemistry at Harvard Medical School in 1972, where he began his research on mammalian fertilization and development and on the proteins involved in fertilization in the mouse. Fourteen years later, he moved to the Roche Institute of Molecular Biology, where he organized and chaired the Department of Cell and Developmental Biology until that Institute closed its doors in 1996. A leading reproductive biologist, Wassarman has served and continues to serve on numerous editorial boards and on many study sections and advisory panels. He has been the chair and organizer of many national scientific meetings and has been a highly sought-after speaker internationally.

While at Harvard, Wassarman and his colleagues, using the mouse model, identified a set of genes that encode three glyco-proteins that make up the extracellular coat (zona pellucida) of oocytes.[46-48] One of the three proteins is the receptor that is recognized by sperm when they bind to unfertilized eggs. The three proteins are also constituents of all mammalian eggs and have been a focus of study for Wassarman's laboratory for more than twenty-five years. The laboratory's current focus is on the surface components of mammalian egg and sperm that account for species-specific binding of sperm to eggs during fertilization. The overall object of the research is to understand the molecular basis of the multiple functions of the glycoproteins during mammalian oogenesis, fertilization, and preimplantation development.[49-51]

The faculty of the Brookdale Department of Molecular, Cell, and Developmental Biology now numbers more than thirty members, the majority of whom have their primary appointment in the Department. Collaboration remains the order of the day. David Sassoon, Ph.D., who also has an appointment in the Derald H. Ruttenberg Cancer Center, investigates the molecular mechanisms that control embryonic patterning and cell differentiation and their applications to cancer cell biology. It is anticipated that the understanding of normal developmental mechanisms that control cell growth and differentiation will have a strong bearing upon the understanding of cancer, where normal regulatory control is lost or misregulated. Sassoon and his laboratory team have recently cloned and characterized a large protein, PW1, that appears to play a role in both cell death and cell survival.[52] The group is also studying the role of the Wnt family of growth factors on female reproductive tract development. Recent observations suggest that the DES syndrome, the propensity of women who received diethylstibestrol (DES) during pregnancy to develop reproductive tract tumors in later life, is due to abnormal regulation of one of the Wnt growth factors.[53]

Reshma Taneja, Ph.D., who also has a secondary appointment in the Cancer Center, has investigated the development of the immune system in the mouse. She and her colleagues have generated mice deficient in a single transcription factor (Stra13), which then go on to develop a systemic autoimmune disorder similar to the human systemic lupus erythematosus.[54]

Scott Henderson, Ph.D., whose research interests concentrate on the functional three-dimensional organization of chromatin in the interphase nucleus, also directs the MSSM Microscopy Center, another institutional shared research facility. Other, more recent faculty recruits include Serafin Pinol-Roma, Ph.D., whose laboratory studies the structure and function of the nucleolus and the exchange of macromolecules between the nucleus and the cell cytoplasm, and Bettina Winckler, Ph.D., a developmental neurobiologist whose major interest is the development of the vertebrate neuron.

The Brookdale Department carries an important teaching responsibility, especially in developmental biology, primarily in the Graduate School. Wassarman directs the Core I curriculum, in which Bieker and Pinol-Roma also teach. Hirsch is the codirector of the Core II curriculum. The Department faculty lecture in the "Molecule and Cells" course in the Medical School and are involved in small-group teaching in the preclinical years.

The Department is currently well funded, with more than $4 million in NIH funding as well as grant support from other sources. Looking to the future, Wassarman hopes to expand further into the field of developmental neurobiology, especially in the area of axon guidance and neuronal plasticity, and to develop the capability to work with other model systems.[55]

CENTER FOR NEUROBIOLOGY

THE ARTHUR M. FISHBERG CENTER FOR NEUROBIOLOGY

THE FISHBERG RESEARCH CENTER FOR NEUROBIOLOGY AND THE KASTOR NEUROBIOLOGY OF AGING LABORATORIES

The Center for Neurobiology was established in 1986 to investigate the nervous system with a multidisciplinary emphasis. The Center's research seeks to understand the anatomy and pathology of the nervous system, as well as the molecular and cellular mechanisms of neuronal function. In keeping with Dean Kase's mandate to create a multidisciplinary environment, it was proposed from the outset that there would be strong interaction and collaboration with neuropathologists and with clinical and basic neuroscientists in Neurology and Psychiatry. These collaborations, along with the Department of Geriatrics and Adult Development, focus on the mechanisms of aging and of psychiatric and neurologic diseases.

In 1986, James Roberts, Ph.D., was appointed the Irving and Dorothy Regenstreif Professor of Neurobiology and the Acting Director of the Center. From the outset, it was the intention to recruit a codirector who would serve with Roberts to lead the new enterprise. While construction of laboratory space was still under way, planning began to develop projects that would investigate the molecular basis of neurodegenerative disease. The major themes that evolved and have been carried forward include molecular neuroendocrinology, the role and control of hypothalamic function, synaptogenesis, and how synaptic circuits are formed and how they change and reform in response to various stimuli to maintain plasticity.

Roberts was recruited to Mount Sinai from the Department of Biochemistry and the Center for Reproductive Sciences of Columbia University, College of Physicians and Surgeons. A molecular neuroendocrinologist, Roberts focused his research efforts on the molecular nature of the neuroendocrine system of the brain, looking primarily at the complex systems of gene expression and gene regulation in the hypothalamus, that area of the brain responsible for the regulation of stress, reproduction, and metabolism. The proopiomelanocortin gene (POMC) and the gonadotropin-releasing hormone gene (GnRH) are central to this regulatory process, and, during his fifteen years at Mount Sinai, Roberts and his associates made major contributions to the understanding of how these genes are themselves regulated and expressed, and how alterations in their expression affect growth and development.[56–63]

Within a year, the nucleus of an outstanding faculty had been recruited and multiple investigations were under way. These included studies of the molecular neuroendocrinology of stress and reproduction; neural growth factor receptor gene expression; and the molecular defects associated with Alzheimer's disease, Parkinson's disease, and schizophrenia. Stephen Salton, M.D., Ph.D., one of the earliest members of the Center, has remained on the faculty. Salton and his colleagues have identified several gene products that are regulated by neurotrophic growth factors and may play an important role in regulating energy balance and hypothalamic function.[64–66]

Mariann Blum, Ph.D., another original member of the Center (and the wife of James Roberts), began an analysis of the relationship of transmitter release and gene expression, work that she continued and expanded on during her tenure at Mount Sinai. An independent investigator and well funded by extramural agencies, Blum was a recipient of the prestigious five-year Irma T. Hirschl Career Scientist Award. In collaboration with Lori Lazar, then an M.D./Ph.D. student at Mount Sinai, Blum was the first to demonstrate that growth factor gene expression remains high in the adult brain, belying the concept that growth factors are important only in early development.[67] She and her colleagues also provided the first report on differential effects of growth factor on the dopamine neurons of the midbrain.[68] In the early 1990s, Blum initiated a successful major collaborative effort with scientists from the Karolinska Institute in Stockholm, Sweden. She also developed a unique murine model of aging and degenerative disease.[69]

The faculty of the Fishberg Center for Neurobiology grew rapidly in size and in 1988 collaborated with the Brookdale Center to establish a core transgenic mouse facility. In that same year, the Center was formally inaugurated as the Dr. Arthur M. Fishberg Center for Neurobiology.

One year later, John Morrison, Ph.D., was recruited to become the other codirector of the Center. Roberts and Morrison led the Center until April 2001, when Roberts and Blum relocated to their home state of Texas. Tragically, Blum's career was cut short by a battle with cancer, which she lost in 2003.

Morrison now serves as the sole Director of the Center. After receiving his Ph.D. from Johns Hopkins University School of Medicine and after a postdoctoral fellowship at Hopkins, Morrison spent two years as a postdoctoral fellow in the laboratory of Floyd Bloom, M.D., at the A. V. Davis Center for Behavioral Neurobiology of the Salk Institute in San Diego. He then joined the faculty of The Scripps Research Institute, where he remained until his move to Mount Sinai in 1989. A leading figure in the field of neurobiology, Morrison has served on numerous study sections, review committees, and panels of the NIH, has held leadership positions in regional and national organizations, and serves on the editorial boards of many neuroscience journals.

The Alfred B. and Gudrun J. Kastor Neurobiology of Aging Laboratories (KNAL), directed by Morrison and jointly sponsored by the Fishberg Research Center for Neurobiology and the Brookdale Department of Geriatrics and Adult Development, were established in 1996. With all of KNAL's faculty members having joint appointments in Neurobiology and Geriatrics, the Kastor Laboratories, which now constitute about one-half of the Center, provide another example of how the linkage of a basic science department and a clinical science department can produce a strong research enterprise.

For more than fifteen years, Morrison's research has focused on the aging brain and Alzheimer's disease. In collaboration with Patrick Hof, Ph.D., who accompanied Morrison to Mount Sinai as a postdoctoral fellow and is now the Regenstreif Research Professor of Neuroscience, he has characterized in detail the neuronal cell type in the human cerebral cortex that is selectively vulnerable to degeneration in Alzheimer's disease and has advanced a cellular hypothesis as to why Alzheimer's disease devastates certain cortical circuits and functions.[70, 71] Morrison and Hof have also presented a comprehensive analysis of the key differences between Alzheimer's disease and normal aging, the most important part of which is that normal aging is not accompanied by neuron death. In this same publication, the authors also proposed that the most important changes coincident with normal aging were synaptic alterations in otherwise intact circuits.[72]

Morrison's laboratory also reported that in one particular circuit that mediates memory, critical receptors are decreased, although the circuit remains intact. In Alzheimer's disease, this circuit deteriorates, suggesting that the more moderate memory decline seen in normal aging is due to molecular changes at key synapses in intact circuits. Since the circuit remains intact, the hypothesis was generated that this molecular alteration linked to aging might be reversible.[73] More recently, members of the laboratory have shown for the first time that the aging brain responds differently to estrogen stimulation than the young brain and lacks the capacity to form new synapses, a capacity that is so robust in a young brain.[74]

All of the members of the Center have been remarkably productive. Andrea Gore, Ph.D., who first came to Mount Sinai in 1991 as a neuroendocrinology postdoctoral fellow in Roberts's laboratory, focused on the changes that occur in neuroendocrine function at both ends of the life cycle. The major interest of her laboratory has been the regulation of gonadotropin-releasing hormone (GnRH) gene expression during development and puberty, and during menopause and aging in female mammals. As GnRH is a key hormone controlling reproductive function, these studies have helped shed light on the timing and onset of puberty and menopause and on the maintenance of the normal ovarian cycle.[75–81] Active in the Women's Faculty Group at Mount Sinai and the recipient of the Faculty Council Award for Academic Excellence in 2001, Gore relocated to Texas in 2003 and joined the faculty of the University of Texas at Austin.

Charles Mobbs, Ph.D., and his associates have devoted their energy to the study of the basic molecular mechanisms by which hypothalamic neurons sense and regulate the metabolic state, including body weight and food intake, and how these mechanisms are impaired in metabolic diseases and during aging.[82–84] The role of the hypothalamic proopiomelanocortin (POMC) gene in obesity and diabetes has been the target of numerous investigative efforts.[85] In animal models, POMC activity is reduced in diabetic and obese mice, particularly those in which the obesity and diabetes are attributable to a deficiency of the hormone leptin. Patients with mutations of this gene may also be obese and diabetic. In an era where obesity has become a national epidemic, this work can have profound clinical relevance. As Mobbs has stated:

> A driving question of our laboratory is what may be called the metabolic mystery. This refers to the fascinating phenomenon that obesity is a risk factor for most age-related diseases and indeed for mortality, and conversely dietary restriction appears to slow down the aging process and extend maximum lifespan. Considering that almost all major pathologies are influenced by caloric intake, the mechanisms underlying the metabolic mystery may be considered among the most compelling in biomedical science.[86]

In November 2003, Mobbs and his colleagues reported a major breakthrough.[87] Utilizing leptin-deficient mice in which POMC expression is diminished, they showed that increased expression of POMC attenuated the overeating response to fasting and reversed metabolic impairments, returning the glucose of diabetic animals to normal. The implications of the work have enormous potential, since drugs that mimic POMC are currently in development.

Synaptogenesis has been approached by Morrison in the context of the aging brain; Deanna Benson, Ph.D., and George Huntley, Ph.D., have investigated the process in the development of the adult brain.

Synapses, the complex wherein a nerve impulse passes from one neuron to another, form in restricted regions and cellular domains and ultimately dictate the flow of information. Benson has studied synaptic assembly during development and the changes that occur in response to injury. Huntley has focused on the mechanisms of plasticity through which brain function is modified by experience.[88, 89] The two have also investigated the role of cadherin and other adhesion molecules in the formation and maintenance of the synaptic junction.[90, 91]

The Center added strength in behavioral neuroscience research with the recruitment of Peter Rapp, Ph.D., and Matthew Shapiro, Ph.D. Rapp is a leading expert on cognitive decline in aging, particularly in nonhuman primate models, and in 2002 published a seminal paper on the effects of estrogen on cognition in aged primates.[92] Using animal models of human memory and amnesia, Shapiro studies how the brain encodes, stores, and retrieves information in memory. Both Shapiro and Rapp have added important elements of behavioral neuroscience to the Center and have forged several active collaborations with other members of the Center and with the Mount Sinai neuroscience community.

The laboratories of the Center's faculty have been popular sites for research participation by the medical students, especially the M.D./Ph.D. students. Faculty members provide substantial input to the teaching of "Brain and Behavior," the main neuroscience course for the Medical School, and they serve as preceptors for a variety of research projects in neurobiology. Center faculty also form a large part of the Interdisciplinary Graduate Training Program in Neurosciences of the Graduate School of Biological Sciences. This program represents one of six training areas that compose the Ph.D. program and the Ph.D. component of the M.S.T. program (M.D./Ph.D. program). In addition, Shapiro and Rapp recently launched a new track in Cognitive Neuroscience, adding an important component to Ph.D. education in neuroscience at Mount Sinai. Moreover, the Center collaborates with other academic departments to deliver comprehensive medical education through such initiatives as joint seminars, instruction at the residency level, and postdoctoral fellowships.

Morrison strongly believes that in addition to the studies in neural repair now being carried out in the Center, the time is ripe for major advances in the area of neural protection and repair.[93] As a consequence, the next decade will see not only greater collaboration in the neuro-science community of the institution but also the inclusion of experts in stem cell research, gene therapy, and transplantation.

THE DERALD H. RUTTENBERG CANCER CENTER

"The study and treatment of cancer at Mount Sinai transcends any one Department or Division, encompassing the efforts not only of oncolo-gists, but also of chemists, microbiologists, cell biologists, pathologists, radiologists, radiotherapists, environmental scientists, and surgeons, among others."[94] Indeed, the study of cancer at Mount Sinai has tran-scended time itself. In 1885, Arpad Gerster, M.D., then the chief of surgery at The Mount Sinai Hospital, was the first to suggest that a sur-gical procedure might hasten the spread of cancer.[95] In 1937, Richard Lewisohn, M.D., upon his retirement as an active surgeon, established Mount Sinai's first designated cancer research laboratory, where he and his colleagues were the first to establish the significance of folic acid in the biology of the cancer cell.[96] In summing up the first five years of his research efforts, Lewisohn commented: "After all, most of us agree that the reputation of a great hospital is made not only by the proper care of the patients—all modern hospitals perform that function admirably— but also by important contributions of its members to medical research. By doing so they may advance the clock of human accomplishments and thus help to diminish the great reservoir of human suffering."[97] After detailing the accomplishments of the research, Lewisohn con-cluded: "It is the charm and the joy of research that the future of any research problem is full of hope and speculation."[98]

But the hope of cancer research lies in the future, as the develop-ments are spread over years of research effort. This was the case with Charlotte Friend, Ph.D., a basic scientist recruited by Mount Sinai in 1965 when the School was in formation. Friend made two pioneering contributions to cancer research, fifteen years apart. The first was her development, in 1957, of a murine model for leukemia, caused by what became known as the Friend leukemia virus; this at a time when it was still not universally accepted that cancer could be caused by a virus.[99] Later, from her laboratory at Mount Sinai, she and her colleagues were the first to note that dimethyl sulfoxide (DMSO) could induce cancer cells to progress or differentiate to a normal pattern of development.[100]

This work created a new field of cancer research, cell differentiation, which continues to this day not only at Mount Sinai but worldwide.

It has been the norm in the Medical Center that whenever discussions have taken place as to what areas of clinical or basic science should be stressed in the institution, cancer is at or near the top of everyone's list. A multidisciplinary cancer center, therefore, was a natural choice for Kase to champion as he created new science entities.

In 1988, the Derald H. Ruttenberg Cancer Center was formally dedicated, with James Holland, M.D., the Jane B. and Jack R. Aron Professor of Neoplastic Diseases, as its Director. An internationally recognized medical oncologist, the author of more than five hundred publications, and the recipient of the prestigious Albert A. Lasker Award, Holland had created the Department of Neoplastic Diseases in 1973 and served with great success as its Chairman for two decades.[101] The establishment of the Ruttenberg Cancer Center provided significant additional support for the more than 125 physicians and scientists who conducted the clinical and research activities in neoplastic diseases. The clinical component of the new Center consisted of a fifty-three-bed inpatient unit; a Chemotherapy Day Unit provided the outpatient services. Multiple clinical trials and epidemiological studies were in place, and research scientists studied the basic controlling mechanisms for the expression of cancer genes, the selective targeting of cancer cells by attaching a cytotoxic molecule to a monoclonal antibody directed at the cell, and the mechanisms by which cancer cells develop resistance to chemotherapeutic agents.

In 1993, Holland reached retirement age for a department chairman but continued both his clinical and research activities. The Department of Neoplastic Diseases was absorbed back into the Department of Medicine as the Division of Neoplastic Diseases, and Stuart Aaronson, M.D., was appointed the Director of the Ruttenberg Cancer Center and the second Aron Professor of Neoplastic Diseases. Under Aaronson's leadership, the Center was given authority to grant primary faculty appointments and was provided with dedicated space to expand cancer research at Mount Sinai.

Born in Michigan and raised in California, Aaronson received his medical training at the University of California, San Francisco, and after an internship he joined the Viral Carcinogenesis Branch of the National Cancer Institute (NCI) in 1967 as a Staff Associate. It was an exciting time to be at the NCI because the "war on cancer" was being mounted, additional funding was being made available, and "the opportunity to do exciting research at a high level was ripe."[102] Rising rapidly through the ranks, Aaronson was appointed Chief of the Laboratory of Cellular and Molecular Biology of the NCI in 1977, a position he held for sixteen years until coming to Mount Sinai in 1993.

At the NCI, Aaronson began studies on the SV40 DNA tumor virus, which had the capacity to transform normal cells. He then noted that RNA tumor viruses were more tumorigenic in animal models and that these viral-like elements were present as endogenous viruses in chickens, mice, and other species. In certain mouse strains, these viruses were the inciting agent for a form of leukemia, while others caused sarcomas. Although many of these viruses were defective, they had all of the information required within the viral genome to cause cell transformation. Aaronson's laboratory subsequently developed important cell lines for studying these phenomena.[103–105] With the advent of molecular cloning, it was then discovered that the viruses had the ability to capture the cellular genes known as oncogenes; the oncogenes were responsible for the transforming activity.

One of the most important contributions at the time was the discovery that the gene sequence of one of the cell-derived oncogenes was identical to a sequence of the platelet-derived growth factor (PDGF).[106, 107] When this sequence was expressed in a cell that had a receptor for that growth factor, that cell could develop a continuing proliferative drive leading to malignancy. Aaronson's work on growth factors led to the identification of a gene related to the epidermal growth factor that was amplified and overexpressed in a primary breast cancer.[108] This gene, designated erbB2 or HER2/Neu, is overexpressed in about 25 percent of breast cancers. When HER2/Neu is present, the tumors are aggressive and have a poorer prognosis than when it is absent. In 1998, the Food and Drug Administration approved a therapeutic monoclonal antibody that, in combination with aggressive chemotherapy, has improved survival of patients with overexpression of HER2/Neu. Aaronson discovered another growth factor, keratinocyte growth factor (KGF), in the late 1980s.[109, 110] This growth factor targets epithelial cells and is under study by industry with the goal of developing drugs to combat the mucositis caused by chemotherapy or radiation.

As the new Director of the Center, Aaronson set out to recruit scientists who would expand the understanding of the molecular basis of cancer development and then devise new diagnostic and management strategies based on emerging insights. Major goals were the identification of genetic markers that might improve diagnosis, prognosis, and therapy and the development of multidisciplinary, organ-specific tumor initiatives.

One of the factors that led Aaronson to decide to move to Mount Sinai was the commitment on the part of the Board and the senior administration to build a new research building that would include the laboratories of the Ruttenberg Center. It would be almost four years from the time of Aaronson's arrival until the laboratories became functional. Aaronson and his biology program moved into the East Building in 1997. The Center occupies almost one and one-half floors (about 40,000 gross square feet) in the building and houses basic scientists of the Cancer Biology Program and investigators in the Center's Cancer Prevention and Control Program. The activities are supported by almost $15 million per year in grants. In addition, the Center has administrative responsibility for a number of federally funded cancer-related training programs.

The Center has had a significant impact on research funding at Mount Sinai. Under Aaronson's leadership, NCI funding has increased tenfold; an institutional initiative to enhance Mount Sinai's infrastructure for cancer research resulted in the awarding of three NCI Shared Resource grants in 2000. Four additional resource grants were awarded in 2002. These five-year resource awards, reflecting more than $7.5 million in total funding, provided important shared support for DNA sequencing, mass spectrometry, imaging, vector production, cancer biorepository, biostatistics, and mouse oocyte microinjection.

In 2000, the Cancer Center launched a new Developmental Research Program to support innovative translational pilot projects in cancer research throughout Mount Sinai. Utilizing an internal peer review process, the program awards pilot research funding to a team of collaborative investigators for the generation of preliminary data that will result in expedited applications for externally funded translational cancer research projects. Funding for the initial ten pilot projects totaled $200,000; several projects subsequently received externally funded cancer grants.

On the clinical side, the multidisciplinary Derald H. Ruttenberg Ambulatory Treatment Center, built with the generous support of the

Ruttenberg family, was dedicated in 2002 to foster multidisciplinary outpatient care for Mount Sinai's cancer patients.

Over the course of his thirty-five-year career as a preeminent cancer biologist, Aaronson has received numerous honors and has served on the editorial board of every major journal in the field of cancer. In addition, he has organized many national and international meetings, served on the advisory boards of a number of cancer centers, and was elected President of the Harvey Society in 1996.

The Cancer Biology Program

The Ruttenberg Cancer Center has about twenty primary faculty members, many of whose research is laboratory focused. The most senior member of the research faculty currently is Ze'ev Ronai, Ph.D., who obtained his graduate degree from The Hebrew University in Jerusalem, Israel, in 1985. He established the Molecular Carcinogenesis Program at the American Health Foundation in Valhalla, New York, and was recruited from there in 1997 to lead the Cancer Center's laboratory research initiatives in cell signaling. Among his current research interests are studies on the important role stress kinases play in cellular responses and in the regulation of protein stability.[111–114] Using melanoma tumors as a model, Ronai and his colleagues are also investigating the reasons why melanoma cells are resistant to most therapies and are seeking ways to render the cells sensitive to chemo- or radiotherapy.[115, 116] Ronai has also conducted research on the role of ubiquitination and proteasome degradation of proteins and the role of RING (Really Interesting New Gene) finger proteins in the process.

Zhen-Qiang Pan, Ph.D., and his laboratory associates have discovered an important RING finger protein family and are researching its function in regulating the abundance of substrate proteins required for the control of the progression of the cell cycle, the activation of signal transduction pathways, and the execution of tumor suppressor activities.[117–121]

The Cancer Center was devastated, in 1998, by the deaths of Eugenia Spanopoulou, Ph.D., along with her husband Andrew Hodtsev, Ph.D., both members of the Center's faculty, and their son, Platon, in the crash of Swissair flight 111. Spanopoulou, recruited in 1994, was a gifted researcher who had been awarded a Howard Hughes Medical

Investigatorship in 1977. In addition to her research on immune system mechanisms and their potential role in the development of leukemia and lymphomas, Spanopoulou is also remembered as an outstanding mentor to her students.[122–127]

James Manfredi, Ph.D., and his colleagues have focused their studies on the determinants of the cellular response to the tumor suppressor protein p53.[128–130] This gene, depending upon particular cellular conditions, has been found to play a role in inducing cellular growth arrest or apoptosis and is abnormal in a majority of malignancies. The work has significant clinical relevance because elucidating the molecular mechanisms that are responsible for the ability of p53 to induce either growth arrest or apoptosis may lead to more effective therapeutic interventions and/or a way to overcome the chemotherapy-resistant phenotype found in many tumors.

Like p53, the Ras family of proteins is mutated in a high percentage of human tumors. Localized to the inner surface of the plasma membrane, the Ras genes are the subject of investigation by Andrew Chan, Ph.D., and his team, which has identified several related members of the gene family and is studying their biological functions.[131–135] The group is also studying the regulation of the human PTEN tumor suppressor protein in normal and tumor cells. Working in close collaboration with Scott Friedman, M.D., Chief of the Division of Liver Diseases, and his laboratory colleagues, the group has identified a tumor suppressor gene that is mutated in more than 75 percent of patients with prostate cancer.[136]

New evidence suggests that the metastatic dissemination of epithelial tumor cells is also strongly influenced by the activity of cell adhesion molecules, in particular members of the cadherin family. Rachel Hazan, Ph.D., has shown that N-cadherin is upregulated in invasive breast cancer cells and promotes invasiveness of non-invasive breast cancer cell lines.[137, 138]

Although the lymphatic system is the most important pathway for tumor metastasis, little is known about the mechanisms by which tumor cells interact with lymphatic vessels. This fertile field is the source of investigation by Mihaela Skobe, Ph.D., who was recruited to the Center in 2001. Current studies in Skobe's laboratory are focused on understanding the mechanisms of VEGF-C-mediated metastasis and on investigating the specific adhesive interactions of cancer cells with the lymphatic endothelium that are critical in the earliest stages of metastasis.[139]

Cell cycle regulation and control is the major interest of Matthew O'Connell, Ph.D., who joined the Center in 2003. O'Connell's research emphasizes the identification of cell cycle checkpoints that respond to DNA damage and delay the onset of mitosis, DNA repair and its coordination with cell cycle progression, and the molecular events that take place during cell mitosis.

Aaronson's current research includes investigations into cancer genes and their signaling pathways, cellular senescence in aging and cancer, the study of the familial breast cancer genes BRCA1 and BRCA2, and the genetics of Wnt signaling in human tumors. The Wnt genes belong to a family of protooncogenes that are expressed in species ranging from *Drosophila* to humans.[140–146] Toru Ouchi, Ph.D., a collaborator of Aaronson's, has also directed his efforts to examine the breast cancer genes and their relationship to p53 and other genes that are involved in the development of cancer.[147–150]

In addition to the primary faculty members of the Center, many Mount Sinai scientists have secondary appointments in the Ruttenberg Cancer Center, and much of their work is cancer related. Many of their accomplishments are documented in other sections throughout this volume.

The Cancer Prevention and Control Program

In 1997, William Redd, Ph.D., was recruited from the Memorial Sloan-Kettering Cancer Center to develop and lead the Cancer Center's Prevention and Control Program. The program's mission is to increase understanding of the role of psychological, social, and biobehavioral factors in the prevention of cancer and in the control of aversive sequelae of cancer and its treatment. The program's efforts focus on three distinct areas: research, education, and outreach. Research currently includes the investigation of factors affecting participation in cancer screening and genetic testing, lifestyle modification to reduce cancer risk, biobehavioral factors in adjustment to cancer and its treatment, and long-term adjustment among cancer survivors and their families. The education effort includes two postdoctoral training programs, one directed toward psychosocial issues in breast cancer and the other directed toward issues in cancer prevention and control. The outreach effort focuses on enhancing the public's cancer knowledge and awareness and on expanding the participation of medically underserved minority groups in cancer research.

A behavioral psychologist, Redd focuses his research on five broad areas: classical (Pavlovian) conditioning during cancer treatment, cognitive-behavioral intervention to reduce distress during invasive medical procedures, psychobehavioral factors in nonadherence to cancer screening guidelines, long-term adjustment in mothers of bone marrow transplant survivors, and posttraumatic stress disorders in cancer patients and their families.[151-153] A major component of his current clinical research is the study of how to increase participation in colorectal cancer screening among medically underserved minority groups. An important outreach is the East Harlem Partnership for Cancer Awareness, a collaboration with the Mount Sinai School of Medicine, Metropolitan Hospital Center, Boriken Neighborhood Health Center, and Settlement Health that seeks to address the disproportionate incidence of cancer among the predominantly Hispanic and African American residents of East Harlem.

Christine Ambrosone, Ph.D., recruited in 2000 to direct the Cancer Epidemiology Program in Cancer Prevention and Control, has research interests that focus on molecular epidemiology, and the role that interindividual variability in metabolism may play in both cancer etiology and prognosis after treatment for cancer. She is interested in the impact of differences in carcinogen metabolism as they affect the relationship between smoking and cancer and the effects of variability in steroidogenesis and hormone metabolism on the risk of breast, lung, and prostate cancer.[154-157]

Dana Bovbjerg, Ph.D., Guy Montgomery, Ph.D., and Heiddis Valdimarsdottir, Ph.D., have investigated not only unique behavioral aspects of patients with breast cancer but also the responses of families and friends. Although each investigator has his or her special interests, they have functioned as a team and published extensively, both together and separately.[158-166] Bovbjerg's expertise is in the interaction of cognitive and emotional factors on behavior and biology that are relevant to cancer etiology, progression, and treatment. His research contributions include the use of animal models of classical conditioning to demonstrate that the immune system is regulated by the central nervous system, studies on classically conditioned reactions in patients receiving chemotherapy for cancer, and biobehavioral investigations of stress in healthy women who have family histories of breast cancer.

Montgomery's research program focuses on the psychological mechanisms that directly contribute to patients' experiences of cancer-related symptoms. His research is based on the hypothesis that cognitive factors (e.g., patients' expectations of specific symptoms) are likely to play an important role in cancer patients' experiences of adverse side effects. Valdimarsdottir is studying the psychological, behavioral, and biological consequences of being at familial risk for cancer, including investigating the psychosocial issues that surround genetic testing for cancer susceptibility. The overall research goal of Joel Erblich, Ph.D., is to better understand how genetic factors impact on smoking behavior. Projects in this area include studies of the effects of dopamine-related genetic polymorphisms on smokers' cigarette cravings induced by laboratory stressors and other manipulations. Other projects have centered on the stress associated with familial risk for breast cancer and genetic counseling.

The Clinical Care Program

On any given day, more than two hundred patients with a diagnosis of cancer are treated in the dedicated cancer units (now called the Oncology Care Center) and other patient care units of The Mount Sinai Hospital. Patients in the Oncology Care Center are cared for by the oncology faculty and a cadre of nurses specially trained to manage the needs of cancer patients receiving chemotherapy and/or immunotherapy for advanced cancer.

In 1998, Janice Gabrilove, M.D., was recruited from the Memorial Sloan-Kettering Cancer Center to head the Division of Neoplastic Diseases of the Department of Medicine and the clinical care program of the Ruttenberg Cancer Center.[167] A graduate of the Mount Sinai School of Medicine (Class of 1977), Gabrilove is a pioneer in the understanding of the biology of hematopoietic growth factors and their clinical utility in promoting hematopoietic reconstitution. Gabrilove played a critical role in the development and design of the Ruttenberg Cancer Treatment Center on the first floor of the Guggenheim Pavilion. Formally dedicated in December 2002, the Treatment Center has fourteen examination rooms, sixteen infusion stations, an on-site laboratory and pharmacy, and an educational resource area. The Treatment Center provides the Medical Center with a state-of-the-art treatment facility for the outpatient care of cancer patients from the newly created Hematology/Medical Oncology Division of the Department of Medicine and in the Division of Gynecological Oncology of the Department of Obstetrics, Gynecology, and Reproductive Sciences.

In addition to the existing Cancer Registry, a critical research resource that documents long-term care of Mount Sinai's cancer patients, the Ruttenberg Center has implemented a Familial Cancer Registry to identify individuals from families at high risk of developing colon cancer. These individuals have access to the cancer genetics testing and counseling program of the Department of Human Genetics.

The Training Programs

Clinical research fellowships in Hematology/Medical Oncology are administered through the Department of Medicine. Designed to produce academic leaders, the fellowship program, codirected by Gabrilove and Jonathan Schwartz, M.D., consists of clinical rotations, a multifaceted educational program, and a career track rotation where fellows choose a research mentor and pursue a research project.

The Ruttenberg Cancer Center sponsors an NCI-funded predoctoral research training program under Aaronson's leadership and codirected by James Manfredi, Ph.D. The research areas include oncogenes and signaling pathways, tumor suppressors and cell cycle regulation, transcription factors, differentiation and development, metastasis and cell adhesion, and DNA replication and repair. The research is combined with a curriculum designed to challenge the trainee to consider the translational aspects of their work in improving the diagnosis and treatment of cancer. Postdoctoral positions are also available in cancer cell biology through an NCI-funded training program under the leadership of Ze'ev Ronai, Ph.D.

The Center's Cancer Prevention and Control Program sponsors a postdoctoral training program in Biobehavioral Breast Cancer Research that is designed to provide trainees with the broad-based scholarly background and the intellectual skills needed to conduct interdisciplinary biobehavioral research in breast cancer.

Thanks to continuous strong leadership, recruitment of an outstanding group of scientists, and the spirit of cooperation so vital to

Mount Sinai's mission, the Derald H. Ruttenberg Cancer Center plays an ever-expanding role in the life of the Mount Sinai Medical Center.

THE CARL C. ICAHN CENTER FOR GENE THERAPY AND MOLECULAR MEDICINE

The concept of gene therapy began to come closer to realization in the 1980s as a result of the rapid strides being made in molecular biology. Mount Sinai, with major programs already in place in cancer, neurobiology, and genetics, was fertile soil for gene therapy to thrive in, and in the early 1990s a decision was made to create a strong presence in this promising field. The search for a founding Director for the Institute for Gene Therapy came to fruition with the appointment of Savio Woo, Ph.D., on September 1, 1996. The Institute was renamed the Carl C. Icahn Center for Gene Therapy and Molecular Medicine in 2000, in honor of Mr. Icahn for his generous support of the Center.

Woo, originally from Hong Kong, was trained as a biochemist and received his Ph.D. in Biochemistry from the University of Washington in 1971. As for the transition from biochemistry to molecular biology and gene therapy, Woo has stated, "I always had a real interest in research in biomedicine. . . . I always wanted to have some application potential of my work in future years. So with that in mind, I went for postdoctoral training in a leading laboratory that was working on phenylketonuria, which is an inborn error in metabolism that predisposes untreated children to severe and irreversible mental retardation."[168]

Continuing his quest to move into the field of biomedicine, in 1973 Woo began a twenty-three-year career at the Baylor College of Medicine in Houston, Texas, studying steroid hormone-regulated gene expression in the laboratory of Bert O'Malley, M.D. Woo rose rapidly through the academic ranks at Baylor, was appointed a Howard Hughes Medical Institute Investigator, and served for twenty years. Woo's career evolved and flourished in tandem with the burgeoning interest and extraordinary advances in the field of molecular biology. He was not only a Professor of Cell Biology and Professor of Molecular and Human Genetics but also the founding Director of the Graduate Program in Cell and Molecular Biology, and in 1991 he was appointed the founding Director of the Baylor Center for Gene Therapy.

In the 1970s, Woo made major contributions to O'Malley's work on the cloning of the chicken ovalbumin gene and demonstrated that the gene is split. It was in the 1980s, however, as an independent investigator, that Woo set the stage for his entrance into the field of gene therapy with his landmark work in cloning the gene for phenylketonuria (PKU), his studies on the molecular basis of the disease, and the description of the first mutation to be identified in PKU.[169–172] PKU is a rare hereditary condition in which the amino acid phenylalanine is not metabolized properly because of a faulty gene for phenylalanine hydroxylase, an enzyme required for the conversion of phenylalanine to tyrosine. The excess phenylalanine can lead to mental retardation; the only available treatment consists of consuming a restricted diet low in phenylalanine. Woo's cloning of the PKU gene was a first in the molecular genetics of metabolic disorders in humans. It paved the way for the discovery of a whole set of disease-causing mutations and served as a paradigm for future research in the field of inborn errors of metabolism in children. Worldwide collaboration on the molecular genetics of PKU ensued; many mutations were identified and mapped, their geographical migration was studied, and it became possible to correlate the genotype with the patient's phenotypic expression. This enabled investigators to predict the outcome of the PKU,[173] which varies in severity, and also to establish a prenatal diagnosis for the disorder in families at risk, leading Woo to comment: "I was finally able to see that what I do has translational medical applications!"[174] It also piqued Woo's interest in the prospect that gene therapy could cure PKU and many other genetic disorders and stimulated him to pursue the field with vigor. Woo also made landmark contributions to the field of gene therapy for hemophilia that led to his selection as a corecipient of the 1999 National Hemophilia Foundation's Researcher of the Year Award.

Woo's work led to many other honors and also brought him to Mount Sinai's attention as a prime candidate to head and develop the institution's gene therapy initiative. The overture came at a propitious moment. Woo was already the Director of the Gene Therapy Center at Baylor College of Medicine, but the ravages of managed care had hit Texas, and the College lacked sufficient funding. The lure of the new East Building, the guarantee of the space and resources, and the forcefulness of the vision of the Mount Sinai leadership and its Boards of Trustees led Woo to move to New York and Mount Sinai in 1996.

At Mount Sinai, Woo and his laboratory collaborators have expanded their efforts on gene therapy in cancer.[175] Gene therapy of cancer has been conducted by attempting to deliver a "suicide" gene to the tumor cell that will destroy it directly, a technique shown to be effective in laboratory mice bearing different tumors. Three Phase I clinical trials for patients with metastatic colorectal, breast, and prostate cancers have been carried out at Mount Sinai.[176] These trials established that there was no toxicity, but unfortunately also little efficacy. Another approach under study by Shu-Hsia Chen, Ph.D., a long-time associate of Woo's, is a combined approach that utilizes a "suicide" gene in combination with a gene that confers antitumor immunity. It has been shown that adenovirus-mediated interleukin-12 gene therapy and T cell activation lead to simultaneous induction of both innate and adaptive immune responses to tumor burdens in an animal model, resulting in cure for animals with metastatic disease.[177] Another promising area of investigation in cancer therapy has been the inhibition of angiogenesis. Woo and his colleagues have demonstrated significant reduction of tumor growth rates and tumor volume after the systemic administration of vector-mediated antiangiogenic gene therapy.[178]

With the support of the institution and Carl C. Icahn's major philanthropic gift, the Center in 2003 had thirteen faculty members with primary appointments. These are independently funded investigators, and others are still being recruited. In addition to the research on gene therapy per se, other major areas of investigation are stem cell research, immunology, and gene vector development.

Leading the group in stem cell research is Gordon Keller, Ph.D., who was recruited from the National Jewish Medical and Research Center in Denver, Colorado, in 1999. One of the earliest pioneers in the field of embryonic stem cell biology, Keller focuses his research on defining and characterizing the early events involved in the establishment, growth, and maturation of the embryonic hematopoietic and vascular systems. He has developed an in vitro model for studying the differentiation of embryonic stem cells into blood cells, which is used for the rapid analysis of genes that are essential for normal embryonic development. Utilizing this model, Keller and his associates identified a novel precursor cell that has the potential to generate cells of both hematopoietic and endothelial lineage and represents the earliest precursor cell of its type described to date.[179–181] The group has isolated a number of novel genes that are expressed at this early stage of development, and it is defining the function of these genes and their role in developmental hematopoiesis and vascular biology.[182, 183] Keller has also collaborated with the Division of Endocrinology, Diabetes, and Bone Diseases of the Department of Medicine to study early thyroid embryogenesis.[184]

Hans-Willem Snoeck, M.D., Ph.D., is continuing studies that he began in Belgium prior to coming to Mount Sinai on the biological mechanisms that determine whether a hematopoietic stem cell will renew itself or undergo differentiation. These cells are vital in bone marrow transplant patients as they reconstitute the hematopoietic system following massive radiation. Since it is known that, at least in the mouse model, the kinetic behavior of this group of cells changes with aging and is genetically determined, Snoeck's group has focused on both ends of the life cycle, investigating the mechanisms responsible for age-related changes and also the genetically determined differences in the size of the stem cell pool.[185, 186]

Jonathan Bromberg, M.D., Ph.D., Professor of Surgery and Surgical Director of the Recanati/Miller Transplant Institute and Chief of Kidney/Pancreas Transplantation as well as a Professor of Gene Therapy and Molecular Medicine, is one of the senior investigators in the Icahn Center studying immunologic aspects of gene therapy. Prior to his recruitment to Mount Sinai, Bromberg was one of the first to show that gene transfer of immunosuppressive cytokine genes to allografts and the subsequent expression of these genes within the allograft would prolong graft survival.[187] The major focus of Bromberg's current research is mechanisms that can be employed to induce tolerance in either grafted organs or cells.[188] The process whereby a graft is either rejected or tolerated is controlled by the trafficking and complex interaction of a variety of receptors and ligands, which include adhesion molecules and chemokines. Bromberg and his associates are studying interleukin-10, a molecule that plays a major role in immunosuppression, and they have identified a unique subpopulation of T cells that express high levels of the IL-10 receptor.[189, 190] Two other areas also under investigation are the ability of gene transfer and gene therapy to deliver immunosuppressive molecules to allografts, resulting in local immunosuppression without the side effects associated with conventional systemic immunosuppression,[191, 192] and the use of stem cells or mature pancreatic progenitor cells to provide a source of pancreatic islet cells for transplantation and the cure of type 1 diabetes.

Gwendalyn Randolph, Ph.D., recruited to the Center in 1999, identified a novel pathway by which monocytes differentiate into dendritic cells in vitro, consistent with a mechanism by which they may convert in vivo. If a monocyte, after migrating into the cellular matrix, remains in the tissue, it becomes a macrophage. If it migrates back into a vessel, it becomes a dendritic cell.[193] The dendritic cells are responsible for transporting antigens to lymph nodes, and as such they can play a vital role in immunotherapeutic approaches to disease, in the design of vaccines, and in the induction of tolerance in transplantation.[194] Randolph and her colleagues have also identified two molecules that mediate trafficking of dendritic cells into lymph nodes.[195–198] Her achievements at such a young stage in her career are globally recognized and led to her selection by the Mount Sinai Faculty Council as the recipient of the Outstanding Young Investigator Award in 2003.

The Center's gene vector development program is headed by Michael Linden, Ph.D., who studies the recombinant adeno-associated virus (AAV), the vector of choice for the delivery of genes to treat genetic disorders. AAV holds great promise for several reasons: it has not been implicated as causing any human disease; it can efficiently infect cells; and the virus's DNA integrates into a specific site within the host genome.[199] The objective of Linden's research is to investigate the biochemical requirements for the targeting of site-specific integration of AAV and to attempt to integrate the AAV into selected loci in addition to the usual target sequence.[200–202] Underlying this attempt is the goal of replacing a defective host gene with the normal gene.

Woo and the faculty of the Gene Therapy Center consider the teaching of graduate students to be of great value because it provides the opportunity for the faculty to attract the students into their laboratories. In addition to the course "Advanced Topics in Gene Therapy," the Center has a significant teaching role in a number of other Graduate School courses including some of the following: "Advanced Topics in Human Genetics," "Cellular Physiology and Ion Channels," and "Genetics and Genomic Sciences." The faculty also participates in the small-group teaching sessions of the Medical School's preclinical years.

In concert with the other Centers and Institutes, the Icahn Center has responsibility for shared institutional facilities. The Vector Core and GMP Production Facility provides viral vectors for laboratory research and clinical translational studies. The Morphology and Assessment Core offers assistance and/or collaboration to investigators whose basic science or clinical research projects require light microscopic and molecular pathological analyses. The Clinical Immunology Core offers a variety of services directed at measuring cancer patients' cell-mediated immune responses to their tumors.

Looking to the future, Woo has no illusions that gene therapy or stem cell therapy by themselves will become a cure-all. On the other hand, he believes that, in the next decade, gene and cell therapeutics will have a major impact on the treatment of cancer and other major diseases in society. A realist, Woo sees these novel therapeutic approaches to disease as another avenue to lengthen human life and to improve its quality.[203]

THE CENTER FOR IMMUNOBIOLOGY

Although immunobiology was one of the original Centers of Excellence conceived of by Kase after he became Dean in 1985, it was not until 1997 that the Center for Immunobiology became a reality. The Center brings together the multidisciplinary interests of basic science researchers and scientists from the clinical departments. Chronologically young in age, the Center, directed by Lloyd Mayer, M.D., has contributed in a significant way to Mount Sinai's research enterprise.

Mayer, a graduate of the Mount Sinai School of Medicine (1976), received his postgraduate training in internal medicine at New York University/Bellevue Hospital. Following a fellowship in gastroenterology at Mount Sinai, he then spent five years at the Rockefeller Institute, where his research centered on cytokine regulation and the maturation of lymphocytes and the roles of these cells in immune disorders, especially immunodeficiency disease. Returning to Mount Sinai in 1985, he was appointed Chief of the Division of Clinical Immunology of the Department of Medicine one year later.[204] In 2003, Mayer was given an added responsibility when he was also appointed Chief of the Division of Gastroenterology of the Department of Medicine.

Mayer and his associates continued their studies on the nature and function of the immune system and the immune deficiency disorders but expanded their focus to study the immune mechanisms of the gastrointestinal tract with the goal of providing insights into the molecular events involved in the pathogenesis of inflammatory bowel disease (IBD). Mayer has also been active clinically, serving as either a principal investigator or coprincipal investigator, representing Mount Sinai on national and international study groups and leading Mount Sinai's participation in a number of clinical trials involving the use of new drugs for the management of IBD.[205–207]

Mayer's laboratory has been extraordinarily productive. He and his colleagues have identified the lack of induction of suppressor T cells in patients with IBD and have characterized the differences in expression of antigenic molecules on intestinal epithelial cells in normal individuals and in IBD patients.[208–210] More recently, the group has focused on the role of CD8+ T cells and their interactions with intestinal epithelial cells.[211–214] Research is also ongoing on the role of lymphokines in the regulation of human B cell differentiation.[215]

There are approximately two dozen faculty members of the Center for Immunobiology. Among the group with primary appointments are Jay Unkeless, Ph.D., Curt Horvath, Ph.D., Sergio Lira, M.D., Ph.D., Adrian Ting, Ph.D., Patricia Cortes, Ph.D., and Huabao Xiong, M.D., Ph.D.

Unkeless, a long-time member of the Department of Biochemistry, has devoted much of his scientific career to the study of the Fc family of receptors and has made seminal contributions to the understanding of their critical role in the immune process.[216–219] The Fc[gamma] receptors are found on neutrophils, macrophages, and natural killer cells and link these effector cells of the immune system to the major product of the adaptive immune system, IgG. Following cross-linking by immune complexes, these receptors can trigger the secretion of superoxide, the release of hydrolytic enzymes, the synthesis of cytokines, and the killing of sensitized target cells. A major aim of the laboratory is to analyze the signaling pathways leading from receptor activation to cell effector activity.[220, 221]

Recruited from The Rockefeller University in 1998, Horvath has had a long-standing research interest in the Signal Transducer and Activator of Transcription (STAT) family of proteins. STAT proteins, of which seven have been identified in mammals, are cytoplasmic proteins that, when activated, move to the cell nucleus, where they bind to specific DNA elements and participate as signal messengers and transcription factors involved in normal cellular responses to cytokines and growth factors. Abnormal activity of the STAT proteins is associated with a variety of human malignancies and immune diseases. These proteins have a causal role in oncogenesis and, as such, have been investigated as a possible target for new cancer drugs. Horvath and his associates have found that STAT protein function can be interfered with and suppressed by a protein expressed by the measles virus.[222]

Mount Sinai graduate and medical students have accomplished exciting work in Horvath's laboratory at Mount Sinai and have become the first authors of their respective papers. One study provides basic insights into the cellular machinery involved in antiviral gene regulation and describes potential targets for antiviral therapeutic intervention.[223] Another study describes a molecular mechanism, by which the mumps virus can evade destruction by the host. An unexpected finding was that mumps has the ability to destroy STAT3, a protein that is often hyperactivated in human cancer. The mumps viral protein, therefore, has the ability to kill cancer cells that are dependent upon STAT3.[224]

Sergio Lira joined the faculty of the Center in 2002 after more than two decades of experience as a senior scientist in both academic environments and industry. In 1994, Lira and his colleagues reported the first genetic experiment showing that chemokines can induce migration of specific leucocyte subsets *in vivo*.[225] More recently, Lira and associates described the generation of the first animal model for Kaposi's sarcoma.[226] Lira's current research employs genetic approaches in mice to study the function of chemokines and their receptors in leucocyte trafficking, angiogenesis, inflammation, and cancer and the role that chemokines play in lymphoid organogenesis.

Patricia Cortes, Ph.D., and Adrian Ting, Ph.D., have continued in the areas of research which they began prior to their recruitment to Mount Sinai. Cortes has studied the mechanisms and regulation of immunoglobulin gene rearrangements, with a recent focus on the specialized DNA rearrangement, V(D)J recombination, that is used by cells of the immune system to assemble immunoglobulin and T cell receptor genes from the preexisting gene segments.[227] V(D)J recombination takes place during lymphocyte development; aberrant recombination has profound effects *in vivo*, leads to abnormal T and B lymphocytes, and has been suggested as the cause of a number of lymphomas and leukemias. The early steps of this recombination are mediated by the products of the recombination genes 1 and 2 (RAG1 and RAG2).[228] Inactivation of either of these genes in the human or mouse results in a severe immunodeficiency state. Omenn's syndrome, a rare genetic disorder caused by a defect in V(D)J recombination secondary to RAG protein abnormalities, has been extensively studied by Sandro Santagata, M.D., Ph.D., a former Mount Sinai MSTP student, in collaboration with Cortes and the faculty of the Ruttenberg Cancer Center.[229] Over the years, members of the Immunobiology Center have collaborated extensively with all of the other Centers and Institutes.

Ting's research, the study of signal transduction in the tumor necrosis factor receptor superfamily and in T cell receptors, may also have important clinical relevance.[230] This work, as well as the discovery by Ting and colleagues of new genes that relate to the mechanism of programmed cell death (apoptosis), can play a role in furthering the understanding of cell responses to viral infection as well as improving the understanding of how cancer cells are destroyed.[231]

Notable research contributions are also being made by faculty members whose secondary appointments are in the Immunobiology Center. The work of those whose primary appointments are in a basic science department or another Institute or Center are described elsewhere in this volume. Immunobiology also plays a substantial role in the research of a number of faculty in the Division of Clinical Immunology of the Department of Medicine and in the Departments of Surgery and Pediatrics.

Charlotte Cunningham-Rundles, M.D., Ph.D., a long-time colleague of Mayer's, in addition to her administrative and clinical positions as Director of the Allergy Immunology Training Program, Director of the Jeffrey Modell Laboratory, and Director of the Immunodeficiency Clinic, has considerable extramural funding for her research in the immunodeficiency diseases and in immuno-reconstitution.[232]

Kirk Sperber, M.D., Associate Professor of Medicine, investigates the interactions between the HIV-1 virus and macrophages to ascertain what causes the loss of immunity in HIV infection, and also what is responsible for the apoptosis of neuronal cells in patients who have HIV-1 infection and AIDS dementia.[233, 234] He and his colleagues have also studied the therapeutic role of chloroquine and hydroxychloroquine, either alone or in combination with other drugs, in the management of HIV-1.[235, 236]

Hugh Sampson, M.D., Chief of the Division of Allergy and Immunology in the Department of Pediatrics, Director of the Jaffe Food Allergy Institute, and Director of Mount Sinai's General Clinical Research Center, has focused his research efforts on food allergy disorders. His work has included studies of the immunopathogenic role of food sensitivity in atopic dermatitis, the pathogenesis of food-induced anaphylaxis, and the characterization of food allergens and of food-induced gastrointestinal hypersensitivites.[237-240] He has developed immunotherapeutic strategies for treating food allergies.[241] In 2003, a major publication reported the protective effect of a monoclonal anti-body against peanut allergy and the immediate anaphylaxis that can accompany the allergic response.[242]

Based in the Division of Clinical Immunology, Karen Zier, Ph.D., interacts with the Center, the Department of Microbiology, and both the Icahn Center of Gene Therapy and Molecular Medicine and the Derald H. Ruttenberg Cancer Center. She also directs the Office of Student Research in the Medical School. Her research focuses on two related areas: the development of effective immunotherapy for solid tumors and the elucidation of the pathways by which different forms of tumor antigen are processed for presentation to T cells, with the goal of defining which treatments induce cell-mediated immunity to autologous tumor antigens.[243-245]

Members of the Center for Immunobiology have major teaching responsibilities. In the Medical School, they direct the teaching of immunology in the preclinical years, and in the Graduate School they offer courses in the fundamentals of immunobiology and advanced molecular and cellular immunobiology and also conduct seminar and journal clubs in immunobiology.

With a young, vibrant, enthusiastic, and growing faculty, the Center for Immunobiology represents the ideal melding of basic and clinical scientists working together to foster the goals of teaching, research, and patient care in an academic medical center.

6

The Department of Community and Preventive Medicine

THE MOUNT SINAI School of Medicine (MSSM) created its Department of Community Medicine (now the Department of Community and Preventive Medicine) in 1965 at the time of the formation of the School. The first such department in an urban setting in the United States, the Department embodies an important component of the "Mount Sinai Concept": interest in the health needs of the community at large and the value in seeing the patient as part of a larger context.[1] From the earliest days of planning for the medical school, Community Medicine was envisioned as having equal status with the basic science and clinical faculties and as being the bridge between the biological sciences, clinical care, and the social sciences.[2]

The importance that the Mount Sinai School of Medicine would place on community medicine was apparent from the beginning with the selection of George James, M.D., as the first Dean and President of the Mount Sinai Medical Center. James, then serving as the New York City Health Commissioner, was already a leader in the field of public health and knew intimately the perils and promise of community medicine. Although community medicine had long been recognized as a public health discipline in Europe, its acceptance in the United States was slow. James wanted to help change this and had the support of the Mount Sinai Trustees to reach out to the community.

In addition to his other duties, James was named the Chairman of the Department of Community Medicine, serving from 1965 to 1968. During these years, he emphasized the importance of preventive medicine. Furthermore, he maintained that the hospital experience, both outpatient and inpatient, should be designed around the needs of the patient as a member of a family and a community and not around the needs of the institution. To institute these concepts would require the Department to have a presence in the Hospital as well as in the School. The initial plan, therefore, included having the Department take over the areas of Ambulatory Care and the Emergency Room so that these areas could become more focused on preventive medicine and more responsive to community needs.

One of the components of the "Mount Sinai Concept" was its emphasis on the study of human behavior. To that end, in 1968 the Division of Behavioral Sciences was established within the Department and given the responsibility for teaching the social sciences, a large segment of which would be integrated into the curriculum of the newly created interdisciplinary course "Introduction to Medicine."[3]

With the support of The City University of New York, the School created a professorship in medical sociology to provide the leadership and a sustainable, stable base for the Division of Behavioral Sciences. Samuel W. Bloom, Ph.D., was appointed the first director, in July 1968. In 1969, utilizing funding from the National Institute of Mental Health, he established a Ph.D. program in Medical Sociology in conjunction with The City University Graduate Center. Sixteen Ph.D. degrees were awarded in the first ten years. Bloom has been highly productive throughout his years at Mount Sinai, publishing on the interrelationships between sociology and medicine and between medical schools and medical education.[4-7] In his most recent work, *The Word as Scalpel: a History of Medical Sociology*, published in 2002, Bloom provides an interpretation of the ongoing search for knowledge about the relationship among illness, medicine, and society.[8]

To assist in curriculum development within the Department and the School, James recruited Bessie S. Dana, M.S.S.A., in 1967. Already well known in the field of social work for her contributions in the conceptualization of collaboration and partnering between social work and physicians, and in medical education for her work on behalf of patients and their families, Dana brought a unique skill set to the new school. She had the uncanny ability to encourage collaboration between disparate elements of the faculty. This came to the fore in the development of the "Introduction to Medicine" course.[9] As Curriculum Development Coordinator for the School, and staff to the "Introduction to Medicine" course committee, Dana brought invaluable talents. Dana continued her career in medical social work and medical education in the Department and the School for more than twenty-five years, earning a well-deserved national and international reputation as an outstanding teacher and leader in the development of interdisciplinary educational efforts.

In 1968, James asked the Trustees for permission to seek a new Chairman for the Department. While he had accomplished a great deal in developing the departmental framework and recruiting key appointments, James felt he could not successfully be both the Dean and Chairman of Community Medicine. Instead, he sought a person with a deep understanding of the concept of the Department of Community Medicine who could help transform the early goals into reality. In 1968, Kurt W. Deuschle, M.D., widely regarded as the founding father of community medicine in the United States, was recruited to MSSM.[10]

Deuschle received his M.D. from the University of Michigan in 1948. After completing an internal medicine residency in 1954, he applied for a position with the U.S. Public Health Service. Expecting an assignment in India to study tropical disease, he was surprised when he was assigned instead to the U.S. Indian Health Service and the Navajo Nation. As a senior assistant surgeon and chief of the Tuberculosis Program at the Navajo Medical Center in Fort Defiance, Arizona, Deuschle advanced clinical care in tuberculosis and was involved in the first field trials of isoniazid. While there, he came to appreciate the intimate links among community, family, work, environment, and health that illuminated his entire career.

Deuschle was beginning to become known for his groundbreaking field work with the Navajo Nation when, in 1960, he was appointed to the chairmanship of the first community medicine department in the United States, at the University of Kentucky. At Kentucky, he worked with local doctors in Appalachia to demonstrate how an academic medical center could promote health in a rural setting. He created appointments at the medical school for several Appalachian doctors, an innovative move that helped improve town-gown relations, and he was able to expand on the model of clinical community medicine that he had begun to develop with the Navajo Nation.

His appointment as Chair of the Mount Sinai Department of Community Medicine in 1968 allowed Deuschle to put his Kentucky model to the test in an urban setting. Again, Deuschle rose to the challenge set before him. By 1988, there were seventy-five full-time faculty members, two part-time, seventy-nine voluntary staff, and fifty-one nonfaculty staff members in the Department, making the Mount Sinai Community

Kurt W. Deuschle, M.D., the first Lavanburg Professor and Chairman of the Department of Community Medicine. The Chair later became known as the Ethel H. Wise Professor of Community Medicine.

Medicine department the largest in the country. Deuschle designed the Department of Community Medicine so that it embodied a balance of teaching, research, and service, based on a foundation of rigorous scientific inquiry. The Department concentrated on creating models of care and delivery systems that might be adapted in other situations or communities. Deuschle also focused on empowering community leaders to change the social and behavioral conditions that cause illness.

Under Deuschle's guidance, the Department conducted several significant studies in community health care and disease. The first major research effort, directed by Louise Johnson, was "The People of East Harlem," carried out between 1969 and 1970.[11] In this investigation, researchers went door-to-door in East Harlem, discovering which health issues residents of the community surrounding Mount Sinai found to be most important. Almost 1,200 households, totaling 3,260 people, were interviewed. In the introduction to Johnson's published report on the study, Deuschle outlined the objectives of the survey, stating, "It was essential that the health problems be identified from the community viewpoint—that a sound, reliable and sophisticated household health study be undertaken. It was considered a departmental responsibility to help see to it that solutions to these health problems be found."[12]

In practice, this meant not that the departmental physicians would treat all the individual patients but, rather, that they would help identify the problems and seek the cooperation of other departments inside or outside the Hospital and/or the community to help resolve them. To aid in this, the Department created the position of Coordinator of Community Relations; the occupant would work with community groups to help assess and answer the needs of East Harlem residents and to identify ways to make the Hospital more responsive and less intimidating. This position was initially held by Marx G. Bowens, a social worker. The role of the Coordinator gradually expanded; staff increased and eventually became Mount Sinai's current Community Relations Department.

In 1979, the Department took this study a step further with the "Report on Primary Care in East Harlem."[13] Information was gathered from residents of East Harlem through a series of forums and town meetings that allowed community members to prioritize healthcare needs in the area. This report later formed the basis for a guide for local health planning.

Unfortunately, not all the original plans for the Department went smoothly. The initial hopes of reorganizing the Outpatient Department and the Emergency Room under Community Medicine never materialized. The same was true of the Division of Nursing Research, created in 1971, when The Mount Sinai Hospital School of Nursing closed. Also in the early years, there was a Division of Nutrition that sought to assess and improve the nutritional status of the community. This division was dissolved when its leader, George Christakis, M.D., left, in the 1970s. In many ways, the Department's experiences mirrored those of the School itself. The poor financial climate of the 1970s led to cutbacks in many of the visionary elements in the School's curriculum. The hope that the School would become a complete "university of the health sciences" had been pinned on cooperative efforts with The City University of New York, but it, too, experienced retrenchment. Like the other departments, Community Medicine was forced to scale back its plans and to concentrate on doing what it could do well and still continue to have a maximum impact on the community.

From the outset, the Department's educational and research perspective stressed the social sciences.[14] The expertise of the Department's faculty was utilized in computer-based, long-term community planning and in the study and facilitation of the effective social organization of health services. In 1967, a Division of Health Care had been formed, as well as a Division of Health Economics, the latter led by Victor Fuchs, Ph.D., a well-known health economist. These divisions continue today as part of the Health Sciences Research and Development Unit (HSRDU).

Deuschle believed that effective social change comes through the actions of those who have the most direct stake in the change. Rather than impose its ideas on the community, the Department has worked to empower community leaders by facilitating their actions to change the social conditions that cause illness. In 1973, a grant from the Commonwealth Fund provided for a development program to study and advance the integration of service, education, and research in prepaid group practice. Through this program, close relationships were established with the Health Insurance Plan of New York for both teaching and research.[15, 16] In that same year, a grant was received from the Robert Wood Johnson Foundation to support a program to evaluate the effectiveness of the five community projects aimed at children and youth in East Harlem that the Department was managing. The goal was to create guidelines for future additional activities either in the East Harlem community or in other neighborhoods.

In 1993, the Department, now chaired by Philip Landrigan, M.D., M.S., began a collaboration with the Health of the Public organization. Health of the Public was founded in 1986 to help academic health centers adapt to the changing demographic and health care environment. The collaboration has developed programs of population research and completed fundamental curriculum reform for the teaching of primary care, prevention, and population health. Through this partnership, the Department has enhanced medical education and teaching of population-based medicine by redirecting the focus of existing undergraduate teaching programs more explicitly toward urban health. The Department has also created a series of model projects in partnership with community-based organizations to improve the health of East Harlem residents. These cooperative ventures require that the partners undertake needs assessments, reach consensus on the priorities of various community health problems, and construct databases for program evaluation and outcomes research.

The Department's approach to undergraduate medical education has focused on helping students to develop the methodological skills and critical reasoning of clinical epidemiology. It is essential that medical graduates master the rudiments of biostatistics and decision theory in order to possess "reasonable proficiency in the logic of scientific inference and the methods of statistics, [the] tools to assess the conclusions drawn in medical articles."[17] The Department's contributions to medical teaching initially took the form of a first-year course in epidemiology and biostatistics and a third-year clerkship. In the latter, students were required to conduct a population-based research project. The work included problem formulation, data collection, analysis, and preparation of a final report. The clerkship provides the opportunity for students to obtain basic community medicine education by becoming seriously involved in the identification and solution of authentic community health problems. The community medicine clerkship eventually became mandatory for third-year medical students, equal in status to other clinical clerkships. The ultimate goal of the School's teaching program is to ensure that every graduate understands that every patient seen in an office, clinic, emergency room, or hospital bed is a member of a community and that his or her patterns of illness and health are profoundly affected by social, economic, and environmental factors within that community. This was one of the basic features of the "Mount Sinai Concept," and it continues to be embodied in the Department's educational efforts.

The Department's educational mission has not focused exclusively on higher education. In 1970, the Department created the Student Health Opportunity Program (SHOP.) Housed off-campus, SHOP had the goal of helping high school students remain in school and to learn about the various careers available to them in health care. The Secondary Education Through Health program (SETH) was established in 1973 by combining SHOP and a Health Careers Program that had been run by the School of Nursing prior to the latter's closing. Directed since 1973 by Lloyd Sherman, Ph.D., SETH, which is based in the Cummings Basic Sciences Building, is a cooperative effort with the New York City Board of Education to encourage high school students to consider careers in the health professions. Students take classes and receive hands-on experience in subjects ranging from environmental health to genomics. The courses are taught by high school teachers, medical school faculty and students, nurses, and senior technical staff. Over the past thirty years, SETH has directed more than eight thousand minority and disadvantaged high school and college students toward careers not only in medicine and science but also in law and other fields. The Center for Excellence in Youth Education was established in 1987 to track the use of resources and to monitor the wide range of new campus and off-campus outreach programs. In addition, more than forty-five high school teachers have been trained through the program in the creation of new curricula and in the use of project and theme-based learning to advance student knowledge of science, mathematics, and language. Beginning in 1993, a process of moving the programs back into the high schools was successfully undertaken and has been achieved at the Life Sciences Secondary School, the Central Park East Secondary School, and the High School for Environmental Studies. In June 2003, 101 students successfully completed their programs both on campus and at the cooperating schools. The program has been eminently successful; more than fifty students have gone on to and graduated from medical school, and several hundred have completed nursing school training.

SOCIAL WORK: AN IMPORTANT COMPONENT OF COMMUNITY MEDICINE

In 1906, under the aegis of Sigismund S. Goldwater, M.D., the first physician administrator of the Hospital, Mount Sinai created the Social Welfare Department, the first such Department in New York City and the fourth in the United States. Jennie Greenthal, R.N., a graduate of The Mount Sinai Hospital School of Nursing, was named the first social worker in the Department. In announcing the creation of the Department, Goldwater commented, "when the social service worker asks not only for more thorough treatment but for a more careful inquiry into the causes of disease, he [sic] touches elbows with the apostles of preventive medicine."[18] In 1915, the Department was renamed the Social Service Department.

In 1916, the Social Service Auxiliary (now called the Auxiliary Board) was created under the leadership of Mrs. Herbert H. Lehman. Originally a committee of The Mount Sinai Hospital, the Auxiliary

Board's prime purpose is to promote the welfare of patients and patients' families by complementing the Hospital's direct patient care activities with a variety of social and auxiliary services. Today the Auxiliary Board is a committee of the Boards of Trustees of the Medical Center, and the President of the Auxiliary Board serves as an ex-officio Trustee. Using its own resources, the Auxiliary Board has played a vital role in the development and continuing support of the Department of Health Education, the Therapeutic Activities Department, the Volunteer Department, and the Department of Social Work Services. Through its Project Funding and Review Committee, the Auxiliary Board has been the initial sponsor of more than twenty major ongoing activities and services, including the AIDS Volunteer Liaison Project, the Breast Cancer Resource Program, the Domestic Violence Prevention and Education Program, the Resource Entitlement Advocacy Program, the Palliative Care Project, and the Employee Assistance Program, among others.

The Hospital's Social Service Department expanded steadily and had remarkable growth under the leadership of Doris Siegel, the Director from 1954 to 1971. In 1977, the Department was once again renamed and became the Department of Social Work Services.

With the opening of the School in 1968, James took the bold step of giving social work a strong position alongside the traditional population-related health sciences by creating a Division of Social Work as a component of the Department of Community Medicine.[19] The mission of the Division was and remains to support the Department of Community Medicine and MSSM through innovative community service programs, research, and participation in medical student education. These activities are carried out with the highest standards of scientific and clinical excellence and social responsibility. An endowed chair, the Edith J. Baerwald Professor of Community Medicine (Social Work), was established in 1969 by the Aron family, the first and only social work chair in a school of medicine in the United States. In what is now called the Division of Social Work and Behavioral Sciences, the Director is the incumbent of the Baerwald Chair. Doris Siegel was the first to be installed in the chair (1969–1971), making her the first woman to hold an endowed chair in the new School of Medicine. Siegel "concentrated her pioneering efforts in the areas of social work administration, education, and public health." She "was deeply involved and committed to bridging the gap between health care administration and the general public."[20]

Doris Siegel, M.S., the first Edith J. Baerwald Professor of Community Medicine (Social Work), the first such chair in an American medical school and the first woman to hold an endowed professorship at the Mount Sinai School of Medicine.

Following Siegel's death, in 1971, her successor, Helen Rehr, D.S.W., occupied the Chair until her retirement in 1984. Rehr's innovations in social work have been unique and lasting. She was one of the first to recognize the needs of the aging population for social work services and was in the forefront in encouraging social work practitioners to take the lead in case finding and to avoid becoming dependent on physician referrals.[21–24] Rehr brought a research component to social work practice at Mount Sinai and was also the first to apply the principles of quality assurance and best practices to social work.[25–29] She also applied public health principles to social work practice, especially in the area of patient screening and risk identification.[30] Rehr's creative teaching and many peer-reviewed publications and books have identified her as one of the leaders in the field over decades. She has received numerous awards for social work leadership and has had two Professorial Chairs in Social Work named for her, one at the Columbia University School of Social Work, where she was also inducted into the

Helen Rehr, D.S.W., (left) at the ceremony where she was invested as the second Baerwald Professor, 1974. With Dr. Rehr are Jane and Jack Aron, both Mount Sinai Trustees, who endowed the chair in honor of Mrs. Aron's mother, Edith J. Baerwald. Mr. Aron also later provided funding to build the student residence, the Jane B. Aron Residence Hall, in memory of his wife, who died in 1983.

school's Hall of Fame, the other at the Hunter College School of Social Work. Rehr has herself provided a lectureship in the Mount Sinai Department of Community and Preventive Medicine. As noted by the Foundation of the National Association of Social Workers: "Rehr's work has touched the lives of many thousands of the sick and frail. . . . Her teaching . . . further extends her impact as she has enabled others to meet the needs of the people they serve. Indeed she has been a pioneer in health related social services through her practice examples, her teaching, and her research."[31]

Since 1986, the Baerwald Chair has been held by Gary Rosenberg, Ph.D. Recruited to Mount Sinai in 1976 as the Associate Director of Social Work Services, Rosenberg has become one of the most respected leaders in the field of social work. He has been the recipient of numerous awards and has lectured throughout the world. He has written or edited nine books and is the author of more than fifty publications in professional journals. Rosenberg has also held important senior administrative positions in the Medical Center since 1989.

Under Rosenberg's leadership, faculty members of the Division of Social Work and Behavioral Sciences have undertaken the creation and management of Medical Center programs in addition to Social Work (e.g., Health Education, Organizational Development, Community Relations, Patient Representatives, The Rape Crisis Intervention Program, and the Employee Assistance Program).

During Rosenberg's tenure, the faculty of the Division has grown. It currently consists of forty-five full-time and thirty-eight adjunct social workers, as well as sociologists, nurses and six behavioral scientists. These faculty members are based at the Mount Sinai Medical Center and its affiliates, Elmhurst Hospital Center, North General Hospital, The Jewish Home and Hospital for Aged, the Bronx Veterans Administration Medical Center, Englewood Hospital, and the Queens Hospital Center.

The Division is internationally recognized as a world leader in the training of professionals in social work and in the creation of innovative patient care services. Faculty members edit two internationally renowned journals. Rosenberg is the Editor-in-Chief of *Social Work in Health Care*, and Rosenberg and Andrew Weissman, Ph.D., are the coeditors of *Social Work in Mental Health*. Weissman, a member of the faculty for a quarter of a century, has been a favorite teacher of the medical students in the "Introduction to Medicine" and in the "Art and Science of Medicine" courses, and he also plays a vital teaching role in the Community Medicine primary care clerkship. Susan Blumenfield, D.S.W., the current Director of Social Work Services, has herself made significant contributions to social work practice. The Division has sponsored an international educational exchange program for social work directors from Australia and Israel since 1989, and the faculty conducts numerous ongoing interdisciplinary research studies. Faculty members teach in two required courses for medical students and also provide educational opportunities for more than twenty-five graduate social work students each year. Established by Rehr, the Division also sponsors the prestigious Doris Siegel Memorial Colloquium every three years, bringing internationally distinguished social scientists as speakers and as participants to the Medical Center to discuss the current issues facing the behavioral sciences in the healthcare arena. Further, the Division has been an active cosponsor of the International Conference of Social Work in Health and Mental Health. Prior conferences have taken place in Jerusalem, Israel, Melbourne, Australia, and Tampere, Finland. The Division also cosponsors the World Wide Web Resources for Social Workers website (http://www.nyu.edu/socialwork/wwwrsw/).

The faculty's research grants include approximately $12 million a year in National Institutes of Health (NIH) funded research grants and $2 million in foundation funded studies. A major bequest has established the Edith Ehrman Center for Health Education in East Harlem as a permanently endowed Center. The focus of the Ehrman Center is on providing services to the elderly and youth in the East Harlem area through programs aimed at cardiovascular health, health education, and pregnancy prevention. It also provides opportunities for qualified high school students to participate in ongoing training to prepare them to enter the health professions.

Judith S. Brook, Ph.D., a psychologist, and David W. Brook, M.D., a psychiatrist, and their collaborators, Martin Whiteman, Patricia Cohen, Steven Finch, and Rosenberg, have made vital contributions in the area of behavioral health. Their work, which has had continuous funding from the NIH for more than twenty-three years, centers on the behavioral aspects of the major scourges in public health today: drug abuse, AIDS, adolescent violence, and smoking. The group has investigated the psychosocial risk and protective developmental factors of etiological significance in the development of these problematic behaviors and their sequelae.[32–35] Longitudinal family studies have enabled the researchers to assess the intergenerational transmission of risk factors for substance use and abuse from parents to children to grandchildren.[36]

The faculty members of the Division have conducted clinical social work and behavioral science practice and research in the Medical Center and its affiliates on a wide range of subjects. The concept of populations at risk has involved social work and behavioral science faculty in such diverse studies as research on the needs of the elderly in East Harlem, factors predictive of child abuse, the relationship of social support to ethnicity and culture, the longitudinal examination of drug use, smoking and violence, and numerous other projects geared to primary and secondary prevention. Through research, the Division contributes new knowledge of the social and psychological costs of illness and the impact of illness on the social functioning of individuals and families.

THE INTERNATIONAL PROGRAM IN
COMMUNITY MEDICINE

In 1979, with many East Harlem programs in place, Deuschle began to expand the reach of the Department by establishing international programs in response to requests from around the world. The Department's first international project assisted in organizing a comprehensive health system for 80,000 people in the eastern region of the Dominican Republic. The success of this project led, in 1983, to the endowment of the Charles G. Bluhdorn Professorship in International Community Medicine and the creation of the International Division of Community Medicine.[37] The Bluhdorn chair was the first endowed chair in a United States medical school devoted to international community health issues. Samuel Bosch, M.D., was invested as the first Bluhdorn Professor and was given the responsibility to organize the International

Samuel J. Bosch, M.D., Bessie S. Dana, M.S.S.A., and Kurt Deuschle, M.D., at the investiture of Dr. Bosch as the first Charles G. Bluhdorn Professor of International Community Medicine, 1983.

Division and its operational unit. Bosch came to Mount Sinai as a post-doctoral fellow in 1969 and remained on the faculty, becoming a full professor in 1978. In 1973, he was appointed Deputy Director of the Department. Bosch retired in 1990 to become Professor Emeritus.

The overall goal of the program is to promote community development outside the United States through improvement in health and health care delivery systems.[38, 39] Since 1983, a multidisciplinary faculty from the International Division has participated in collaborative efforts with universities and medical institutions around the world. The Division has been involved in the Pan-American Consortium for Health -Policy Development since 1986. The Consortium is a network within which health professionals collaborate to promote education, leadership, and the development of community-based healthcare delivery systems throughout the Americas. The Consortium's activities have led to collaborative programs in several countries, including Argentina, Brazil, Chile, the Dominican Republic, Guatemala, and Venezuela. International relationships established through the activities of the Consortium are evolving into long-term agreements of cooperation and linkages between and among institutions from different parts of the Americas. Most important, the Latin American universities with which the Consortium has worked have translated their projects into funded programs that are now being implemented, thereby enhancing the capabilities of the able, local faculties and strengthening their institutions.

In 1996, the Department began offering individualized international fellowships in occupational and environmental medicine. Funded by the Fogarty Foundation, the fellowships are codirected with Queens College. Trainees spend most of their time in their own countries, where their research training is prioritized by the local university. The fellows make several brief visits to Mount Sinai, where Department faculty are intimately involved with precepting the trainee's field projects. The international program has worked with nine collaborating institutions in three countries—Brazil, Mexico, and Chile—and has trained twenty-one scientists to date. The research projects of the fellows vary widely in scope and degree of complexity. Major topics include studies of chronic lung disease and the damaging effects of pesticides and heavy metals. Not only do the fellows benefit from the program, but also their home institutions and countries have benefited from the enhanced research capacity that has led to changes in health policy. For example, a fellow from the Ministry of Health in Chile

Irving J. Selikoff, M.D.,
Professor of Community
Medicine and Director of
the Environmental Sci-
ences Laboratory, 1973.

established the first national surveillance system for occupational
trauma fatalities in Chile. This system now serves as a demonstration
project for the Pan American Health Organization, which has designated occupational trauma fatalities as a top priority for occupational
health surveillance activities in Latin America.

THE MOUNT SINAI–IRVING J. SELIKOFF CLINICAL CENTER
FOR OCCUPATIONAL AND ENVIRONMENTAL MEDICINE

When Kurt Deuschle came to Mount Sinai, he was fortunate to find
Irving J. Selikoff, M.D., already at work in the Division of Environmental Medicine. Selikoff's Environmental Science Laboratory had been
incorporated into the Department of Community Medicine in 1967,
when George James was Chair.

Selikoff first joined Mount Sinai as a volunteer assistant in Morbid
Anatomy in 1941. He then moved to Sea View Hospital, on Staten
Island, where he and Edward H. Robitzek made a breakthrough in the

treatment of tuberculosis by establishing the value of isoniazid therapy through clinical trials.[40] In 1955, they received the prestigious Albert Lasker Award in Medicine for Outstanding Achievement in Public Health for this work. In 1947, Selikoff returned to Mount Sinai, where he soon noted the unusually high incidence of lung disease among his patients who worked with asbestos.

In 1961, Selikoff created the Mount Sinai Environmental Sciences Laboratory to conduct in-depth research into the connection between asbestos and disease. After conducting systematic studies in New York and New Jersey, Selikoff was the first in the medical community to recognize the occupational dangers of asbestos exposure and to support this claim with hard scientific proof. In 1964, Selikoff presented his study on the deaths that occurred between 1943 and 1962 among 632 asbestos workers; he reported that their death rate from lung cancer was 6.8 times that reported for the general white male population.[41]

Between 1963 and 1967, Selikoff, E. Cuyler Hammond, M.D., and Jacob Churg, M.D., investigated the correlation among asbestos exposure, cigarette smoking, and cancer. In a classic epidemiologic study, published in April 1968 in the *Journal of the American Medical Association* (*JAMA*), they reported that men who both smoked and had been exposed to asbestos had about fifty-five times the risk of dying of cancer as men who neither smoked cigarettes nor worked with asbestos. The authors concluded "that asbestos exposure should be minimized, that asbestos workers who do not smoke should never start, and that those now smoking should stop immediately."[42]

In 1970, with the support of the Asbestos Workers' Union, Selikoff began to conduct epidemiological studies throughout the United States and Canada. By monitoring the health of 17,000 workers for many years, he amassed a body of evidence strong enough to convince physicians and lawmakers that asbestos was the leading cause of lung and digestive tract cancers and mesotheliomas among those workers, even though some persons did not manifest their asbestos-related disease for two to four decades or more after exposure.[43]

In 1973, after the National Institute of Environmental Health Sciences awarded the Environmental Sciences Laboratory funding as a core center (funding not tied to any specific project,) the laboratory was renamed the Environmental Health Sciences Center. This allowed the Center to enhance its infrastructure and to hire new researchers.

Under Selikoff's able leadership, Center researchers discovered fundamental etiological associations between environmental toxins and human disease. Their pioneering research into the health hazards not only of asbestos but also of lead, solvents, vinyl chloride, and other dangerous chemicals helped the Center gain national and international recognition.

In 1985, Selikoff retired as director of the Center, and Deuschle recruited Philip Landrigan, M.D., M.S., to be the next director of the Division of Environmental and Occupational Sciences. Landrigan and Selikoff worked side by side until the latter's death in 1992. In 1987, the Mount Sinai–Irving J. Selikoff Clinical Center was established to diagnose and evaluate work-related disease. Today, a new generation carries on the work in occupational medicine at the Center and at the Irving J. Selikoff Asbestos Archives and Research Center, which maintains an extensive database of Selikoff's research documents.

Philip Landrigan received his M.D. from Harvard Medical School in 1967. After completion of pediatric training, he joined the Epidemic Intelligence Service (EIS) at the Centers for Disease Control in Atlanta. As an EIS officer, Landrigan was sent around the world to study epidemics and the effect of toxic chemicals on human health. He served in Nigeria and El Salvador. He also obtained a Master of Science degree in occupational medicine from the London School of Hygiene and Tropical Medicine.

After accepting the post of director of Mount Sinai's Division of Environmental and Occupational Sciences in 1985, Landrigan worked to build the scientific base of the Division, creating links to basic science departments in order to study the biochemical and cellular mechanisms that cause disease. In 1986, Landrigan went to the New York State legislature to request funding to establish a network of public clinics throughout the state. He received $100,000 from the legislature to conduct a study in collaboration with Steven Markowitz, M.D., on the extent of occupational disease in New York. The results were staggering: 5,000 to 7,000 people were dying from occupational disease each year in New York State, and tens of thousands more became ill.[44] Based on these findings, the New York State Department of Health established a unique network of clinical centers in occupational and environmental medicine across the state, with the Mount Sinai–Irving J. Selikoff Clinical Center serving as the flagship facility.

After having served as chairman for twenty-two years, Kurt Deuschle retired in 1990, and Philip Landrigan was appointed Chairman of the Department of Community and Preventive Medicine. Stephen Levin, M.D., and Robin Herbert, M.D., are the current co-directors of the Selikoff Center for Occupational and Environmental Medicine. As Chairman, Landrigan continued his interest in environmental and occupational sciences. With his background in pediatrics, Landrigan concentrated on studying the impact of environmental factors on the health of children, the section of the population most vulnerable to environmental disease. In 1998, with a grant from the Pew Charitable Trusts, he established the Center for Children's Health and the Environment (CCHE) as an academic research and policy center organized to examine links between exposure to toxic pollutants and childhood illness. CCHE is committed to promoting children's health and to conducting environmental health research. It assists federal agencies in establishing health standards based on the unique health needs of children, serves as an expert resource to the media and policy-makers, conducts public education campaigns, and works to make environmental pediatrics a core component of pediatric practice.

THE DIVISION OF ENVIRONMENTAL HEALTH SCIENCE

Prevention research in environmental health science has been a key mission of the Department of Community and Preventive Medicine since its inception. Under the leadership of Mary Wolff, Ph.D., a long-time member of the Department and a collaborator with Selikoff, a separate research division was created with the ultimate goal of protecting the public's health by understanding, elucidating, and preventing diseases that arise from environmental exposures. The Division has had a distinguished history in identifying and quantifying cancer hazards, reproductive dysfunction, and neurological disease associated with exposures to toxic agents in the environment.[45–47] Current directions in research include study of the molecular epidemiology of cancer, lead exposures and their effects on reproductive and neurological health, the neuropathology of environmental agents, and the environmental health of children. Laboratory programs are in place to provide quantitative exposure assessment of lead in bone and of organochlorines and dietary micronutrients in body tissues. A major area of current research is the assessment of exposures to lead and organochlorines in the New York Harbor.

THE DIVISION OF EPIDEMIOLOGY

Directed by Gertrud Berkowitz, Ph.D., the Division established a major program in Women's Health Research in 1994. Current studies include the effects of psychosocial stress during pregnancy on cardiac responsivity and on the risk of preterm delivery; an investigation of vaginal douching as a risk factor for bacterial vaginosis among African American women; participation in a multicenter investigation of semen quality among the partners of pregnant women; estimation of historical exposures of women in the Long Island Breast Cancer Study, using geographic modeling techniques; and collaboration in a study of gender-specific risk factors for recurrent stroke.

A pilot study has been carried out to test the feasibility of a longitudinal study of the health benefits of intentional weight loss among obese and overweight women. The cohort will be followed for psychological well-being and quality of life, in addition to disease endpoints. In collaboration with a major New York City union, the Division has developed health promotion activities for a population of older, largely immigrant Hispanic and Chinese women, who have been little studied up to now.

THE DIVISION OF BIOSTATISTICS AND DATA MANAGEMENT

Critical to the analysis of the Department's research activities, the Division of Biostatistics and Data Management, led by James Godbold, Ph.D., studies investigator choices regarding sample size, power calculations, and data collection instruments. In the data collection phase, the staff is involved in database design, development of data tracking systems, and the generation of interim reports to monitor data quality and to study progress and safety. In the final phase of a study, Division members are heavily involved in data analysis and interpretation, using sophisticated statistical packages. The Division of Biostatistics also plays an active role in the classroom teaching and curriculum development of the School of Medicine.

GRADUATE EDUCATION

Today, under the aegis of the Department of Community and Preventive Medicine, more than six hundred Mount Sinai physicians, scientists, and staff serve annually as instructors and mentors for medical students, medical professionals, and young people from inner-city schools. The Department has been especially active in training and encouraging medical students to participate in community service and conduct applied health services research. In 1988, a curriculum that had been offered informally for many years was augmented to allow the granting of a Master of Science degree in Community Medicine to selected undergraduate medical students, residents, and fellows.[48]

Since the early 1970s, the Department has offered residency programs in General Preventive Medicine and in Occupational and Environmental Medicine. Each of the two-year programs, currently directed by Elizabeth Garland, M.D., and Jacqueline Moline, M.D., respectively, incorporates the Master of Science in Community Medicine degree into the curriculum. These residencies educate physicians in the application of population-based medicine for positions in environmental medicine, healthcare planning, clinical epidemiology, industrial medicine, clinical preventive services, and health services research. In 1991, Wykoff Medical Center, in Brooklyn, New York, became an affiliate of the Department's General Preventive Medicine residency. Wykoff residents enroll in the Department's Master of Science degree program, and Department faculty members provide oversight for the residents' practicum.

In 1972, the Medical School, in partnership with Baruch College of The City University of New York, began offering a Master of Business Administration (MBA) degree in Healthcare Administration. In 1991, the MBA program moved into the Department of Community and Preventive Medicine. Raymond Cornbill, M.B.A., a member of the faculty for more than twenty years and the codirector of the program, has been pivotal not only in its organization but in the teaching and mentoring of students. Matriculants in the program receive a strong grounding in basic management disciplines applicable to the unique problems of the health care industry. As will be seen later, Cornbill has also played a vital role in the Department's outreach to the local community that the Medical Center serves.

THE DIVISION OF FAMILY MEDICINE

One of the Department's most recent projects has been the creation of a Division of Family Medicine, which conducts family practice research. This also provides Mount Sinai students with an opportunity to learn primary care from general practitioners and to participate in a family medicine clerkship. Family Medicine programs have developed across the country since the early 1970s, typically finding their initial home in departments of Community Medicine.[49] In 1998, in response to a California regulation requiring any physician seeking licensure in that state to have had training in family medicine, the Department received permission from the Dean and the Primary Care Working Group of the Curriculum Committee to proceed with the development of an academic family medicine program. This included a new family medicine–oriented primary care clerkship, evolving out of the existing community medicine third-year clerkship and the clinical programs at affiliated institutions.

In 2000, Francis Kohrs, M.D., a physician trained in both family practice and public health, was recruited to head the new Division of Family Medicine. Within a year, Kohrs received two major federal grants, the first to support the development of the primary care clerkship and the second to support the development of the new academic unit. At the program's inception, the Family Practice Program at Jamaica Hospital Medical Center, in Queens, became the primary clinical affiliate. In 2002, Heather Sealy, M.D., became the program director, and Maimonides Medical Center and Jersey City Medical Center became additional training sites.

SERVING THE COMMUNITY TODAY

Since its inception, the Department of Community and Preventive Medicine has instituted many local community programs that continue today. In connection with the Division of Social Work, the Health Services Research and Development Unit (HSRDU), directed by Cornbill, conducts projects that create and test new approaches to the delivery of health care with a strong emphasis on the participation of the community. The Unit also provides technical assistance in community advancement, health planning, program initiation, implementation, and evaluation and offers a "systems" approach to community problem solving. A prime example of the result of these activities is the Boriken Neighborhood Health Center. Established in 1975, the Center provides quality and affordable outpatient health care to East Harlem's diverse community. The HSRDU calls on the resources of faculty, residents, students, staff, and community membership, melding research, education, and service delivery.

The HSRDU has also demonstrated the strength of its commitment to the local community by partnering with the New York City Department of Health's Immunization Registry Project. The project relied heavily upon Mount Sinai School of Medicine students and faculty to gather information about East Harlem residents from medical records and community health providers and through door-to-door canvassing. East Harlem was chosen as the pilot site for the project because earlier Sinai-facilitated partnerships in this neighborhood had been so successful.

The Department's outreach has also extended to playing a major role in the East Harlem Community Health Committee, Inc. (EHCHC). Founded in 1976, EHCHC is a membership organization of community health centers, hospitals, consumers, social service providers, schools, government agencies, and other community-based agencies. Cornbill is the cochair of the general membership. To fulfill its mission of improving the health status of East Harlem residents, EHCHC provides a forum for providers and consumers to collaborate, plan, exchange ideas, and build networks; to identify gaps and inefficiencies in health service delivery; and to recommend improvements in access, quality, accountability, coordination, education, and training. In response to the needs of the community, EHCHC has established Pediatric/Child Health, AIDS, Substance Abuse, Older Adults, Housing, and Mold subcommittees. A number of these are chaired by Mount Sinai personnel. All of the groups are committed to improving community health by assessing the issues, developing and implementing interventions, and reviewing the results.

THE EFFECTS OF SEPTEMBER 11, 2001

Within months after the attack on the World Trade Center (WTC), there was growing evidence that upper and lower respiratory illnesses, musculoskeletal injuries with potentially chronic sequelae, and persistent psychological distress were prevalent among those who had been heavily exposed to a wide range of hazardous conditions at or near "Ground Zero." With the support of the Centers for Disease Control and Prevention (CDC), the Mount Sinai–Irving J. Selikoff Center for Occupational and Environmental Medicine (COEM), led by Levin and Herbert, has established a comprehensive medical surveillance program to provide medical assessments, diagnostic referrals, and occupational health education for more than 8,500 exposed workers and volunteers. Systematic medical evaluations and follow-up for approximately 7,000 workers are carried out at the Center. In order to provide geographically accessible medical services, another 1,500 workers are followed at other facilities in the greater New York/New Jersey metropolitan area.

Conducted over a twelve-month period, the program has the following goals: to identify individuals who sustained exposures at or near Ground Zero during rescue and recovery activities; to provide clinical assessments for exposed individuals; to identify those with persistent WTC-related medical conditions; to coordinate referral for follow-up clinical care for affected individuals; to educate exposed individuals about their possible associated health risks and to advise them about benefit and entitlement programs available; and to establish "baseline" clinical status for individuals exposed at or near Ground Zero for purposes of comparison with future clinical assessments.

In 2002, the National Institute of Environmental Health Sciences (NIEHS) awarded a Superfund Basic Research Program World Trade Center Supplement to the Department. With Landrigan as the Principal Investigator, and in collaboration with the Lamont-Doherty Earth Observatory of Columbia University, the NIEHS WTC program includes four projects. Two are epidemiologic investigations at Mount Sinai: a prospective clinical and epidemiologic study of ironworkers (Stephen Levin, M.D., and Alvin Teirstein, M.D., Investigators) and a study of pregnant women and their infants (Gertrud Berkowitz, Ph.D., Investigator). The third project, directed by Columbia University faculty, will study the WTC emissions utilizing imaging spectroscopy. The fourth project, directed by Joel Forman, M.D., will provide guidance concerning the health risks of the disaster to children and their families.

Were George James, the founder of the Department of Community and Preventive Medicine, alive today, he would take great pride in what he created. Outstanding leadership, innovative research, and pioneering programs have enabled the Department to be a major asset in the fulfillment of the Mount Sinai Medical Center's stated mission: "Mount Sinai will be ever sensitive to the social and health care needs of the many different communities it serves. The Center will be a participant in efforts to define and solve health problems in population groups and communities through its capability in developing scientific knowledge, education and service."[50]

7

The Department of Human Genetics

IN THE EARLY planning for the opening of the Medical School, it became evident that Mount Sinai required a stronger presence in the field of genetics, a discipline that would be essential in the teaching of both medical and graduate students. Horace Hodes, the Chairman of the Department of Pediatrics, took on the task of creating a Division of Medical Genetics within his Department that would not only have teaching and clinical care missions but also engage in meaningful research. In 1966, Hodes recruited Kurt Hirschhorn, M.D., an internist and world-renowned human geneticist, then at the New York University School of Medicine, as the Chief of the new Division.

Successful from the outset, the Division was designated as one of five NIH-funded genetics centers and also received an NIH training grant in human genetics. A named professorship, the Arthur J. and Nellie Z. Cohen Professor of Pediatrics and Genetics, was created in 1968 to support the work of the Division Chief. The program flourished, particularly in the areas of cytogenetics, somatic cell genetics, and biochemical and clinical genetics. Hirschhorn described new chromosome abnormalities, studied the genetics of lymphocyte response related to transplantation and cellular immunity, and established one of the country's first prenatal diagnostic laboratories. When Hirschhorn succeeded Horace Hodes as Chairman of the Department of Pediatrics in 1977, he recruited Robert J. Desnick Ph.D., M.D., to become Chief of the Division and Cohen Professor of Pediatrics and Genetics.[1]

A native of Minneapolis, Desnick received his M.D. and Ph.D. degrees from the University of Minnesota and, following residency training in pediatrics, joined the faculty of his alma mater. Desnick's research focused on basic and applied studies of inherited metabolic diseases, and in particular the lysosomal storage diseases (e.g., Tay-Sachs, Fabry, Gaucher, Schindler, and Niemann-Pick diseases) and the

The investiture of Kurt Hirschhorn, M.D., as the first Arthur J. and Nellie Z. Cohen Professor of Pediatrics (Genetics), February 1968. From left to right are Mrs. Nellie Z. Cohen, a Mount Sinai Trustee; Milton Steinbach, President of the Mount Sinai School of Medicine; Dr. Hirschhorn; and Dean George James.

inherited disorders of heme biosynthesis, the porphyrias (e.g., acute intermittent porphyria and congenital erythropoietic porphyria). These investigative efforts initially led to the clinical and biochemical characterization of these disorders and, subsequently, to the isolation of the disease-causing genes and identification of their mutations. These advances led to the improved diagnosis and, ultimately, to the development of specific therapies for Fabry and Niemann-Pick diseases.

Desnick has authored or coauthored more than five hundred papers and edited nine books. He has been the recipient of numerous awards, including the Ross Award in Pediatric Research, the E. Mead Johnson Award for Research in Pediatrics of the American Academy of Pediatrics, an NIH Research Career Development Award, an NIH MERIT Award, and the Mount Sinai Outstanding Faculty Award. Recognized worldwide, Desnick is a past President of the Society for Inherited Metabolic Diseases and was the President of the Fifth International Congress on Inborn Errors of Metabolism. He served as a Director of the American Board of Medical Genetics and is a Founding Diplomate of the American College of Medical Genetics. In addition, he is a founder and Past President of the Association of Professors of Human and Medical Genetics, the organization of chairs, chiefs, and directors of human and medical genetics departments or units in North America. In 1990, Desnick was appointed the Program Director of Mount Sinai's General Clinical Research Center, a position he held for ten years. He also founded and directs Mount Sinai's Center for Jewish Genetic Diseases.

Under Desnick's leadership, the Division continued to grow, attracting significant extramural research funding and expanding the range of its clinical and research activities, especially in the areas of biochemical genetics, gene mapping, molecular genetics, and the development of treatments for inherited metabolic diseases. This was the era of molecular genetics, and gene cloning was now possible. Spearheaded by senior investigators Edward H. Schuchman, Ph.D., and David F. Bishop, Ph.D., Division researchers isolated and sequenced the genes for multiple diseases, notably the lysosomal storage diseases[2-7] and porphyrias,[8-10] and identified the specific disease-causing mutations in these and other genetic diseases.[11-14]

This was also a time when the field of medical genetics was becoming recognized nationally as a medical specialty. The American Board of Medical Genetics was established in 1980, and, in 1991, this Board became an official member of the American Board of Medical Specialties. Board certified geneticists were recognized as having expertise in clinical, biochemical, and molecular genetics and in cytogenetics. In 1991, the American College of Medical Genetics was founded. As preeminent geneticists, Hirschhorn and Desnick played important roles in the establishment of these organizations. Human and medical genetics were brought to the forefront with the initiation of the NIH-sponsored Human Genome Project; the latter vastly accelerated research in gene discovery, genomics, and gene therapy. These accomplishments led to the rapid expansion of DNA-based diagnostics and "translational genetics," the development of molecular- and cellular-based therapies for genetic disorders.

CENTER FOR JEWISH GENETIC DISEASES

In 1982, Desnick founded the Center for Jewish Genetic Diseases at Mount Sinai, the first Center in the world devoted to the study of genetic diseases that are prevalent in the Jewish population. The mission of the Center was and remains to provide expert patient care and to devise improved diagnostics and new therapies for these diseases. The Division had established a Tay-Sachs prenatal screening program in the 1970s, utilizing a biochemical test to identify couples who had a one-in-four chance of conceiving a Tay-Sachs baby. These couples were offered genetic counseling and various reproductive options to prevent the birth of an affected child. In the 1980s, the ability of DNA testing to identify the gene defects that cause several recessive diseases in the Ashkenazi Jewish community led to the expansion of the Division's prenatal carrier screening program. The Division established one of the first academic DNA diagnostic laboratories and inaugurated the first "multiplex" prenatal screening programs for Jewish genetic diseases.[15] This carrier screening program has been extraordinarily successful and currently screens for nine severe Jewish genetic diseases.

In addition, a unique premarital genetic screening program, Chevra Dor Yeshurum, was founded in 1993 by Desnick and Rabbi Josef Ekstein to meet the special needs of the Hasidic and Orthodox Jewish communities. Large numbers of single men and women were screened and the results entered into a database. When an engagement was proposed, parents could contact the central office and ascertain whether the proposed couple was genetically "compatible." To date, this program has identified more than four hundred proposed matches that each carried a one-in-four risk of conceiving a child with a severe Jewish genetic disease.[16]

The Mount Sinai Comprehensive Gaucher Disease Treatment Program, instituted in 1985 as part of the Center for Jewish Genetic Diseases, today represents the single largest clinic in the world devoted to the diagnosis and treatment of Gaucher disease. More than five hundred patients and their families with the disease have been enrolled and managed in the program since its inception. At the outset, the Program was directed by Desnick and Gregory Grabowski, M.D., who had been recruited from the University of Minnesota. When enzyme replacement therapy for Gaucher disease was approved by the Food and Drug Administration (FDA), in 1991, this Program was among the first to offer the new treatment.[17] The Comprehensive Gaucher Disease Treatment Program has proved to be an outstanding model for providing expert multidisciplinary care to patients. In fact, in 1998, a similar program was established at Mount Sinai for the coordinated care of patients with Types A and B Niemann-Pick diseases, another of the group of rare Jewish genetic diseases.

ESTABLISHMENT OF THE DEPARTMENT OF HUMAN GENETICS

As noted elsewhere, a major expansion of the science base of the Medical School occurred during the tenure of Dean Nathan Kase.[18] In 1991, an evaluation of the School's research strengths and needs was carried out by a Dean's Committee, headed by Paul Lazarow, Ph.D., then Chairman of the Department of Anatomy and Cell Biology. The Committee's recommendations included the establishment of new Centers in Molecular Biology, Neurobiology, Structural Biology, and Immunobiology and the creation of a Department of Human Genetics. These entities would be housed in a new research building that would be constructed on the northeast corner of 98th Street and Madison Avenue. Following a nationwide search, Desnick was selected as the first Chair of the new Department of Human Genetics.

From its inception, the Department of Human Genetics was a combined basic science and clinical department. Its goals were clear: to assemble an internationally recognized faculty capable of obtaining significant extramural funding; to initiate high-quality training programs at both the degree-granting level (M.S., Ph.D., M.D./Ph.D.) and the postdoctoral level; to create state-of-the-art core laboratories to facilitate basic and clinical research; and to provide the highest-quality clinical care and clinical laboratory services. In addition to the ongoing clinical, educational, and research activities in the Division of Medical and Molecular Genetics, the new Department would also undertake investigative efforts in genomics and gene discovery, use animal models to elucidate disease pathogenesis, and direct translational efforts to develop molecular-based therapeutics, including enzyme and gene replacement.

When Desnick suggested that he would recruit in the area of gene therapy, an institutional decision was made to establish a separate program for gene therapy, thus giving rise to the Institute for Gene Therapy and Molecular Medicine. Desnick subsequently chaired the search committee that recommended Savio L. C. Woo, Ph.D., of the Baylor College of Medicine, who became the Institute's first Director.

FACULTY RECRUITMENT

With the establishment of the Department of Human Genetics, the faculty of the previous Division of Medical and Molecular Genetics transferred their primary Medical School appointments to the new Department; the M.D. and M.D./Ph.D. faculty retained secondary appointments in the Department of Pediatrics. Early faculty recruitment was directed to the enhancement of the clinical program.

The first recruit to the new Department, as the Vice Chair for Clinical and Educational Activities, was Margaret McGovern, M.D., Ph.D., then at the State University of New York at Stony Brook. McGovern spearheaded the expansion of the Department's Inherited Metabolic Disease program with the recruitment of Selma Snyderman, M.D., and Claude Sansaricq, M.D., Ph.D., from the New York University School of Medicine. This program, which includes the disorders identified by the New York State Newborn Screening Program, now treats more than five hundred infants, children, and adults with phenylketonuria, maple syrup urine disease, galactosemia, and other inherited metabolic disorders. In 1991, Bruce D. Gelb, M.D., a pediatric cardiologist interested in gene discovery, was recruited from the Baylor College of Medicine to the Department of Pediatrics. In addition to his genetic research interests, he, together with Judith P. Willner, M.D., the Medical Director of Clinical Services, established a cardiac genetics clinic.

To enhance the efficiency, capacity, and volume of the Department's diagnostic genetic testing laboratories, Nataline Kardon, M.D., and Ruth Kornreich, Ph.D., were recruited to direct the Cytogenetics Diagnostic Laboratory and the DNA Diagnostic Laboratory, respectively. Christine Eng, M.D., a clinical and research geneticist and a long time member of the Genetics faculty, in collaboration with Kornreich, significantly expanded the Department's diagnostic menu and volume.

The laboratory has become one of the busiest academic molecular diagnostic laboratories in North America. In addition, Jose Abdenur, M.D., and then Kimiyo Raymond, M.D., were recruited to direct the Biochemical Genetics Diagnostic Laboratory.

With the imminent opening of the new research facility, the East Building, in October 1995, recruitment efforts focused on research faculty. Douglas Forrest, Ph.D., and Lilly Ng, Ph.D., experts in the generation of "knockout" mice and in the use of animal models to delineate the biology of the thyroid hormone receptors, were recruited in 1996 from the Roche Institute of Biology as Assistant Professors. Dieter Bromme, Ph.D., is an authority on lysosomal proteases and a researcher with particular interest in the structure and function of cysteine proteases who had already cloned and characterized several of these proteases, including five novel cathepsins. Peter Warburton, Ph.D., was recruited from the University of Edinburgh as an Assistant Professor in 1998. Warburton, a molecular cytogeneticist with expertise in centromere and neochromosome structure and function, also has a major interest in using artificial chromosomes for gene transfer. Jin-Qiang Fan, Ph.D., a glycobiologist and molecular biologist with expertise in chemical chaperones, was recruited in 1999 from the Tokyo Institute of Medical Sciences.

As envisioned by the planners of the Department, collaboration has been the order of the day. Many secondary appointments have been granted to faculty of other departments. These have included, among others, George Atweh, M.D., an internist and hematologist with expertise in the hemoglobinopathies and other inherited hematologic disorders; Jon Gordon, M.D., Ph.D., a reproductive biologist who, in collaboration with Frank Ruddle of Yale, was the first to create a transgenic mouse; Max Levitan, Ph.D., an anatomist and *Drosophila* geneticist who wrote one of the early textbooks on human genetics; and James Wetmur, Ph.D., a microbiologist and DNA expert who published a *Current Contents* "classic" on DNA hybridization.[19] Many primary faculty members have joint appointments either in other clinical or basic science departments or in one or another institute or center.

Several of the Department's outstanding M.D. and Ph.D. postdoctoral fellows remained on the faculty and have made meaningful contributions to the Department's effort in both research and patient care. These include Calogera Simonaro, Ph.D. (1996), John

Martignetti, M.D., Ph.D. (1998), Melissa Wasserstein, M.D. (1998), George A. Diaz, M.D., Ph.D. (1999), Xing-Xuan He, M.D. (2000), Ainu Prakash-Cheng, M.D., Ph.D. (2001), Maryam Banikazemi, M.D. (2001), and Brynn Levy, Ph.D. (2001). As is described later, Martignetti and Diaz were both involved in genomics and gene discovery, each having successfully identified the genes for genetic diseases. Prakash-Cheng became Director of the Comprehensive Gaucher Treatment Center in 2001, and Banikazemi was responsible for the clinical trials of enzyme replacement therapy for Fabry disease. Wasserstein has clinical and research responsibilities in the Inherited Metabolic Disease Program and has also been involved in the clinical trials of enzyme replacement therapy for Fabry disease and the efforts to develop enzyme therapy for Niemann-Pick disease.

CLINICAL PROGRAMS

The Clinical Genetics Program of the Department provides services that range from preconception genetics counseling to management of adults with genetic disorders. The Jewish Genetic Disease Screening Program identifies couples at risk for nine recessive genetic disorders that are prevalent in their ethnic group and serves as a model for other prenatal screening programs nationwide. In concert with the Division of Maternal and Fetal Medicine in the Department of Obstetrics, Gynecology and Reproductive Sciences, the Prenatal Diagnosis Program provides state-of-the-art prenatal diagnostic services. The faculty in this program has also carried out NIH-funded research initiatives to evaluate new diagnostic techniques, such as chorionic villus sampling. More recently, this faculty has evaluated multiple biochemical markers in maternal serum for the identification of fetuses at increased risk for chromosome abnormalities.

The Department's Inherited Metabolic Disease Program, one of the largest in the country, provides multidisciplinary care for patients with inborn errors of metabolism, including those identified by the New York State Newborn Screening Program. Recently, the Department established a Cancer Genetics Program that offers counseling and testing to patients at increased risk for genetic forms of cancer. The Department also continues to expand the menus of its clinical diagnostic laboratories, which provide expert cytogenetic, biochemical, and molecular genetic testing.

EDUCATIONAL PROGRAMS

The Department plays an important role in the teaching of genetics to Mount Sinai medical students, residents, and fellows. Departmental faculty have responsibility for the didactic course in Medical Genetics for the first-year medical students. They lecture on genetics to each clerkship group in Medicine, Obstetrics, Gynecology and Reproductive Sciences, and Pediatrics and provide elective courses in Clinical Genetics for third- and fourth-year students. Medical students also have the opportunity to participate in basic genetics research during summer electives; a number of students have graduated with "Honors in Research" based upon their studies in genetics.

Responsibility for the clinical, diagnostic laboratory, and clinical research training of the M.D. postdoctoral fellows is shared by Willner and McGovern. In 1995, the Department was the first in the nation to have a residency in Medical Genetics approved by the Accrediting Council on Graduate Medical Education (ACGME). More recently, a five-year ACGME-approved joint residency program with Pediatrics was established for those individuals interested in joint certification; a combined program with Pathology in Molecular Genetics was also initiated.

The Department has a long history of training predoctoral students, Ph.D. postdoctoral fellows, and M.D. postdoctoral fellows in clinical, biochemical, and molecular genetics and in cytogenetics. A training grant in Mental Retardation and Developmental Disabilities (MRDD), with a focus on genetics, was first obtained by Hirschhorn and has been funded continuously by the National Institutes of Health for more than thirty-five years. The training grant has been directed by Desnick since 1977. The Department has trained more than one hundred predoctoral students, most of whom have earned doctoral degrees in human genetics. In 1998, Schuchman was instrumental in structuring the Genetics and Genomic Sciences training program in the Mount Sinai Graduate School of Biological Sciences and served as its first Director. This interdisciplinary training program draws on faculty throughout the institution who have research interests in genetics and genomics. Departmental faculty also have trained more than one hundred Ph.D. postdoctoral fellows in cytogenetics, biochemical, and molecular genetics. Many of these individuals are certified by the American Board of Medical Genetics, and many have gone on to occupy tenured faculty and administrative positions at prominent universities.

As a major resource for the training of genetic counselors, in 1995, the Department instituted a highly successful Master's Degree program in Genetic Counseling, directed by Randi Zinberg, M.S.

RESEARCH HIGHLIGHTS

Funded by both NIH and private agencies, Departmental faculty have pursued independent basic and clinical research in programs focused primarily on (1) the discovery of genes responsible for inherited diseases, (2) the further understanding of the fundamental biology and pathology of genetic diseases, particularly involving the use of naturally occurring mammalian and genetically engineered mouse models, and (3) the development of therapies based on protein, gene, or stem cell replacement. Schuchman serves as Vice Chairman for Research.

Disease Gene Discovery

Departmental researchers have isolated the genes for more than a dozen genetic diseases. Early successes by Desnick, Schuchman, Bishop, and Yiannis Ioannou, Ph.D., included the isolation and characterization of the genes for lysosomal disorders, including Fabry,[20] Farber,[21] Niemann-Pick types A, B,[22] and C,[23] Schindler,[24] and Maroteaux-Lamy[25] diseases. The genes and/or genomic sequences for several of the porphyrias were also isolated and characterized by Desnick, Bishop, and Wetmur. These included ALAD-deficient porphyria,[26, 27] congenital erythropoietic porphyria,[28,29] acute intermittent porphyria,[30] porphyria cutanea tarda,[31] and X-linked sideroblastic anemia.[32] Other Departmental researchers were responsible for the isolation and characterization of numerous other genes that are essential for human function. These include the thyroid hormone receptor genes[33-37] and the genes for cathepsins F, K, V, W, and other cystine proteases.[38-42]

With the gene mapping and gene sequencing advances of the early 1990s, accelerated by the Human Genome Project, it became possible to identify the genes responsible for diseases in which the nature of the underlying defect was unknown. Gelb, with Desnick, began "gene discovery" research in the Department. These efforts were focused on the identification of causative gene defects for diseases in which there were no clues as to their etiology. The first success, in 1996, was the identification of the gene for a rare bone-thickening disease, pycnodysostosis, that causes dwarfing.[43] The discovery that the cathepsin K gene is defective in this very rare disease suggested cathepsin K inhibitors as potential drug targets for the treatment of the common disease osteoporosis. If one could inhibit the action of cathepsin in bone reabsorption with specific small molecule inhibitors of this protease, then one might retain bone, particularly in older women. This approach is being pursued by industry.

Subsequently, "positional cloning" techniques were employed by departmental researchers to identify the genes responsible for other genetic diseases, including Noonan syndrome (a common congenital heart defect),[44] WHIM syndrome (an immunodeficiency that creates susceptibility to the papilloma virus),[45] May-Hegglin syndrome (a coagulation disorder due to defective platelets),[46, 47] MONA syndrome (a new form of juvenile arthritis),[48] Char syndrome (an inherited disorder involving patent ductus arteriosus),[49, 50] Rogers syndrome (a thiamine transporter defect that causes diabetes mellitus, megaloblastic anemia, and sensorineural deafness),[51] Kenny-Caffey syndrome (a disorder that involves skeletal dysplasia and hypothyroidism),[52–54] and Niemann-Pick C disease (a neurodegenerative disease due to abnormal cholesterol transport).[55, 56]

Animal Models of Genetic Diseases

Advances in embryonic cell manipulation made it possible to introduce gene defects into mice so that the consequences of a defective gene could be investigated biochemically, pathologically, and clinically. Forrest and Ng have produced mice with a variety of thyroid hormone receptor defects. Studies of these mice have led to the elucidation of the unexpected, but essential, role of the thyroid hormone receptor in hearing and color vision.[57] Departmental investigators have also created mouse models for Fabry,[58] Schindler,[59] Niemann-Pick types A and B,[60,

[61] and Farber diseases,[62] as well as for Rogers syndrome[63] and pycno-dysostosis. These mice have been useful in helping researchers understand the basic pathophysiology of these diseases and are useful models for preclinical studies of various therapeutic strategies.

Novel Treatments of Genetic Diseases

The ultimate goal of translational research and a major focus of the Department's research has been the elaboration of novel therapies for inherited metabolic disorders. Efforts to formulate enzyme replacement therapy for Fabry disease have been successful; a drug was approved in Europe in 2001 and in the United States in 2003. This is the first therapeutic drug designed by Mount Sinai genetic researchers (Desnick, Bishop, and Ioannou). To do this required the isolation of the causative gene,[64, 65] the development of a novel method to produce normal recombinant enzyme (human alpha-galactosidase A),[66] and the ability to generate a Fabry disease mouse model for pre-clinical testing.[67, 68] The enzyme, which was created in collaboration with the Genzyme Corporation, underwent clinical trials in Europe, the United States, and at Mount Sinai (conducted by Eng, Wasserstein and Banikazemi) and has been found to be safe and effective.[69, 70]

With successful therapy for Fabry disease, efforts have been undertaken to produce an enzyme replacement therapy for Niemann-Pick type B disease. Schuchman's laboratory was responsible for isolating the gene for acid sphingomyelinase[71] and the generation of the Niemann-Pick mouse model.[72] Preclinical studies have demonstrated the effectiveness of the replacement enzyme,[73] and clinical trials, sponsored by the Genzyme Corporation, began in 2003 under the direction of McGovern and Wasserstein.

A new approach to the treatment of genetic diseases due to misfolded proteins, particularly enzymes, was pioneered by Jian-Qiang Fan, Ph.D., who joined the faculty in 1999. This strategy involves the use of small molecules, called chemical chaperones, that are specifically designed to bind to the misfolded protein and to rescue it for delivery to the appropriate subcellular compartment.[74] Proof of concept for this exciting approach has been demonstrated,[75] and a private biopharmaceutical company, Amicus Therapeutics, has been established with venture capital to create drugs for genetic diseases using the Department's novel technology.

CORE FACILITIES

The Department established several core facilities that provide state-of-the-art techniques and are available to the benefit of all faculty throughout Mount Sinai. These include a DNA synthesis and sequencing core, directed by Bishop, a tissue culture core, directed by Warburton, and a proteomics core, directed by Rong Wang, Ph.D. The DNA core also provides genomic scans for gene discovery, loss of heterozygosity testing, and mutation/polymorphism detection services. In addition, the Department has cores for gene mapping and production of knockout/knockin animal models.

PHILANTHROPY

Generous and sustained philanthropic support has played a pivotal role in the achievement of the Department's goals. The Center for Jewish Genetic Diseases was supported for the first decade of its existence by the generosity of Florence and Theodore Baumritter. In 2002, the National Foundation for Jewish Genetic Diseases, under the leadership of George Crohn, transferred its funds to the Center for Jewish Genetic Diseases to support combined educational programs and create a unified website. This provides investigators, counselors, patients, and their families with learning resources and up-to-date information on the Center's efforts in research, prevention, and treatment of diseases that affect primarily the Ashkenazi Jewish population.

In 1997, families affected by genetic diseases founded the nonprofit Genetic Disease Foundation to support the Department's research and education missions, primarily by the purchase of state-of-the-art research equipment such as mass spectrometers, DNA sequencers, and other sophisticated and expensive equipment. The generosity of the Foundation and its devotion to the Department and institution have greatly enhanced research capabilities and accelerated the research progress of all Mount Sinai investigators.

THE DEPARTMENT OF HUMAN GENETICS IN 2003

In its first decade, the Department has not only met but exceeded its goals of becoming a preeminent Department in both clinical and basic research. As a hybrid basic science and clinical department, it offers a broad-based program of instruction, research, and clinical services in human and medical genetics. The faculty, which has grown from the original ten to more than thirty, has expertise in the application of molecular, biochemical, immunologic, cytogenetic, and somatic cell approaches for the study of genetic diseases.

Ten years after the creation of the Department of Human Genetics, its leadership and faculty can point with pride to its accomplishments. As the Department moves into the future, its unique initiatives and its collaborative efforts with so many of Mount Sinai's Institutes, Centers, and Departments have set the stage for dramatic advances in the field of human genetics.

8

The Department of Health Policy

THE DEPARTMENT OF Health Policy was conceived by John W. Rowe, M.D., then President of the Mount Sinai Medical Center, in the early 1990s. The Department became a reality in March 1995 when a letter of agreement was signed by Mark R. Chassin, M.D., M.P.P., M.P.H., the newly appointed Chairman, who would also serve as Senior Vice President for Clinical Quality at the Mount Sinai Medical Center and Health System. The plan was to develop a new initiative in quality of care, one that would integrate an academic research program with operational clinical quality improvement. The new effort would encompass a "program in quality which includes state-of-the-art measurement capabilities, system-wide accountability, and the ability to bring to all participating physicians the latest knowledge regarding the effectiveness of health-care services through the use of practice guidelines and the latest technologies."[1]

Before coming to Mount Sinai, Chassin served as Commissioner of the New York State Department of Health. Elected to the Institute of Medicine of the National Academy of Sciences in 1996, he cochaired its National Roundtable on Health Care Quality. A board-certified internist, Chassin practiced emergency medicine for twelve years. He also served as a member of the Board of Directors of the National Committee for Quality Assurance and the Association for Health Services Research. In addition, he was a senior project director at the RAND Corporation, where he led several major health services research studies; Senior Vice President and cofounder of Value Health Sciences, a private-sector firm that developed software and systems for quality assessment and utilization review; and Deputy Director and Medical Director of the Office of Professional Standards Review Organizations of the U.S. Health Care Financing Administration.

Chassin's research has focused on creating measures to quantify the quality of health care, using those measures to improve quality and to understand the relationship of quality measurement and improvement to health policy.[2–8] As Health Commissioner of New York State, Chassin was able to achieve a significant statewide reduction in mortality following coronary artery bypass grafting.[9–11]

As enunciated by Chassin and his colleague Albert Siu, M.D., M.S.P.H., the Department was to be a full academic department in the School of Medicine but was also to function as a resource for the management of the Medical Center. It was to combine an academic mission of research and teaching in health services and health policy with an administrative mission to improve quality for the patients and populations served by the Medical Center and its growing health system.[12]

Originally housed in rented space at The New York Academy of Medicine, Chassin undertook the task of recruiting faculty and staff and discussed the Department's vision with the various constituencies within the Medical Center. A conceptual framework for measuring and improving quality was designed, and a strategy for bringing quality improvement to the Mount Sinai Medical Center was constructed. A series of research and quality improvement projects was initiated, and the first of many grant applications for external funding was submitted. As the Senior Vice President for Clinical Quality of The Mount Sinai Hospital and Health System, Chassin was able to gain the support and collaboration of the clinical leaders of the institution.

The physicians and staff of The Mount Sinai Hospital had long been involved in quality assurance activities. These activities, however, were for the most part departmentally driven and directed mainly at morbidity and mortality reviews. The earliest institutional attempts at continuous quality improvement (CQI) had been in the administrative arena and were primarily focused on reducing length of stay and on cost containment. Chassin's definition of "quality" was that adopted by the Institute of Medicine Committee to Design a Strategy for Quality Review and Assurance in Medicare of which Chassin was a member: "Quality of care is the degree to which health services for individuals and populations increase the likelihood of desired health outcomes and are consistent with current professional knowledge."[13]

In Chassin's view, healthcare quality problems may be usefully categorized as overuse, underuse, or misuse.[14] Overuse is the provision of ineffective services; underuse is the failure to provide an effective service. Misuse occurs when an effective service is provided badly and patients are exposed to an increased risk of avoidable complications. Solving these problems depends critically on research that pinpoints when particular health services improve patient outcomes and when they do not. Research must also design, deploy, and evaluate robust improvement strategies and systems to achieve and sustain substantial gains in quality.

The Department's physician faculty members are all accomplished health services researchers and represent the clinical disciplines of general internal medicine, cardiology, geriatrics, pediatrics, obstetrics and gynecology, and surgery. Along with Chassin, they have been extremely successful in achieving external funding for their projects. Indeed, by early 2003, just prior to the Department's eighth birthday, its faculty had secured a total of $30 million in research funding, about two-thirds from federal sources. As practiced by the Department, health services research focuses on interdepartmental and multi-institutional collaboration. Almost all of the major projects that have been instituted have involved one or more departments in the Hospital or multiple hospitals both within and outside the Mount Sinai–New York University Health System.

Albert Siu, who accompanied Chassin from Albany to Mount Sinai, has engaged in a series of projects on the impairments and diseases that contribute to poor functional status and disability in the aging. The patient cohort for this project was drawn from a large number of patients with hip fracture who had been admitted to four hospitals in the Health System.[15–17] In 1998, Siu was appointed the Chief of the Division of General Internal Medicine in the Department of Medicine, and in 2002 he became the third Chairman of Mount Sinai's Department of Geriatrics.

The research interests of the other members of the faculty have encompassed a broad spectrum of clinical problems. Nina A. Bickell, M.D., M.P.H., an internist and the first faculty member recruited to the Department in 1995, has focused her research efforts on the management and coordination of multidisciplinary care in patients with early stage breast cancer[18–20] and directs a study on the effects of time delays on the outcomes of patients with appendicitis, intestinal obstruction, and ruptured ectopic pregnancy.[21] In 2001, in a program project to reduce disparities in health, she became project leader for a federal grant designed to reduce underuse of effective breast cancer treatments among minority populations.[22] In 2003, she became codirector of an NIH Comprehensive Center grant similarly aimed at reducing racial and ethnic healthcare disparities.

The current research interests of Carol R. Horowitz, M.D., M.P.H., who joined the Department in 1996, center about the quality of care rendered to underserved and ethnically diverse populations. She has explored interventions that aim to decrease racial and ethnic disparities in health care.[23] Patients with congestive heart failure, adult-onset diabetes, and community-acquired pneumonia[24] have been investigated in an attempt to identify patterns of patient behavior that can be changed for the better through patient education and patient self-management. Horowitz is the principal investigator of the East Harlem Diabetes Center of Excellence, funded by the New York State Department of Health. The Center aims to develop simple, sustainable interventions to improve diabetes care and patient-centered outcomes in the East Harlem community, one of Mount Sinai's major service areas.

Ethan Halm, M.D., M.P.H., a member of the faculty since 1997, is the principal investigator of a study to develop and validate measures of instability on discharge in patients with pneumonia, asthma, and hip fracture.[25] He has established evidence-based rules for decision making in the admission, management, and discharge of patients with community-acquired pneumonia,[26] and he is also a coinvestigator of a New York statewide study to measure risk-adjusted outcomes of carotid endarterectomy and to assess the effectiveness of various surgical techniques on these outcomes.[27, 28] His health policy research includes a comprehensive analysis of the relationship between volume and outcome in health care.[29]

Elise Becher, M.D., M.A.P.P., a pediatrician, has studied firearm injury prevention and the appropriate use of tympanostomy tubes in children with frequent ear infections.[30, 31] More recently, Becher has made important contributions to the health policy debate on medical errors and how to prevent them from doing harm, barriers to quality improvement, and the physician's role in improving quality of care.[32-35] Becher collaborated with Elizabeth Howell, M.D., M.P.H., an obstetrician/gynecologist, in conducting studies on prematurity. Howell is also involved in the investigation of postpartum disability. Mary Ann McLaughlin, M.D., M.P.H., a cardiologist and geriatrician, has undertaken studies on the management of heart failure and hypertension and on issues of women's health.[36]

With the recruitment of Bruce Vladeck, Ph.D., in 1998 to head the newly created Institute for Medicare Practice, another dimension was added to the Department. Vladeck came to Mount Sinai after ten years as President of the United Hospital Fund of New York City and immediately after serving four years as Administrator for the United States Health Care Financing Administration (HCFA) in Washington, D.C. In that position, he directed the Medicare and Medicaid programs, with combined annual expenditures of more than $300 billion. Elected to the Institute of Medicine (IOM) of the National Academy of Sciences in 1986, Vladeck chaired the IOM's Committee on Health Care for the Homeless.

The Institute for Medicare Practice was designed to bring together members of the Mount Sinai faculty in projects to study and improve care for elders in the community. Vladeck and his colleagues partnered with other faculty in Health Policy and in the Department of Geriatrics, where he holds a joint appointment. Monthly "Medicare Seminars" attracted a significant following from the metropolitan area. In October 2001, Vladeck became acting Chairman of the Department of Geriatrics, a position he held until Albert Siu was appointed to the Chair. Unfortunately, as a result of Mount Sinai's fiscal crisis, the Institute closed in June 2003; Vladeck remains on the faculty of both Health Policy and Geriatrics.

Jane Sisk, Ph.D., a senior health economist, joined the Department in 1999 after being recruited from the Columbia University School of Public Health. Her current research focuses on interventions to improve the quality of care and to reduce disparities among population subgroups; the evaluation of Medicaid managed care; and the cost-effectiveness of healthcare interventions such as pneumococcal vaccination in the elderly.[37] Elected to the Institute of Medicine of the National Academy of Sciences in 2001, Sisk has served on numerous study sections and has also served as President of the International Society for Technology Assessment in Health Care.

The burgeoning research agenda of the Department highlighted the need for additional faculty. Mary Rojas, Ph.D., was recruited in 2002 from IPRO (Island Peer Review Organization), where she had served as the Senior Director for Data Analysis and Epidemiology. Her

236 TEACHING TOMORROW'S MEDICINE TODAY

contributions were immediately apparent as she collaborated with other faculty in developing the data analysis for several of the major projects. Paul Hebert, Ph.D., a health economist with expertise in the analysis and management of large databases, was recruited in 2002 from the University of Minnesota. He has developed an independent research portfolio and has provided major analytic, programmatic, and statistical support to the Department.

In August 2001, and in collaboration with the General Electric (GE) Corporation's Medical Systems' Healthcare Solutions business unit, the Department of Health Policy launched a "Six Sigma Quality Improvement Program" to explore the effectiveness of GE's quality improvement methods in addressing a variety of important quality problems within the Medical Center.[38, 39] From the outset, this program has had the enthusiastic support of the Boards of Trustees and senior Medical Center administration.

Invented at Motorola in the 1980s, Six Sigma is a rigorous method of reducing the frequency of serious errors or defects to extremely low levels. In statistics, "sigma," the eighteenth letter of the Greek alphabet, is the mathematical symbol for standard deviation, a measure of how far something is from the average. Achieving Six Sigma quality results in a process that produces no more than 3.4 defects per million opportunities. A "defect" is defined as a product or service that fails to meet agreed-upon expectations, while an "opportunity" is a chance for something to go wrong. In percentage terms, Six Sigma performance yields a defect rate of 3.4 ten-thousandths of 1 percent.

Six Sigma Quality Improvement methods, as refined by GE, are a set of tools and methods for achieving and maintaining unprecedented levels of excellence across a wide variety of complex technical, business, administrative, or service processes. In the first phase of its initiative, Mount Sinai trained a group of sixteen managers, analysts, nurses, and physicians to implement the Six Sigma methods. Known as "Green Belts," these individuals spent at least one-third of their working time leading the various projects. Kathryn Colson, M.P.H., of the Department of Health Policy, herself a Green Belt, has been the project's coordinator.

Four Six Sigma projects, two in the Hospital and two in the Faculty Practice Associates (FPA), were initially undertaken with a view toward achieving substantial and measurable benefits. The first Hospital project was designed to reduce medication errors, the second to enhance revenue collection. In one FPA project, the team helped to generate a noteworthy improvement in billings. In the second project, service quality and efficiency were greatly improved by reducing delays between orthopedic surgical procedures in the operating rooms.

As 2002 came to a close, the first four Six Sigma projects were nearing completion, and a second phase of training and improvement projects was begun. Sixteen additional Green Belts were being trained, four additional improvement projects were launched in the Hospital and School, and two Black Belts (expert Six Sigma practitioners) were being trained. These initial exposures to GE's Six Sigma training programs and quality improvement methods provided Mount Sinai with invaluable experience with which to assess the potential role of these tools in addressing problems in a broad array of clinical, administrative, and patient service areas. Should these programs prove their worth, a final phase will train Mount Sinai personnel in Six Sigma training techniques so that the entire effort can be sustained by the Medical Center itself without a continuing dependence on outside consultants.

In 2002, the Department was awarded an NIH Comprehensive Center grant by the National Center on Minority Health and Health Disparities. This $6 million grant created a Center with the goal of improving health and reducing disparities among patients in East and Central Harlem by enhancing patients' self-management skills. The core components of the Center will enhance existing disparities research in the Department, foster community outreach, and establish new training programs for minority health services researchers. North General Hospital, the only other not-for-profit hospital in the East and Central Harlem community, and already an affiliate of the Medical School, will be a major partner in the Center's activities.

This Center builds on a previous grant award to Department researchers. In 2000, the federal Agency for Healthcare Research and Quality funded a $7.4 million Program Project Grant, "Improving the Delivery of Effective Care to Minorities." In four related studies, researchers are measuring the underuse of selected, proven-effective medical and surgical interventions in the large and ethnically diverse communities of East and Central Harlem in New York City. The underlying causes for underuse will be assessed and interventions will be devised, implemented, and evaluated to eliminate underuse by targeting the underlying causes. This program represents an extension of previous work and includes the control of hypertension, the prevention of recurrent stroke, the correction of underuse of early-stage breast cancer treatment, and the assessment of variations in the management of prematurity. Patricia Formisano, M.P.H., is the program manager.

In early 2003, shortly after his appointment as Dean of the School of Medicine and President of the Medical Center, Kenneth L. Davis, M.D., appointed Chassin to the position of Executive Vice President for Excellence in Patient Care of the Mount Sinai Medical Center. In June 2003, Mount Sinai, under Chassin's leadership, launched a major new initiative to establish unprecedented excellence in all aspects of patient care, including clinical outcomes, safety, the experiences of patients and families, and the working environment of the institution's care-givers. The program was designed to support the Medical Center's overarching mission: to provide the highest-quality, most compassion-ate care possible to patients and their families. To sustain a central focus for the new quality enterprise, the Joseph F. Cullman Jr. Institute for Patient Care, which had been created by a gift from the Cullman family, the office of Organizational Development and Learning, the Survey Center, and the Six Sigma program, are all now administered through the new office of the Executive Vice President. The first activi-ties of the initiative focused on three areas: developing a broader understanding of the concerns of the Hospital's caregivers regarding quality and safety and their ideas for improvement; designing and implementing focused clinical improvement efforts in selected patient care areas, beginning in obstetrics; and engaging in an ongoing educa-tional and communication program to improve quality and safety throughout the Medical Center.

The Cullman Institute, directed by Mary Dee McEvoy, Ph.D., R.N., of the Department of Nursing, works in partnership with all segments of the Mount Sinai community to promote a culture of patient-centered care that emphasizes service and compassion from the moment a patient or family member enters the doors of the Hospital until dis-charge. One of the Institute's first initiatives is the Ambassador Pro-gram, created by Jean Crystal, a member of the Board of Trustees and a former President of the Auxiliary Board. Volunteer "Ambassadors" meet patients and family at the Hospital entrances and escort them per-sonally to their destinations. The program has grown rapidly, and a number of medical students have joined the corps of Ambassadors.

Thanks to its strong and innovative leadership and highly productive faculty, the Department of Health Policy, although chronologically young, has already become an essential academic resource for health services research and policy throughout the nation.

9

Graduate and Postgraduate Education

IN THE JANUARY 15, 1852, Articles of Incorporation of The Jews' Hospital, the founding fathers stated that they wished to associate themselves into a "benevolent, charitable and scientific" organization. The primary purpose of the Hospital was to provide "medical and surgical aid to persons of the Jewish persuasion."[1] Dedicated to patient care, the Hospital did not consider education to be a priority at that time. The first evidence of an educational effort occurred in December 1855, six months after the opening of the Hospital, when Mark Blumenthal, M.D., the Resident and Attending Physician, requested permission from the Board of Directors to perform an autopsy to "justify himself and his reputation concerning the cause of death of the patient."[2] As best as can be determined, an occasional autopsy was the only educational effort in the Hospital throughout the fifteen years that it was located on West 28th Street.

When the Hospital moved to its second site, in 1872, its 105 beds more than doubled the previous complement. At that time, the professional staff created the Medical Board, one of whose first recommendations was the establishment of a house staff, presumably for physician education as well as for service duties to patients. Appointments to the house staff would be made by recommendation of a three-member examining committee. The three original members of this examining committee consisted of Abraham Jacobi, M.D., a pediatrician; Ernst Krakowiczer, M.D., a surgeon; and the internist, Samuel Percy, M.D. In addition, in 1872, an Outdoor Dispensary was established, and it was noted that the institution's mission was expanded to "greatly extend the sphere of usefulness by the establishment of a clinic contributing thereby to the advancement of medical science, and aiding the student in the study of his profession."[3]

240

One year later, and ninety years before the founding of the Mount Sinai School of Medicine (MSSM), Percy appeared before the Board of Directors to recommend the development of a medical college. The Board, already faced with paying for almost 100 percent of the cost of patient care from either donations or their own pockets, deemed it "not expedient."[4]

The year 1877 proved to be a watershed year for the training of house officers. At that time, the Hospital divided the beds into medical and surgical divisions and established a formal two-year training program for each division. Continuing to be appointed by competitive examination, successful recruits to the house staff were able to choose their primary service. They would spend six months on each of the services in their first year and were known as "Juniors." They would then spend the entire second year on their chosen service, at which time they were called either "house physicians" or "house surgeons." A diploma was issued following each recruit's successful completion of the two years. The Hospital records note that in the years prior to 1884, only seventeen physicians were "graduates of the house staff." In 1877, Howard Lilienthal, M.D., became the first house officer to designate surgery as his field of choice. From 1884 until just prior to the Hospital's move to the 100th Street site in 1904, one or two house surgeons and house physicians completed training each year. At that time, the number of house staff appointees increased dramatically. Josephine Walter, M.D., in 1885, was the first woman to graduate from the house staff, and it was believed that she was the first woman to graduate from a formal general hospital training program anywhere in the United States.

Postgraduate education was also ignored for the better part of the first half-century of the Hospital's existence. John Wyeth, a surgeon who joined Mount Sinai's staff in 1882, was so disappointed at the lack of postgraduate training, not only at Mount Sinai, but throughout New York City, that within the year he founded the New York Polyclinic Hospital and Medical College, an institution that provided short postgraduate courses for physicians, primarily surgeons and surgical specialists, until its closing in 1976.[5]

In December 1880, Abraham Jacobi asked the Board of Directors if he might do clinical teaching on the Pediatric service. He did not receive a positive response, but less than two years later, the Board approved a resolution of the Dispensary Committee stating "that the practice of each attending physician of the Dispensary of inviting two medical students during the hours of their service, for assistance and clinical experience be permitted."[6] Education of house officers flourished. A major boon to the educational process was the creation of the library by Adolf Meyer, M.D., in 1883. As Meyer would later note: "I had recognized the need of one when I was an interne only five years before."[7] Specialty training, however, lagged far behind, despite the fact that both medical and surgical specialties were being created in the Hospital. For the most part, those interested in a specialty or laboratory training went abroad for further training; Germany and Austria were the sites most frequently chosen. Clinical instruction of medical students on the inpatient units was formally approved in 1888.

Publications by members of the Hospital Staff have been collected from as early as 1874.[8] In 1899, thirty-five years before Mount Sinai published its own journal, the Hospital began to publish annual medical reports detailing scientific statistics from several departments and articles contributed by members of the attending staff. Regarding this publication, the President of the Hospital noted, "It has proven of great value to the medical profession here and abroad, and its dissemination is well calculated to be of advantage to the general public."[9]

With a new hospital containing more than four hundred beds under construction at 100th Street and Fifth Avenue, the house staff was enlarged in 1902 to include twenty interns. When the buildings opened, there was a separate building for private patients. As a consequence, a separate house staff was created to care for the private patients. Although the private pavilion house officer positions were not considered to be nearly as prestigious as those on the regular house staff because the Attendings allowed the private residents less autonomy, many medical graduates availed themselves of these positions, hoping to be appointed later as house physician or house surgeon on the ward services.

It is clear that the opening of the new Hospital led to the expansion of physician education at both the graduate and postgraduate level. Allusions to the educational role of the Hospital were heard during the ceremonies celebrating the laying of the cornerstone of the new building on May 22, 1901. Among the many eminent speakers present at the time, Seth Low, the President of Columbia University, noted: "Neither a physician nor a surgeon deserving the patronage of an intelligent people can be educated without the clinical privileges afforded by the hospitals."[10] Abraham Jacobi then stressed the views of the staff: "a hospital is a school for doctors who learn and profit in the interest of mankind. . . . It is a school for nurses . . . it is a school for the patients and their families. . . . Finally, it is a school for the medical world."[11]

The first Clinical Pathological Conference (CPC) was held in January, 1905; Sir William Osler was the invited speaker. The Directors, mindful of Osler's worldwide reputation as an educator, availed themselves of the opportunity to discuss the Hospital's future educational plans with him. Osler's recommendations led to the expansion of the educational effort. Plans were put in motion to develop fellowship funds for postgraduate education, primarily abroad, for young physicians; the establishment of formal postgraduate training programs sponsored by the Hospital; and the inclusion of medical student teaching on the ward services. In 1919 the CPC became an established entity that attracted physicians throughout the area and added greatly to Mount Sinai's reputation among physicians.

One of the most outspoken of the Hospital's leaders in regard to the educational mission of the Hospital was Isaac Wallach, who served as President of the Board of Directors from 1896 to 1907. In his President's Report for the year 1905, he made a major statement relating to education when he wrote: "In even a broader sense has our hospital been an educational force. . . . Emblematical of its threefold mission, may its banner with the motto 'Benevolence, Science, Education' wave over Mount Sinai for all time to come."[12] Two months before his death, in March 1907, Wallach's report for the year 1906 established the rationale for Mount Sinai as a teaching hospital:

> Twenty-five or thirty years ago, from the lay standpoint, it was regarded as heresy to consider the patient as furnishing material for education, and the doctor was expected to view him only from the benevolent side. Not so now. It is well understood that the hospital must furnish the facilities and opportunities for education, research, and experience-not alone for the benefits of those within, but also for those outside of such institutions.[13]

The ensuing decade saw Mount Sinai bring to fruition the goals set following the discussions held with Osler. Fellowship funds were developed for both postgraduate clinical and research training, and medical students were accepted on a regular basis for training on the wards. In 1910, the Carnegie Foundation issued the "Flexner Report," a devastating critique of the medical schools then extant. In his discussion of New York schools and their lack of teaching hospitals, Abraham Flexner noted that Mount Sinai was one of four hospitals in the city that, with an appropriate medical school affiliation, "might under such conditions be familiar names in medical science; as well known to the progressive clinicians of St. Petersburg, Vienna, Edinburgh, St. Louis, and San Francisco, as they are to the stricken widows of the East Side of New York City itself."[14] In that same year, Mount Sinai established a formal training program with Columbia University's College of Physicians and Surgeons (P&S), and several members of the staff received faculty appointments. Within a year, the President of the Hospital would state: "The clinical instruction of medical students is permitted and encouraged in the belief that the opportunity to study clinical phenomena is one which hospitals owe to medical students as a matter of public duty."[15]

In 1923, larger specialty departments and divisions were being established in the Hospital, creating the need for graduate specialty training at the Hospital for the first time. Although the exact date of the addition is not known, the Hospital's Constitution was amended during this period so that, in 1918, it stated that one of the Hospital's objectives was "to afford to students in Medicine the opportunity to acquire a practical knowledge of the art and science of Medicine."[16] The stage was now set for the next growth phase in Mount Sinai's undergraduate, graduate, and postgraduate educational activities.

The medical school affiliation with P&S spurred an increase in postgraduate educational activity. Mount Sinai faculty taught in P&S-sponsored courses, and postgraduate courses were developed at the Hospital. The wards and operating rooms of the Hospital had always been meccas of informal instruction for physicians from around the world; the new expanded facilities at 100th Street proved to be a great improvement for both graduate and postgraduate teaching. For the next six decades, postgraduate programming increased slowly but steadily. This activity, however, was directed and coordinated through the individual departments.

The launching of the *Journal of The Mount Sinai Hospital*, in 1934, was a most important factor in authenticating Mount Sinai as a major force in education. Sponsored by the Hospital, the *Journal* was originally devoted to case reports by the staff. Peer review of all articles was instituted from the outset. From the beginning, the *Journal* also published the papers from named lectures given at Mount Sinai. One unique aspect was the publication of abstracts of articles published by the members of the staff in other journals. It also listed all of the postgraduate courses, many of the regularly scheduled conferences, and occasionally Hospital news, as well. Published bimonthly, the *Journal* had initial print runs of 750 copies; there were six hundred paid subscribers, and 150 copies were sent to "libraries and universities in the United States and abroad."[17] The *Journal* was selected for inclusion in the *Cumulative Index Medicus* after one year of publication. Over the years, the focus of the *Journal* expanded. Festschrift issues were dedicated to many of Mount Sinai's great physicians, special issues celebrated landmark events in the Hospital and then the School, theme issues appeared with greater regularity, and gradually more and more papers were accepted from outside the institution.

With the January 1970 issue, the *Journal*'s name was changed to *The Mount Sinai Journal of Medicine*. In its almost seventy years of publication, the *Journal* has had only five editors-in-chief, all physicians. The founding editor, Joseph Globus, the neuropathologist, held the post until just before his death in 1952; Solon Bernstein, an internist, succeeded Globus but died within a year. The internist, Lester Tuchman, served from 1953 until 1974, when he was succeeded by David Dreiling, then vice chairman of the Department of Surgery and the associate editor since 1960. In 1990, Sherman Kupfer, the internist and former Deputy Dean of the Medical School, joined Dreiling as the Coeditor-in-Chief. One year later, Kupfer took over as Editor-in-Chief, a position he held until his death in late 2003.

In addition to the many courses that were given on a regular basis, The Mount Sinai Hospital established itself as preeminent in several areas of postgraduate teaching. For example, the Hospital was asked on numerous occasions to conduct the annual courses sponsored by the American College of Physicians and the American College of Cardiology, and several surgical specialty courses drew participants worldwide. As will be seen later, the opening of the Mount Sinai School of Medicine had a profound effect upon the further development of postgraduate education.

A major impetus for the development of resident training programs, especially in the surgical arena, was the founding of the American College of Surgeons (ACS) in 1913.[18] As criteria for membership were established, it became apparent that the qualifications of the applicants would be of concern, and that standards for surgical education and practice had to be codified. One of the early requirements was the submission of case records, and since most hospital records were sparse or nonexistent, this indirectly led to the establishment of hospital standards. The hospitals were originally certified by the ACS's Hospital Standardization Program, the forerunner of today's Joint Commission on Accreditation of Healthcare Organizations. The leadership of the ACS was diverted from its activities by World War I, but in 1919, William J. Mayo, M.D., the President of the College, stated that henceforth applicants for Fellowship would have to obtain specialty training in the area of their choice. This, coupled with the incorporation of the first specialty Board, the American Board for Ophthalmic Examinations in 1917 (the name was changed to the American Board of Ophthalmology in 1933), provided the stimulus for the beginning of residency training in multiple hospital departments. The Specialty Boards, with their rigid training requirements, proliferated, and the corresponding departments created or revamped their own programs to conform with the new standards of training.

The Department of Pathology had graduated its first resident in 1911; in the 1920s the Departments of Gynecology, Neurology, Ophthalmology, Otolaryngology, Pediatrics, and Radiology initiated formal training programs. They were joined during the 1930s by Dentistry and Orthopaedics, and in the 1940s by Anesthesiology, Dermatology, Physiatry (Rehabilitation Medicine), Psychiatry, and Urology. With the formation of the Boards in Surgery and Medicine, these Departments dropped the title of "House" and adopted the term "Resident" for their house officers. With the opening of the Klingenstein Pavilion in 1952 and the creation of an obstetrical service, the Gynecology residency was renamed Obstetrics and Gynecology. Neurosurgery graduated its first resident in 1950, and a new training program in Oral Surgery was begun in 1952.

Another significant change in graduate education occurred in the early 1940s. The rotating internship was created, stimulated by the advent of World War II and the need to have recently graduated physicians trained rapidly in numerous areas prior to call to military service. The rotating internship—one year of multiple rotations through the major departments—existed for many years and then lapsed into

Anesthesiology residents discuss a paper in their department's library, 1992.

disfavor. Even the term "intern" has essentially disappeared from the lexicon and has been replaced by the term "postgraduate year one" (PGY-1).

After World War II, there was a progressive increase in subspecialty training in the Department of Medicine. This, coupled with the large number of returning servicemen, caused a sizable augmentation in the house staff cadre. Over the decades, existing programs continued to grow, and new ones appeared. In 1961, in an affiliation agreement with New York City, Mount Sinai took over patient care at Greenpoint Hospital, a municipal institution in Brooklyn. Residents from participating departments were assigned there, necessitating a further increase in the total house staff complement. Two years later, the affiliation was shifted to City Hospital Center at Elmhurst, in the borough of Queens. The program was an immediate success; it enabled the City hospitals to provide improved physician services; it accorded a remarkable opportunity for both resident and student training; and it became an anchor in Mount Sinai's educational efforts. Later in the 1960s, Mount Sinai began an affiliation with the Bronx Veterans Administration Medical Center, making available another major institution for student and resident training but also requiring an expansion of existing residency programs.

After the Mount Sinai School of Medicine opened, other institutions became affiliated with the School as venues for teaching undergraduate medical students. In some instances, an interchange of residents occurred for specific rotations not available at one or the other hospital. Although most resident training programs throughout the United States were under the aegis of individual hospitals, both the Liaison Committee on Medical Education, the accrediting body of medical schools, and the Accreditation Council for Graduate Medical Education, the accrediting body for residency programs, were encouraging medical schools to play a larger role in graduate education. At Mount Sinai, in 1994, Barry Stimmel, M.D., who had served as Dean for Academic Affairs and also Dean for Admissions and Student Affairs, was appointed Dean for Graduate Medical Education. Shortly thereafter, the School became the overseer of all graduate training programs in the Medical Center and the affiliates.

The current Bronx Veterans Administration Medical Center. The Bronx VA has been an important affiliate for graduate and undergraduate students of the School of Medicine since 1968.

THE MOUNT SINAI SCHOOL OF MEDICINE CONSORTIUM FOR GRADUATE MEDICAL EDUCATION

With support from a grant from the New York State Health Department in 1995, Stimmel formed the Mount Sinai School of Medicine Consortium for Graduate Medical Education. With the approval of the chief executive officers of all of the hospitals affiliated with the School, the Consortium was "dedicated to centralize, enhance, and monitor the quality of education provided to house staff at all participating institutions as well as to meet the new demands and responsibilities inherent in maintaining graduate medical education programs."[19] As of 2002, the consortium included thirteen institutions and more than one hundred residency programs training 2,022 residents. Mount Sinai Hospital's 801 residents and fellows in fifty-seven different programs constituted forty percent of the Consortium.

The Consortium, which is one of the few actively integrated consortia in the United States, conducts internal reviews of all programs at all member hospitals on an ongoing basis and works with the program directors to prepare for site visits by the accrediting bodies. It provides a core graduate medical education curriculum for all entering first-year house staff (PGY-1), addressing a broad range of medical issues including but not limited to multicultural diversity, alcohol and substance abuse, biomedical ethics, care of the elderly, Acquired Immunodeficiency Syndrome, the dynamics of the physician relationship, preventive medicine, physician impairment, pain management, professionalism, leadership qualities, managed care, and the teaching of medical students. Utilizing standardized patients in the Morchand Center,[20] the Consortium provides a clinical assessment to all PGY-1s prior to the start of their residency training so as to identify any weaknesses that can be remediated during the initial months of training. It also organizes a retreat for upcoming Chief Residents that stresses leadership qualities; promotes minority recruitment; organizes job fairs and career advisory programs; and maintains an extensive demographic database. The Consortium has also fostered the development of computerized programs to facilitate resident evaluation, has expanded the Academic Educational Network linking the participating hospitals, and has helped create uniform policies for residents in all the member institutions.

THE PAGE AND WILLIAM BLACK POST-GRADUATE SCHOOL

As plans were moving forward with the establishment of both the Medical School and the Graduate School, it was only natural that the Hospital's postgraduate activities would be consolidated into a post-graduate school. The latter became reality in 1965 and at that time began to offer Mount Sinai courses through its affiliation with Columbia University's College of Physicians and Surgeons. A major gift from William Black later that year led to the renaming of the School in 1967 as the Page and William Black Post-Graduate School of Medicine of the Mount Sinai School of Medicine. M. Ralph Kaufman, M.D., Chairman of the Department of Psychiatry, was named the first Dean, and Minerva Brown, M.S.S.W., took over the post of Registrar, a position she held until her retirement in 1985.

In its first academic year (1967–1968), the School offered sixty courses varying in length from one day to the full academic year and given at both The Mount Sinai Hospital and City Hospital Center at Elmhurst. Three hundred sixty-four students received instruction from 118 faculty members. Twelve courses were offered by the Department of Dentistry, which had the largest dental faculty of any school not affiliated with a recognized dental school.[21] Under the direction of Charles Friedberg, M.D., the American College of Physicians course in cardiovascular disease was again sponsored. Within two years, the number of courses increased to seventy-seven and the number of students served to more than 1,250.

Kaufman was named Dean Emeritus and a Consultant to the School in 1972, and in 1974 Carter Marshall, M.D., was named Associate Dean for Continuing Education and Director of the Post-Graduate School. Marshall, a member of the Department of Community Medicine from 1969 to 1975 where he played a major role in teaching, also served Mount Sinai as University Dean for Health Affairs from 1972 to 1974, interfacing with The City University of New York to coordinate all health science programs within the system.

In 1977, the scope of the continuing education effort was greatly strengthened by the establishment of the Brookdale Center for Continuous Education, created with support from the Brookdale Foundation. The cardiologist Samuel Elster, M.D., was named Dean for Continuous Education and Director of the Center, positions he held until 1985. The Center administered all educational programs of the Medical School

This was taken at the first course offered by the new Page and William Black Post-Graduate School of Medicine, November 1966. The course was held in the old Blumenthal Auditorium, later torn down to make way for the Annenberg Building. Dean George James is at the podium.

other than the undergraduate and graduate curricula. Divisions were set up to handle continuing education for Nursing, Social Work, Consumer/Patient Education, and Allied Health Education. The curriculum of the Post-Graduate School was restructured and expanded to include an increasing number of courses in various aspects of geriatrics; satellite programs were established at community hospitals; and a new Division of Self-Assessment Tests for Physicians was activated. Courses were also given in association with the Graduate Center of City University of New York. At the time when documentation of continuing medical education (CME) activities became necessary, the Post-Graduate School assumed responsibility for the administration of CME at all of the participating institutions. One of the most highly successful innovations of the new educational offerings was the "mini-residencies." These programs, lasting one to four weeks, were custom-designed for the individual physician, allowing intensive updating of knowledge and skills in a given field at either Mount Sinai or one of its affiliate hospitals. Forty physicians availed themselves of this opportunity in the first year. The 1979 catalog also listed courses in thirty-four separate disciplines.

During Elster's tenure as the Director of Continuous Education, the response to the expanded programming was most gratifying. Registration for courses more than tripled between 1976 and 1981. The satellite program grew to include five hospitals. An annual Brookdale Medical Conference on Aging drew outstanding leaders in the newly emerging field of geriatrics as speakers, and attendance grew rapidly through the years.

Following Elster's retirement in 1985, little change occurred in the Post-Graduate School until 1995, when Mark Swartz, M.D., assumed the position of Dean of the Page and William Black Post-Graduate School of Continuing Medicine Education. Swartz, a graduate of the Mount Sinai School of Medicine and a cardiologist, has been an innovative force in medical education for many years and was responsible for developing the Medical School's Morchand Center for Clinical Competence.[22]

Swartz rejuvenated Mount Sinai's postgraduate activities. In the five academic years ending in June 2002, the number of CME events more than tripled, from 73 to 244; in that time span, the total number of attendees increased from 6,300 to more than 15,200. Over the years, as computer-assisted materials were developed for the School and telemedicine and teleconferencing became more popular, a number of Web-based programs were developed that were applicable to CME. In 2001, Swartz and his colleagues created MSSMTV.org, a Web-based distance learning CME program of the School of Medicine, with the goal of centralizing all of Mount Sinai's ongoing activities in this arena, and also developing software and computer-based teaching products. At least eight on-campus sites, seating up to 260 attendees, are now available for telemedicine and teleconferencing use, and programs are sent worldwide. With substantial support from numerous hospitals and medical schools, medical societies and journals, and multiple corporations, many departments have produced their rounds and conferences online, and the growth of the program has been rapid. By early 2003, the CME website contained 412 lectures totaling 213 course hours. The site was accessed by more than 32,000 visitors in 2002, including 4,000 medical professionals who viewed more than 380,000 pages, with CME credits available for online self-study.[23]

Formal graduate and postgraduate educational activities have now been available at Mount Sinai since 1872 and 1905, respectively. A reputation for excellence, innovative programs, and outstanding leadership will ensure Mount Sinai's preeminence in medical education for decades to come.

PART III

10

The Faculty Practice Plan

AS POINTED OUT by Kenneth Ludmerer in his book *Time to Heal*, a history of medical education in the United States in the twentieth century, clinical practice by full-time faculty is today an essential but problematic component of American medical schools.[1] Finding the right balance of time for teaching, research, and clinical care has long been complicated but has become even more difficult over the past decades by the changes in healthcare financing that encourage shorter hospital stays and increased ambulatory services. These same forces have also led to markedly lower reimbursement rates from insurers, thereby necessitating more patient appointments to generate equivalent income, resources that medical schools have become dependent upon to make ends meet.

Mount Sinai's use of salaried physicians to further its patient care and clinical teaching missions predated and perhaps presaged the Mount Sinai School of Medicine (MSSM) by almost a quarter of a century. George Baehr, M.D., and Isidore Snapper, M.D., were appointed in 1944 as Chiefs of the two teaching services in the Department of Medicine and overall Director of Education and Director of Research, respectively. They were provided with salaries for their services, and they were also able to maintain their extensive practices. Although the laboratory Chiefs had been salaried at Mount Sinai since the 1920s, Baehr and Snapper set the stage for what today is a full-time, school-based clinical faculty that numbers almost eight hundred.

In 1945, M. Ralph Kaufman, M.D., was appointed as the Chief of the newly created Department of Psychiatry. The first full-time clinical chief, Kaufman was, however, "geographic full-time," and he was, indeed, expected to practice to supplement his hospital-based salary.[2] Shortly after beginning his tenure, Kaufman requested that he be made "full full-time," since he felt that the need to practice interfered with his ability to develop his new department.

Over the next seven years, full-time chiefs of service were appointed in Medicine, Pediatrics, Radiology, Obstetrics and Gynecology, and Surgery.[3] With a mandate from the administration and the Board of Trustees to develop research programs and oversee the expanding residency programs, the chiefs began to recruit additional full-time physicians, many of whom wanted time to practice. An immediate issue of concern was the source of salary support for these individuals. Other concerns included the development of a mechanism to handle billing and collections, support for running the practices, and the cost of malpractice insurance.

In 1961, and concomitant with early planning for the Medical School, a Subcommittee of the Hospital's Research Administrative Committee initiated discussions on the need for and support of career research investigators and endeavored to define the relationship of these individuals to the Hospital. It was also recognized that a cadre of full-time clinicians would be required to carry out the patient care and teaching mission of the new School and of the Hospital. Although the Subcommittee recommended groupings of physicians based upon their source of salary support, few appointments were made because of the lack of funds.

Within a month of the granting of the provisional charter incorporating The Mount Sinai Hospital School of Medicine (the original name of the School) in June 1963, a "Policy on Full-time, Geographic Full-time and Part-time Physicians" was drafted that included clinicians as well as career investigators. Voluntary staff who received funds for research or who were unpaid research investigators were also included. Salary scales were proposed, and it was stipulated that, except for voluntary staff, salaried physicians were required to utilize office suites provided by the Hospital. Later in 1963, the Trustees designated funds for support of full-time associates to the chiefs of the clinical services.

In the years leading to the opening of the School, it was recognized that policies regarding the handling of salary and fee income would require uniformity across departmental lines. Shortly after George James, M.D., was appointed Dean in 1965, he appointed Allan E. Kark, M.D., Chairman of the Department of Surgery, to chair a subcommittee of the Executive Faculty to establish such policies, and outside consultants were hired to assist in the task. A number of issues were evident from the outset. Irrespective of the 1963 guidelines, some members of the salaried faculty were not abiding by the "rules." Billing and collection systems were poorly organized and fragmented, and there were major differences of opinion between the full-time and geographic full-time physicians regarding distribution of fee income.

Although the recommendations of the Kark Committee were accepted, it was recognized that many of the proposals were honored more in the breach than in compliance. Faculty who had been in practice before the formation of the School felt that they were being too severely limited in their income because of the restrictions on supplementation. For the first time, the issue of bed availability was raised by the chairs and their colleagues on the salaried faculty. Noting that the voluntary faculty could use other hospitals, several of the chairs requested that beds be set aside for the use of geographic and full-time faculty. As might have been expected, "town-gown" issues suddenly appeared.

In an attempt to address a number of these issues, in 1969, James appointed another committee, this one chaired by Louis Wasserman, M.D., Director of Hematology. In June 1970, the Wasserman Committee issued its report, proposing the establishment of a faculty practice plan (Medical Service Plan [MSP]). The report also outlined the relationship of members of the clinical faculty to the School of Medicine. The plan would be institutionwide, departmentally based, and inclusive of full-time and geographic full-time physicians and would have a broadly representative governing council. Guidelines for determination of base salary sources and supplementation were established. The report was approved in principle and distributed to the chairs and faculty for discussion. Over the ensuing three years, the proposal was modified on numerous occasions. Alternative plans were submitted by Sherman Kupfer, M.D., Associate Dean, and S. David Pomrinse, M.D., the Director of the Hospital. The unexpected and untimely deaths of George James and Solomon Berson, M.D., Chair of Medicine, within a few weeks of each other in 1972 interrupted forward momentum for a time. Hans Popper, M.D., Ph.D., who became the Acting President and Dean, pushed ahead, noting that, with the opening of the Annenberg building and the increase in School class size, there would be a need to expand the salaried faculty in a significant way.

Finally, in February 1973, the Board of Trustees approved a plan to take effect July 1, 1973. That plan contained a number of basic components. There would be two categories of faculty: full-time and part-time/voluntary. Salary ranges for full-time faculty would be established by the Board and negotiated by the chairs with the faculty member. Each participant would have a fund established in the School for depositing fee and consultative income. Depending on fee income after expenses, faculty could supplement their base salary by an amount not to exceed 25 percent of the base, with any overage accruing to the department. Part-time/voluntary physicians would have to devote ten hours per week to School or Hospital activities and would have no income ceilings on the retention of their practice income. This group would have no institutional salary support unless approved by the Dean or Hospital Director. The earlier category of geographic full-time physician would disappear. Those individuals so designated would move to either full-time or voluntary status. Interestingly, clinical chairmen were not included as participants in the plan, and no governance structure was created. A year later, in July 1974, the chairs were allowed to participate.

On October 1, 1973, Thomas C. Chalmers, M.D., was appointed President of the Medical Center and Dean of the Medical School. Two of his early appointees to vacant chairs, Richard Gorlin, M.D. (Medicine) and Arthur H. Aufses Jr., M.D. (Surgery), would have a profound influence on the future of the MSP. Aufses, previously the Director of Surgery at the Long Island Jewish Medical Center, was no stranger to Mount Sinai, having been in private practice at the Medical Center from 1956 to 1971. The two new Chairs were given a mandate to expand the full-time faculty. Keenly aware that the disparity in income between the full-time and the voluntary faculty would preclude meaningful full-time faculty recruitment, Aufses and Gorlin pressed Chalmers to consider again a restructuring of the practice plan. The Board of Trustees had established a Clinical Excellence Committee that was charged with conducting a major institutional self-study.[4] Chalmers created a subcommittee, chaired by Aufses, to review and evaluate the Medical Service Plan.

Chalmers also created an Advisory Council of MSP members, chaired by Aufses, to act as a governance mechanism and to advise him on practice-related issues. An administrative office within the Medical School structure was designated to implement faculty practice activities; Milton Sisselman, an administrator at Mount Sinai since 1952, was assigned responsibility for oversight of the MSP.

On July 1, 1976, a revised MSP structure was implemented that was based on the recommendations of the Clinical Excellence Committee and approved by the Board of Trustees. A major change was the creation of a Dean's Fund, with a 5 percent assessment on all gross revenue going directly to the Dean for support of the School. A further 5.75 percent of gross revenue (the "indirect" overhead) was assessed for services purchased by the Plan, including, but not limited to, space costs, malpractice insurance, support of the MSP administrative structure, and a reserve fund for renovations and upgrading of practice space. Actual practice costs (direct overhead), along with a portion of the participant's base salary and fringe benefits, an expense allowance, and a fund for incidental expenses, were then deducted. The residual balance was divided between the participant and his or her department. The participant's share was allocated for a salary supplement. Finally, and most important for recruitment and retention of full-time faculty, the supplement limit was raised to 100 percent of base salary. In 1983, the 100 percent ceiling on supplementation was removed for selected individuals capable of generating large amounts of practice revenue.

In the first fiscal year (July 1, 1973–June 30, 1974) of MSP operation, the seventeen faculty participants generated $516,000 of gross revenue. The following year, with many of the chairs now participating, income was slightly over $2 million. The advent of the revised plan in 1976 had the desired effect; the number of new faculty increased dramatically, and the School's clinical programs grew at a rapid pace, as did the revenue generated by these activities. By fiscal year 2001–2002, there were 782 participants and a total income of $200,552,000 (see Figure 10.1).

Thus, in a relatively short period of time, the MSP became an important initiative in the fulfillment of the teaching and patient care missions of the School and Hospital, as well as integral to the support of the research programs.

Although the number of participants and clinical activities increased substantially, there was little change in the basic operation of the MSP from 1973 until 1983, when James F. Glenn, M.D., was appointed as the President and Chief Executive Officer of the Mount Sinai Medical Center.[5] Glenn's appointment marked a sea change in the leadership organization of the Medical Center in that the positions of President and Dean would now be occupied by different individuals. In the new structure, both the Dean of the Medical School and the

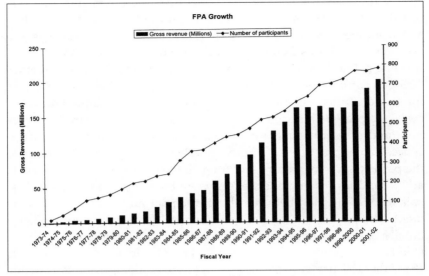

FIGURE 10.1. Gross Revenue ad Number of Participants, Faculty Practice Plan
(MSP, FPA), 1973–2002

Hospital Director reported to the President and CEO of the Medical
Center. Glenn had spent many years as Chief of Urology at Duke Uni-
versity Medical Center and in the three years prior to his arrival at
Mount Sinai had been the Dean of the Emory University School of
Medicine.

From the outset, Glenn was a strong supporter of the MSP.
Although the MSP was part of the School, he insisted that the Chair of
the Advisory Council report directly to him, rather than to the Dean.
Glenn recognized the need to emphasize the unity of purpose of the
practice activities of the full-time faculty and the need to have the
faculty think more along the lines of a group practice. He therefore
polled the faculty regarding a name change, and in May 1984, the MSP
became the Faculty Practice Associates (FPA). Glenn also appreciated
the need for a single site for the practice of the full-time faculty and
was, in fact, the individual who identified 5 East 98th Street as the ideal
practice location. This site was desirable because it was located on
campus, which greatly facilitated the ability of faculty to fulfill their

multiple roles in the furtherance of the institution's clinical, teaching, and research missions.

Five East 98th Street, formerly Guggenheim Hall, had been the home of The Mount Sinai Hospital School of Nursing. When that school closed in 1971, the building was converted to administrative office space and medical student housing. With the approval of the Board of Trustees, the building was leased to the School of Medicine in 1984 for use by the FPA with the understanding that the FPA would help pay for renovations to convert the building to practice space. At the outset, FPA would use four floors and then add floors as other hospital services could be relocated. To accomplish this, a bond issue was authorized, and the indirect overhead assessment to the FPA participants was increased from 5.75 percent to 10.25 percent (over a period of four years) to cover the debt service on the bonds. After extensive renovations were carried out, the first practices took over four floors of the building in November 1987.

Sweeping changes were occurring in the healthcare marketplace during the early 1980s. In a nationwide attempt to help control rising healthcare costs, "managed care" became the order of the day. Both not-for-profit and for-profit health maintenance organizations (HMOs) were created to manage patient care using capitation as the payment method of choice. Patients paid a monthly, all-inclusive fee to the HMO, which then contracted with hospitals and physicians to render care at discounted prices. In most of the HMOs, the patient's primary-care physician functioned as a "gatekeeper," with the responsibility for deciding what studies or consultations were appropriate for each patient. Originating for the most part on the West Coast, the HMO concept moved eastward at a rapid rate.

There was no one more cognizant of the approaching impact of managed care in the New York metropolitan area than Glenn. In 1984, capitalizing on work on HMOs and capitation initiatives done previously by the School's Department of Community Medicine,[6] Glenn created an interdepartmental HMO Planning Task Force that included both full-time and voluntary faculty. The goal was "to explore the feasibility of developing a Mount Sinai sponsored comprehensive prepaid health program for its employees, their families and other population groups."[7] The task force soon reported that Mount Sinai employees and their families alone would not provide an adequate patient base for a viable HMO. The task force also noted that there were not enough primary-care providers on the faculty to provide such services.

At about this time, the Federation of Jewish Philanthropies (FOJP) approached five of its member organizations in the metropolitan New York area, including Mount Sinai, to consider forming a consortium to deal with HMOs. As expected, FOJP and the involved hospitals were approached by investor-owned, for-profit HMOs that had gained entry into the New York market; the major one was the California-based Maxicare. Eventually, FOJP reached agreement among the five centers and Maxicare. As projected, the physician services component would be a network of Individual Practice Associations (IPA) based in each center. It was the birth of the Mount Sinai IPA, Inc.

Sanctioned in New York State by Article 44 of the Public Health Law, an IPA was "an entity formed for the limited purpose of arranging by contract for the delivery or provision of health services by individuals, entities and facilities licensed or certified to practice medicine."[8] From the outset, Mount Sinai's IPA was open to any and all members of the full-time and voluntary staffs, a stipulation that holds true to the present day.

The ensuing months of 1984–1985 were a time of strife within the Medical Center. Hospital census was low, revenue was falling, and Glenn and the Board of Trustees felt that Maxicare and other HMO patients might help solve the problem. The physicians had grave concerns about joining HMOs, believing that it would compromise their ability to render the best medical care and would also adversely impact their incomes. For the most part, they believed that the institutional support of HMOs would result in physicians having to accept unduly low fees for the care of patients and that patient care would suffer. Although many members of the full-time faculty also felt that they were being forced to join the IPA, they complied and did participate. In the context of the proposed capitation arrangements, a structure was established to provide appropriate physician incentives and to protect physician income. A significant number of both full-time and voluntary physicians did join the IPA, but, with the passage of time, IPA membership declined, and, in 1988, Mount Sinai and Maxicare parted company. Many physicians had never seen a Maxicare patient. Unlike managed care (which is discussed later), the concept of capitation was slow to take hold in New York. As a result, the IPA went into a period of dormancy that lasted for almost five years.

As the HMO battle developed, it had become evident that a new system of governance for the FPA was necessary, not only because of its growth but also because of the evolving healthcare system, with its need for a unified response on the part of the participating physicians. Following extensive analysis, a new governance was put into place in July 1986. The major changes were the creation of an eleven-person Executive Council, including a President and Vice-President, and the formation of an Assembly, in which all clinical departments were represented. Arthur Aufses was elected the first President, with a three-year term. In July 1987, Victor Kazim was appointed a Vice President of the Medical Center and Executive Director of the FPA. Sisselman, who had directed the activities of the practice plan since its origin, was appointed Executive Consultant to the FPA.

In 1989, after thirteen years as the physician leader of the practice plan, Aufses stepped down, and Joel Kaplan, M.D., Chair of the Department of Anesthesiology, was elected President of the FPA. Three years later, Kazim resigned, and, after an extensive search, Sloane Elman, Associate General Counsel of Mount Sinai's Legal Department, was chosen as the new Executive Director. Elman directed the day-to-day affairs of the FPA for almost six years, a period in which many changes occurred in medicine and in the FPA.

A new Medical School research building (the East Building) had been approved, and, in July 1993, the FPA pledged $10 million toward the construction. This would be accomplished by an additional 1 percent Dean's Fund assessment until the full amount was achieved. This commitment was fulfilled in fiscal year 1999–2000.

In 1993, the IPA was resurrected, and Douglas Present was named the Executive Director, a position he held until December 1995. During that period, he arranged for contracts with various HMOs and, in fulfilling the main role of the IPA, provided physicians with guidance relative to the benefits and drawbacks of the various contracts and assisted in enrolling physicians in the HMOs.

Managed care was now making major inroads into the New York area, and it became evident that the FPA governance structure created in 1986 was unable to deal with the complexities of the rapidly changing medical marketplace. The existing Council of eleven individuals did not represent all departments and included many non-Chairs. There-fore, in 1995, the Executive Council appointed an Ad Hoc Committee on Restructure of FPA. The new goal was to move the FPA toward a more integrated multispecialty group practice structure as the best way to participate in the School's and the Hospital's clinical and training programs.

In July 1997, the FPA implemented a revised governance that placed the clinical department chairs at the fulcrum of the FPA decision-making process. A Board of Governors composed of all the clinical Chairs, the Dean, and nine non-Chair FPA participants became the controlling authority of the FPA. Kristjan Ragnarsson, M.D., Chairman of the Department of Rehabilitation Medicine, was elected as the first Chair of the Board of Governors. In the same month, Arthur H. Rubenstein, M.B.B.Ch., former Chairman of the Department of Medi-cine at the Pritzker School of Medicine, University of Chicago, was appointed Dean of the Mount Sinai School of Medicine and Executive Vice President of the Mount Sinai Medical Center. He assumed the post later in the year on a part-time basis and moved into the position full-time in early 1998. Later that year, the FPA by-laws were amended to stipulate that the Dean of the School of Medicine, as opposed to the President of the Medical Center, would serve as "the Chief Executive Officer of FPA and will have authority to represent the FPA in all man-agement matters."[9] Although such a move a decade earlier would have been contentious, to say the least, it was almost universally believed that it was appropriate for the time. James Lewis, Ph.D., was appointed Deputy Dean and represented Rubenstein at many of the Board of Governors' meetings.

In the mid-1990s, the Office of the Inspector General (OIG) of the U.S. Department of Health and Human Services had instituted audits of physicians in teaching hospitals (PATH audit.) These reviews cen-tered on compliance with documentation and coding requirements of the Medicare law. In 1996, FPA established a Billing Compliance Unit within the FPA administration. Headed by Jane Whitney, R.N., this entity, an integral component of the Mount Sinai Medical Center Corporate Compliance effort, was part of the FPA's continuing com-mitment to a program aimed at promoting compliance with medical record documentation and coding requirements of third-party payers. In June 1998, the FPA was subjected to a sixteen-month PATH audit. Thanks to the prodigious efforts of Whitney and of Sally Strauss of the

Legal Department, the financial settlement was very favorable to Mount Sinai and the FPA. As part of the final settlement with the OIG, a five-year Institutional Compliance Agreement (ICA) was negotiated. The ICA, which took effect in November 1999, required that FPA execute a compliance program of various key elements for five years. The 1 percent assessment of gross receipts, originally initiated to cover the FPA capital donation for the new East Building, was continued to cover the ongoing cost of the Billing Compliance Unit.

As managed care continued to make inroads into the New York area, albeit more slowly than in many other cities, the IPA began to assume a greater role in the life of the Mount Sinai physicians. By the end of 2002, contracts were in place with twenty-eight HMOs, and the IPA staff had grown; Doreen Nelkin-Warantz was appointed Executive Director of the IPA, and Mari Enrique remained the Director of Managed Care. By now, all FPA participants had joined the IPA, as had 150 voluntary physicians. Despite the overwhelming number of full-time physicians (almost eight hundred) relative to the number of voluntary physicians, the sixteen-member governing board was equally divided. There were co-Presidents, Jack Adler, M.D., representing the voluntary faculty and Paul Goldiner, M.D., representing the full-time faculty. In September 2002, the FPA executed an agreement appointing the IPA as the FPA's agent in nonrisk managed-care contract negotiations.

Before the end of 1998, Elman resigned as Executive Director of the FPA, and Lewis assumed the position of Executive Director until the arrival of Robert Chassin in early 1999. Chassin remained in the position for ten months, and then, once again, Lewis assumed the helm until Stephen Selby, formerly the President and Chief Operating Officer of the University of Texas Southwestern Health Systems, became Executive Director in July 2000, a position he held until 2003.

Kenneth Berns, M.D., appointed as President and CEO of the Medical Center in 2002, recruited Louis Russo, M.D., to become the first full-time physician CEO of the FPA. Russo, who had trained in Neurology at Mount Sinai in the early 1970s and had been a colleague of Berns at the University of Florida, joined Mount Sinai and the FPA in July 2002. At Russo's first meeting with the Board of Governors, in July 2002, he spoke of his beliefs about "the importance of patient-centered health care services, strong physician leadership within Faculty Practice Associates, and a good working relationship between the FPA and the Hospital."[10]

For more than a quarter of a century, the FPA has remained a stable force in the School and the Medical Center and has been critical to the School's and the Hospital's ability to carry out their respective clinical programs, as well as an integral part of their teaching and research activities. The number of participants and revenue continues to grow. Office visits are increasing, and FPA physicians are responsible for more than 55 percent of Hospital discharges.

With the increasing complexity and size of the FPA, there is now in place an administrative structure sufficient to support this expanded enterprise. In addition to the FPA Board of Governors, the FPA Executive Committee, and five active FPA standing committees, the FPAs Senior Administration now includes Louis S. Russo, M.D., Chief Executive Officer; Debra Lunburg, Vice President for Patient Financial Services; Carolyn Bernard, Director of Business Operations; Angus Ramadeen, Director of Finance; and Jane Whitney, Director of Compliance and Practice Support Services. To Russo and his leadership team falls the task of leading the FPA to a new plateau of success in the first decade of the twenty-first century.

11

The Mount Sinai Alumni

Alumni are the guardians of the rich historical traditions of Mount
Sinai, and their voices command respect throughout the country.
Actually, the performance of our physicians—both of the Hospital
and the School—is a true reflection of Mount Sinai and the yardstick
by which we are judged.

—Hans Popper, M.D., Ph.D.[1]

POPPER WAS NOT alone in believing that the work of a body of
alumni can be used to judge an institution. Every school looks to the
quality of its graduates, their impact in their chosen fields, as well as
their devotion to their alma mater, as important indicators of success
or failure. Since the Mount Sinai School of Medicine grew from a hospital with an already existing alumni group, its alumni organization
has had a history and shape different from those of other medical
schools. Most medical schools are part of universities that have an
extensive alumni organization, staffed by full-time administrators, that
oversees outreach to the graduates, gathers data about their careers,
and performs vital fund-raising efforts to support the school. Since the
Hospital alumni group's initial goals were the exchange of scientific
information and fellowship, with services for the individual members
and support for the Hospital added later, the Mount Sinai Alumni
office has always been and remains somewhat different from most
other alumni organizations.

There has been an alumni association at Mount Sinai since 1896,
when graduates of the Hospital's training programs (fellowships,
internships, and house positions) founded the Associated Alumni of
The Mount Sinai Hospital. At this time, there were about sixty-eight

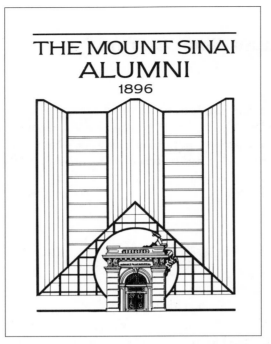

THE MOUNT SINAI ALUMNI
1896

The logo of the Mount Sinai Alumni uses representations of the Annenberg Building, the Guggenheim Pavilion, the sculpture *le sfera*, and the old Administration (Metzger) Pavilion. This image combines the Hospital and the School, the old and the new, into one powerful symbol.

graduates of the Hospital programs, and many of them were still connected to the Hospital in some way, primarily by serving in the clinics.[2] The early years of this group were devoted to sponsoring meetings on scientific topics, followed by a banquet. In 1900, the annual meetings became more focused on entertainment and fellowship. After any pressing business was conducted and speeches given, the meeting was given over to funny skits and songs. Publications with titles such as *The Medical Gewalt* appeared at each dinner. They contained humorous advertisements, articles, and notices, all intent on skewering the senior staff of the Hospital.

These dinners continued for many years, but, at the same time, the Association attended to more serious business. It created a fund to aid sick alumni who needed care but did not have the resources. It took an active interest in the Hospital, since many alumni were on the clinical staff, and wrote letters and sent recommendations to the administration and Trustees on numerous occasions. The group cultivated a close relationship with the house staff, the Association's future members, and tried to serve as a mediator between the Hospital and house staff when asked. In 1935, the Associated Alumni established an Alumni Library Fund to provide formal support for the Mount Sinai library, a tradition that continues to this day. As the 1939 *Alumni Directory* stated, "In the past five years we have graduated from a social organization into one that is taking an active interest in hospital affairs . . . we are growing up."[3]

The active membership of the Association was limited to graduates of the Hospital's training programs, but included one-year externs and fellows. Hospital attending staff who had not trained at Mount Sinai could join as honorary members. Other categories of membership were created and eliminated over the years and dues were assessed. By 1944, there were 255 members, and the number continued to grow.

There is not room here to note in detail the history of the Association, but highlights that must be mentioned include the adoption in 1941 of the tradition of the Gold-Headed Cane as an award to the physician who best exemplifies the personal and professional traditions of Mount Sinai; the creation of the first *Alumni News*, in 1948, edited by Beatrice Aufses, the wife of Arthur H. Aufses Sr., M.D., which has continued off and on to the present day; and the establishment of the Jacobi Medallion in 1952 to honor alumni for service to the Association, to Mount Sinai, or to medicine. By the 1960s, the goals of the organization were to provide member services, such as insurance, and to hold alumni reunions each year. It must be noted that for many of these years, the Association business was coordinated by Bella Trachtenberg, the retired former Secretary to the Mount Sinai Board of Trustees. Bella, as she was known to one and all, was joined later by Helen Wilson.

The 1960s were relatively quiet as the Associated Alumni awaited the results of the efforts to establish the School. The question of the relationship of the School's graduates to the Association was, of course, an important topic. There were many voices that wanted the Association to remain limited to graduates of the Hospital programs and warned that the young School graduates would take over the Association. There were others who argued that all graduates of Mount Sinai should be joined in one organization and that having separate groups would marginalize both. Discussions of this issue began appearing in the 1966 minutes of the Association, culminating at the meeting of March 22, 1969, when the members voted to make graduates of the School eligible for active membership in the Association by a vote of better than three to one. The official name of the organization was changed to The Associated Alumni of The Mount Sinai Hospital and Medical School. The constitution was amended to read: "one object of this Association shall be . . . to promote cooperation with the parent Hospital and Medical School and their staffs, acting and assisting in all relationships affecting the institutions and their graduates."[4]

During the 1970s, the Association struggled to get graduates of the School to join the Association and become involved. Changes were made to accommodate and encourage School alumni participation. The Association instituted a morning of scientific presentations by reunioning School classes at the annual Alumni Day, and in 1986 a family component was added to the weekend to encourage alumni to attend. The traditional dinner dance has remained a fixture of Alumni events.

A milestone was reached in 1987, when Mark H. Swartz, Class of 1973, was elected president of the Association, the first graduate of the School to lead the group. It was during his tenure that it was decided that the Association needed the efforts of a full-time, paid Director to run the day-to-day affairs of the Association. Cynthia (Cyndi) Gruber, the former registrar of the School, was selected for this position and continues in it to this day. Another milestone was achieved in 1992, when Janice Gabrilove, Class of 1977, became the first graduate of the School to receive the Association's Jacobi Medallion.

The pace of the Association quickened in the 1990s. Membership grew from 900 in 1990 to 3,000 in 1995. Samuel Guillory, a 1975 graduate of the School, was president from 1991 to 1993. He reaffirmed the willingness of the organization to change with the times and to work actively for a better Mount Sinai School and Hospital. In 1993, the Association once again amended its name, adopting the more streamlined The Mount Sinai Alumni as its official title. That same year, a unique logo for the Association was adopted that was based on a design by Guillory. The logo combines elements from the Hospital tradition (the Metzger and Guggenheim Pavilions) with a symbol of the Medical School (the Annenberg Building). The logo is coupled with a depiction of the sculpture *le Sfera* by the artist Pomodoro, which is housed on the Mount Sinai campus. As Guillory described it in 1993, "This sphere represents Mount Sinai's world of medicine in which students, residents, physicians and faculty come together from all over the world to learn medicine at Mount Sinai and take what they have learned here to care for people all over the world. The ball is incomplete as science and the study of medicine is never complete."[5]

As the Association reached its centennial in 1996, it had become much more proactive in its support of the Medical Center, increasing its donations to the Levy Library and creating new programs that directly affected the medical students and house staff. This included increased scholarship funds, house staff dinners, an exercise room for medical students, a Match Day event for graduating students, and mentoring opportunities between alumni and students.

As the Association moves into its second century, it is grappling with the issue of how to become a more efficient fund-raising organization in support of its own programs and those of the Medical Center. A professional development officer has been placed in the Association office to help the Alumni meet its potential in this important area. This represents another step toward making the Alumni office more like those at other universities. Still, the Association strives to remain vital and to carry out its traditional mission of uniting the graduates of Mount Sinai for the greater good of the institution. The Mount Sinai Alumni remains "the one [organization] most closely tied to the essence of belonging at Mount Sinai, of feeling a part of a larger good and feeling good about it."[6]

12

Student Voices

In Their Own Words

A school is a living entity that changes over time and is infused with the spirit of the people that work and study there. Mount Sinai School of Medicine is no different. There is much to history that is ephemeral, or that can only be told by someone who was there. In the summer of 2002, Emily Falk, then an undergraduate student at Brown University, worked with the authors accumulating information about the early years of the School. She also embarked on a project interviewing alumni and current students of the School (thirty-four in total), asking them about their experiences at Mount Sinai. Each was asked the same set of basic questions so a comparison over time could be made. What follows is a brief synopsis of the graduates' comments. We are indebted to Ms. Falk for her efforts, which took her up and down the Northeast corridor, as well as the many alumni who took the time to provide their thoughts on the School.

WHY SINAI?

I graduated from college in 1968, and the years before that were tumultuous on college campuses, and Sinai came out with statements in the press about how it was going to be a new school, an innovative school, integrating social concerns with science. And so I thought that was just terrific. . . . I could remain adventurous and still go to medical school.
—Peter Lang, M.D., '72

My connection to Mount Sinai and this medical school at that time was a very emotional one. This was the hospital that my family came to.

This was, in my mind, always regarded as a special place, as a place for great care. . . . I had applied to the hospitals and medical schools that you would expect. . . . I just thought I'd apply to Mount Sinai because it was a great hospital and it was a new medical school and what they wrote in their catalogue seemed very exciting: only forty students in a class, individual attention. This could be more of the kind of education that I thought I wanted than what I was hearing about from other medical schools.

—Kenneth Davis, M.D., '73

I decided to go to Sinai because it was an experiment.

—Terry Maratos-Flier, M.D., '76

I had spoken to someone who had just finished their medical school training, and they spoke highly of [Mount Sinai] in terms of providing a very good education and also the atmosphere and the camaraderie amongst students; that it was an atmosphere that was, I'd say, benevolent.

—Alfio Carroccio, M.D., '96

THE MOUNT SINAI "TYPE"

It was a terrific class, but it had a lot of strange people in it. You might say that that was predictable. You'd have to have certain personality types that would be interested in being in the first class.

—Jeffrey Flier, M.D., '72

They took a chance on students here. I think that they were looking for students that were not necessarily just premed oriented students. They wanted a varied student body.

—Loren Skeist, M.D., '72

I think Sinai does a really good job of picking people who have really diverse interests and want to do other things. We have great musicians in our class, very athletic people. . . . That's a lot to do with it. They pick people who want to do other things—like all sorts of community service.

—Bethany Slater, M.D., '04

EARLY IMPRESSIONS

We all got telegrams accepting us to medical school that year. . . . On arrival, it was really pretty remarkable. There was a sense that probably exists only at the beginnings of new schools in terms of the faculty's interest in having us be there. There were far more professors than there were students and everybody was interested in having students there. . . .

The very first lecture at the medical school was given as an honor, as it should well have been, to Dr. Hans Popper, who was both Chairman of Pathology and had been the first acting Dean of the School. He was followed by Dr. Katsoyannis, the Chair of the Biochemistry Department, followed by Dr. Barka, who was then Chair of the Anatomy Department. So we had an Austrian, a Greek and a Hungarian giving the first three lectures, all in their rather thick accents. . . . I remember somebody wandering around after the third lecture that first morning saying, "I knew medical school would be tough, but I thought I would be able to at least understand the professors."

—Arthur Frank, M.D., Ph.D., '72

We lived at 5 East 98th Street, what's now the faculty practice, and that was our dorm building. It was pretty bad when we first came here. I remember walking into my room, and I had the best room on the floor, and it was characterized by a platform bed, a chest, a small chest, and a little sink in the corner with a broken mirror. It was essentially nine by twelve or fifteen, and this was supposed to be home. There were no showers in the rooms except down the hall, the communal shower. We called it Lake Guggenheim because they would flood every time more than one person would take a shower. . . .

And then when they built this wonderful residence hall across the way [Aron Hall], we were all excited by it, but then when it came time to move, I remember the cost: that the cheapest apartment when I went was $325 [per month]. That doesn't sound like much now, but in the early eighties, to a medical student and with all the loans that were mounting, everyone was up in arms.

As soon as we moved into the new facility, which was truly lovely, at that point, everyone stopped complaining, found the money. . . . It was a much more pleasant experience for the next three years.

—George Raptis, M.D., '87

It just struck me that everyone was warm and friendly and wanted to teach. . . . I never knew that you could do so much as a doctor, rather than just do research or just see patients.

—Bethany Goldstein, M.D., '99

I learned that it was a very good and well-respected medical school, but as I arrived I also felt that they *did* really care about the type of students they took, not just about getting the highest grade point average.

—Peter Klatsky, M.D., '03

SIGNS OF THE TIMES

We had many and impassioned political arguments, night upon night in the lab. . . . There was very little factionalism within our class; I think it was fairly cohesive. What we really factionalized along was political lines. . . . There was really significant sentiment. Outside of that, the overall feeling was closeness.

—Marlene Marko, M.D., '72

We were part of the anti-Vietnam sentiment. . . . I remember making signs and carrying them. . . . We were not striking for more money or better hours. We were mainly doing things for better patient care. The funny thing about that, we actually did walk around outside with signs and picketed, but we didn't just walk out on our patients. We'd come in, we'd care for our patients, we'd write their orders, check up on them, and then we'd go out and picket for a while and then we'd come in in the afternoons, see our patients and take care of them.

—Arthur Frank, M.D., Ph.D., '72

We went out on strike with [Local] 1199. There was a Committee of Interns and Residents, and we went out on strike in sympathy with them. We would picket outside, and it was a little surprising that we couldn't come into the hospital to use our check cashing privileges.

—Loren Skeist, M.D., '72

From '69 to '73 was the period of probably the greatest political activism among students in the last fifty years, so students felt concerned with what was going on in Vietnam and what they had to do about it. They were also concerned about the bomb and what should be physicians' responsibility. . . . Poverty in America was a very important issue. What were the responsibilities of physicians to address the political needs of people who were in great poverty was a major concern. Those were burning issues.

—Kenneth Davis, M.D., '73

I went to school during a very boring time in history . . . no major wars or terrorism.

—Deborah Marin, M.D., '84

In my year, you sensed a lot of cynicism and irreverence. We didn't feel particularly cohesive as a group. There was a lot of sense of being like a second-tier medical school rather than an alternative medical school, a medical school with a different viewpoint. Most people were happy here and got along and liked the teachers and liked the doctors, especially the clinical side, we liked a lot better as a whole than the basic science side. . . . I think the eighties were a pretty cynical time for doctors. Those were the first times where you started to hear that there was going to be a glut of physicians, that there might be too many physicians, that doctor's incomes were going to go down, and a lot of people were very disillusioned to get to medical school and hear that those were the things.

—Anthony Weiss, M.D., '88

We were very worried about the M-D [multidisciplinary] labs . . . something that people are really drawn to at Mount Sinai . . . everyone has their own desk, and so it's nice. . . . Some people keep their whole life there. We also use it for lab. They said that all that space would be gone—the space was being redone. So that was a big issue my first and second year. . . . That idea has now gone by the wayside because of money.

Then this year, obviously the biggest issue has been the increase in tuition . . .the Hunter group coming in and cut everything out. . . . They made a lot of cuts . . . part of that was raising the tuition at the school. . . . Not only did tuition go up by $7,000, but just little perks that we used to get, we don't get anymore—like lunch at this activity or free breakfast and little things like that that boost morale have been cut.

—Stefani Wedl, M.D., '03

THE CURRICULUM AND ACADEMIC ENVIRONMENT

The curriculum was a good balance between the first part of the year, which was traditional lecture style, and then we went into the organ systems. I think that that was a very good way, and I think it took a large amount of faculty input. . . . I imagine today that the faculty would bitch and moan about that, especially the clinicians coming over and having to use up their time with all the pressures they have in terms of generating dollars and seeing patients, to come and teach. But we were doing basic anatomy and doing dissection, and we had the surgeons in there with us.

—Arthur Frank, M.D., Ph.D., '72

I felt that it was everything that I wanted it to be in terms of medical school letting me spend a little more time in the area that I knew I wanted to be in. . . . There was a strong research bias, but still mainly clinically focused . . .

Sometimes [the curriculum] was wonderful, and sometimes it took a while to hone it down. . . . If you have a faculty person who really themself has integrated the information and is working with the information and is working with this new way of looking at it, it's great.

—Loren Skeist, M.D., '72

To me, the greatest thing about Sinai teaching was a certain pragmatism and trying to understand the problem without much pretense.

—Ernst Schaefer, M.D., '72

There was this sort of tug, although I'm not sure I recognized it as that at the time, between the basic science faculty and the clinical faculty for our hearts and minds.

—Peter Lang, M.D., '72

I was extremely happy with my education, although there were some rough edges at the beginning because it was a new curriculum. But I liked the idea that it was a new curriculum and there were people interested much more than usual in how we were doing. We were like this little experimental class. . . . The first lecture, I believe, was in biochemistry. Dr. Katsoyannis talked about the nature of the peptide bond, and I actually just remember that.

—Jeffrey Flier, M.D., '72

The reality of what they described, what they hoped to teach, was breaking down by the time I was here. The "Introduction to Medicine" course was becoming a class that didn't have much content and that students didn't feel very good about. The core parts of the medical school curriculum were being taught unevenly. . . . Some people came in and, in retrospect, taught what they knew, which was their research area. Other people came in and actually taught what they thought you should know. But this was a place that was trying to figure out what's the curriculum and what's the best way to communicate that curriculum. What's the best way to evaluate the people learning the curriculum? So it was bumpy.

The tension for the students was always whether education needed to be more individual. The class was so small, why were we being educated in a way as if the class were 120? Why couldn't there be more independent learning, more tutorial learning, more seminar learning, more graduate school–like education? Because the impression was that we would be treated more like graduate students, but, in fact, we were treated more like the traditional medical student. And that was the main tension. Shouldn't faculty know us well enough that we didn't have to have rote exams and that we wouldn't have to recite memorization, that they would be able to interact with us well enough to know if we really knew the information or not? That people could have more independent learning? I don't think the school was ready for that.

—Kenneth Davis, M.D., '73

They really tried to have the integrated curriculum. I think it was one of the first schools to try to integrate organ systems and physiology with pathology, and they also tried to get us contact with patients relatively early on. . . . They did much more than was usual. . . . The fact that they gave us some time off in the afternoons so that we could study and occasionally do electives was nice. . . . The thing I remember, the most striking thing, is how good the clinicians were when we got to the third year. . . . As a group, the clinicians were amazing. They were the most kind of dynamic and impressive clinicians that I encountered.

—Terry Maratos-Flier, M.D., '76

In retrospect, there were some aspects of the curriculum that were not leading edge. A lot of the curricular issues were still being worked out or worked through or optimized. . . . In the seventies so much more of the education was classroom based. Not even classroom based, but sort of lecture based. So it was a little bit, sort of monotony. We'd always sit in our self-assigned seats; the irreverent ones in the back would occasionally be reading a newspaper and be admonished by the lecturer, the ones in the front would have tape recorders and ask all the questions, so there were almost rituals. . . .

Most of the clinical faculty when I was a student here were voluntary faculty. They were in practice. . . . One of the unique strengths of Mount Sinai is that, because it's a medical school that grew out of a hospital, the tradition of translational investigation, meaning patient based, or discoveries that lead to treatment, is something that is absolutely ingrained in the culture here in a way that isn't necessarily in other institutions.

—Scott Friedman, M.D., '79

I was very well trained. It was a very good medical school. Going to Baylor [for later training], I saw that there was a difference in the way that there are approaches to medicine. Whereas Baylor was a much more procedurally oriented medical school, Sinai was more of a cognitive and academic medical school, but it trained me very well.

I think that the school is successful in taking a raw undergraduate and training them in all of this book knowledge and how to deal with other human beings and how to perform a physical exam and take a great history and think clearly and logically to come up with a diagnosis or train their hands and their mind and their eyes to be great surgeons.

—Jeffrey Mechanick, M.D., '85

There was a sense that [the faculty] had a stake in guiding the direction of medicine in the future, that there was a greater overall mission than just getting us through the four-year process. . . . It was the perfect balance between hands on and didactic training and I thought that the fact that there was such a strong voluntary faculty in this institution in addition to the full-time faculty, and the fact that there were three attendings on internal medicine rounds as opposed to one that did chart rounds in the back . . . was really the strength of the program. I thought that the voluntary faculty brought a richness.

—George Raptis, M.D., '87

When we started, the medical school preclinical years got to be very top-heavy with class hours. And there was not a lot of what they call problem-based learning, which was something that came into vogue later. . . . It was not uncommon to have seven to eight hours of didactic lecture on very dry subjects such as microbiology, followed by biochemistry. And that was cumbersome, and a lot of students reacted by not going, just studying the syllabi and learning what they had to do to pass because it wasn't very relevant. I remember that there were some issues when the new dean came, Dean Kase came and made some changes and started reorganizing some of the departments, and I remember at that time being concerned as a student. We voiced concern that some of these changes were being made without input from students.

One thing that I think most people feel good about is our affiliation with Elmhurst. . . . That's a place where I think students and residents feel more like doctors because they're in a poorer community. They're in a hospital where they really feel like they're friends, and the volume is very high, the patients come from all over the world, they have unusual arrays of diseases. . . . We all did work there.

—Anthony Weiss, M.D., '88

There was a sense of money, but there was and still is a sense of frustration that we never learned the real dollars and cents of medicine. I think it's a shame that medical school doesn't include some kind of finance courses also because, really, yes, it's a humanistic thing, and we're doing an amazing thing taking care of people, but we are sort of in the grip of everyone else around us who is looking at the bottom line and looking at the dollar. You learn about tests that you're going to do in school to diagnose something; you don't learn the cost of them. You don't learn how people's lives are affected by their insurance. You don't know how the hospital works. . . .

I think that everyone understands that [research] is important. In class you learn about new discoveries every day. It wasn't shoved down our throats like I think it is in some places, where you have to absolutely do research. It was encouraged, but it wasn't so easy to do. . . . It was mostly lecture based. In physiology we had smaller groups. They were experimenting with smaller groups. . . .

I think when you're a medical student and you're sitting there in that lecture room for two years and then you're out following a crowd of people for two more years, you feel very small.

—Ram Roth, M.D., '92

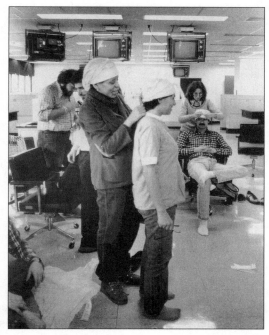

Medical students practice bandaging in the Annenberg student labs.

I don't really agree with or know what the benefits are of introducing clinical work earlier than [the clinical diagnosis class] . . . I actually think it's better to have some more basic knowledge and have spoken to more people, to know a little bit more about medicine before you actually go and sit with a patient.

One of the great things about Mount Sinai's program is that you get a very different experience at the different hospitals that you go to, and being at Mount Sinai was completely different from being at the VA from being at Elmhurst and if you're only at one place, you don't get as much exposure.

—Elise Brett, M.D., '94

I think that everybody would mention the anatomy class. Dr. Laitman turns it into a very huge production. He likes to scare the first years, but he makes it fun. . . . I know now they're trying to integrate more independent learning and free time, which I think is great, but, at the same time, being together as a class for a number of hours really solid-ifies your relationships with them. You really get to know people very well. I think always one of the strengths of Sinai has been the whole environment here has really been geared towards teaching and the students. . . .

The interest that attendings take in both resident and student edu-cation is phenomenal. . . . They're very busy, but they also sign on at Mount Sinai knowing that they're going to have a lot of teaching to do. . . . I think that the quality of the hospital and the quality of people that they're able to attract really trickles down to the students. . . . It's a chal-lenging way to go through medical school. . . . I think the general idea is excellent—to have everything in smaller blocks and try to integrate things better. It's going to be a challenge because the departments themselves don't integrate themselves very well. A lot of the theory sounds good, but I don't know how practical it's been.

—Philip Mulieri, M.D./Ph.D., '03

They have a new curriculum now . . . it's now finished its second year, and I wasn't part of that. Starting from day one . . . they get to go to the hospital and meet patients right away. As far as my class, we had a section of first year for about six weeks at the beginning that was like that . . . that was great. After that it sort of dropped off and then didn't start again until the second half of second year . . . but that's revised now . . . and I think that's a good thing. . . . Now seeing the second years who are a year below me . . . they look more prepared to go into the clinics. . . . I feel like they know more than we did about clinical medicine.

—Stefani Wedl, M.D., '03

I feel like the way that the curriculum change was instituted was top down. I don't know that students asked for more small groups. . . . I think that to really make radical curriculum reform, you have to have the professors themselves that give the lectures, that lead the small groups, want to change it. . . . But I don't think it was like that because the new curriculum, in a lot of ways, feels like a rearrangement of the old curriculum. You take two classes like biochemistry and cell biology, with defined lectures . . . and you change that into a class called "Mole-cules and Cells," which is biochemistry and cell biology all combined into one. You didn't change the lectures you just changed the scheduling of the day.

—Robert Greenberg, M.D., '03

We were the first class for the new curriculum, so we gave a lot of feedback and they've completely changed the model. . . . The "Molecules and Cells" class . . . was really bad for us, and they've completely changed it for this year. They liked it a lot better this year. Other than that, we had histo and physio together. They don't repeat things, and that was really good. All the case integration things, and much more the clinical aspect of things, I liked that a lot.

—Bethany Slater, M.D., '04

I think that the general atmosphere here is an approach to medicine that is integrative and requires an approach to a whole human and sort of the understanding of the whole human body in the context of society and culture as opposed to just an organism or just a physiologic being.

—Adam Kawalek, '05

THE MEDICAL SCIENTIST TRAINING PROGRAM (MSTP OR M.D./PH.D.)

It was a wonderful learning environment with a great deal of care and concern from the faculty. . . . Things were not organized. I was an M.D. student. I was also a Ph.D. student, but there was no organized M.D./Ph.D. program.

—Arthur Frank, M.D., Ph.D., '72

It's similar content but a different scope. The scope in the medical school is much more direct and fact oriented, whereas in the Graduate School, it's much more methods based. . . . The combination of those two things, the graduate school curriculum and the medical school curriculum, made the first year really an enormously multifaceted educational experience that I know that myself and the other members of my class really appreciated, because it wasn't just that we were being told, "this is what we know, here, learn this." But it was, "this is what we know, here, learn this, and here's how you find out more." The group is obviously different in their goals and what they want out of their education, and they're really interested in asking questions and figuring out how to find the answers.

It's neat to have this collaborative feeling that I know I didn't have before coming here. . . . In the first year, because we don't take that one class with the medical students, there's a little bit of a separation between the medical students and the MSTPs that was noticeable, but what happens is that after that semester, everything else is together with the medical students, and where we might not have been as unified at the beginning, we became, and certainly by the end of the second year, it was hard to tell that there was a difference. . . .

Part of the program at Sinai is that the medical students and the Ph.D. students, the MSTP students, are taking classes and interacting with both first-year medical students and first-year Ph.D. students. There are a lot of programs that separate them. . . . I think it's healthy for us to have this interaction with medical students and Ph.D. students because in the future we're going to be colleagues of both these groups of people, and so we should be colleagues with them now.

—Talia Swartz, M.D./Ph.D. candidate

THE GRADING SYSTEM

There was . . . a grading system that had been put into place when we got there: a traditional A, B, C system. . . . We, as a class, coalesced very quickly . . . we asked if we could have a modified grading system of just pass/fail. And not only that, we asked for an anonymous pass/fail system. That meant that we could all help each other. There would be a minimum of any kind of competition. We brought that up with the head of the Student Affairs Committee . . . and something that lasted for many, many years after was the comment we got back from him. He started a response to us by saying, "Now kiddies . . . " and that didn't sit too well with the thirty-six of us, given the time and era of the late sixties.

—Arthur Frank, M.D., Ph.D., '72

We had a lot of debates . . . and eventually it came to a head with one particular exam where a lot of people, either they refused to take it, or they refused to sign their names. . . . The idea was to take it but only put numbers on it, so that [the professor] could list what was a passing grade and figure out who had passed, who hadn't. So the teacher, who was Kurt Hirschhorn, a very illustrious geneticist and pediatrician, he was very offended by this, but what happened in essence was that anybody who showed up for the test, he passed. . . . It tended to reduce the degree of competition and enhance the sense of being able to work together.

I think that the idea of trying to grade people for what you can memorize in the first two years of medical school bears no resemblance to what they'll do whether they end up as clinicians or as researchers or as public health people or whatever they're going to do.

—Terry Maratos-Flier, M.D., '76

Most people passed. Fail was not an option. . . . So if you did a really good job and you did well but you just missed the grade on honors, then you were no different in terms of your ranking than the guy who would just barely pass. So, pass/fail/honors, in some respects, I thought was even worse than a grading system.

—Michael Marin, M.D., '84

I liked the idea that it was a pass/fail system. . . . I didn't like the idea that people would be cutthroat or not working together. I mean medicine is supposed to be a community team approach, or you'd think it should be, and everyone should be learning from each other and not stabbing each other in the back, and that was the impression I got when I interviewed.

—Adam Levine, M.D., '89

I think in my class the students were always very skeptical about if it was really just a purely pass/fail system. I think we learned at the end when we were applying for residencies that they do rank the class, 'though I still never understood how they rank the class. On some level it relaxes people, and on some level they are anxious and nothing is going to change that. Sinai kind of diffuses that competitiveness that is inherent in a premedical student. . . . The atmosphere is one of cooperation.

—Philip Mulieri, M.D./Ph.D., '03

They have what's called a pass/fail system here and what used to be a pass/fail/honors in the second two years. They've since added high honors . . . medical education is a very repetitious process where you just gain more and more as you go along, so the pass/fail system allows you to slowly absorb more and more without the stress, and it does in that way allow you to get out into the community and allow you to do a lot of fun things.

—Robert Greenberg, M.D., '03

The pass/fail system in the first two years is phenomenally important because it creates a system of real cooperation during the first two years. Students really do help each other learn and get through stuff. . . . It changes in your third year because people are more competitive, but it's not as bad as if you had had it the whole time because people are used to not being competitive. It sets a precedent. It's about learning.

—Peter Klatsky, M.D., '03

STUDENT REPRESENTATION

We were involved in all kinds of roles that were not true in most medical schools. There was a committee that I served on for all four years called the Executive Faculty or something like that, which was the Dean, the Associate Deans, and all the department chairmen, and then there were one or two first- and second-year students. And you could just speak your mind to Dean James and Christakas and Sol Berson [Chairman of Medicine]. It was kind of mind boggling.

—Peter Lang, M.D., '72

What was particularly interesting that first week was the meeting that we had in the Dean's office. George James was the Dean . . . but he took time and invited literally all of us, there were only thirty-six and he had a very big office, for coffee and danish on Friday morning from 8 to 9 before our classes started at 9 A.M. We all sat in the office. We had a wonderful visit with him. He listened to our concerns about what we wanted to have by way of a different grading system. . . .

He was a very insightful and witty and thoughtful man. . . . The other speech he made was what was . . . called afterwards the famous "cookie jar" speech. He basically said we could have pretty much everything we wanted. We had a number of requests; we wanted to be on all the committees. He said, "That makes sense. You can be on all the committees except the Faculty Promotions Committee. . . . That's left for full professors who judge the other faculty . . . and it isn't right for students to be judging if faculty ought to be promoted or not." And somebody came back and said, "Well, why can't we be on that committee, too?" And he said, "Look, it's like being offered ten cookie jars and saying you can have cookies from nine of them but not the tenth. Given the fact that some things are not appropriate for students but everything else seems appropriate and you can be on all the other committees." So that was the spirit.

The other thing that was funny about that day was that this took place from 8 to 9, and by midday the word around Mount Sinai was that the students had had a sit-in at the Dean's office and that we had taken over the Dean's office.

—Arthur Frank, M.D., Ph.D., '72

There was a big group of people [in the Dean's office], and it got a little contentious after a while, and it was almost like this feeling that some of the people were going to announce, "We're not leaving until you do this." And maybe that's why I'm remembering it as occupied it. . . . There really was this strong component of kids who were not afraid to confront authority.

—Jeffrey Flier, M.D., '72

The impact of all our agitating was very minimal, and that might have been the right result. . . . But I know that we were regarded as important. We felt important.

—Kenneth Davis, M.D., '73

I think that the faculty, of which there was very little at that time, was very receptive to the students. . . . They knew that they had a brand-new school . . . and I think that there was a significant amount of dialogue between the students and the faculty to see how we were doing, because we knew that . . . in some respects, we were guinea pigs.

—Mark H. Swartz, M.D., '73

I always felt very, very well respected. In fact, as I think about it, knowing that it was a new school, I think that they were particularly receptive to our input. If there was a scale from being disrespected to being included, it was much closer to being completely included and valued for our opinions than being at the other end of the spectrum.

—Scott Friedman, M.D., '79

When I was a medical student I was very interested in development of a rape counseling service here, and it turned out that I and another medical student . . . were actually empowered to hire the coordinator for the rape crisis program that's now called SAVI, and she's still here. So I was very lucky; I had a tremendous opportunity to bring [my idea] to fruition.

—Deborah Marin, M.D., '84

There was no student council. The students could go to the Academic Council meeting. They didn't really have much of a voice. If we organized ourselves and made a protest of something . . . we would be listened to, but no one actively sought us out. We'd have to make ourselves seen.

—Anthony Weiss, M.D., '88

I think that the Mount Sinai administration was very open to student ideas and to their concerns. . . . I think that they were very welcome to hear what the students had to say, within reason.

—Jason Brett, M.D., '94

I was somewhat impressed that, despite the class being 120, in terms of medical schools, it wasn't small, yet if there were any concerns, I know frequently there were concerns, they had meetings and they would present them to the deans or what have you. And I think that a lot of that, percentagewise, was accepted well, and a lot of changes were made.

—Alfio Carroccio, M.D., '96

[My roommate] was involved a lot with the curriculum changes, which I know kind of went on the year after I graduated. . . . A lot of changes started happening, and I know that they took a lot of interest in what the students thought.

—Bethany Goldstein, M.D., '99

A proud marker of their identity is being a school that really allows the students to get involved in their education and guiding and molding future classes. I was always impressed with the speed of the integration of things . . . seeing things implemented for the class below you right after your class was done. They were able to make the changes that you had suggested or other people in your class had suggested within the next year.

—Philip Mulieri, M.D./Ph.D., '03

I have gone to these student committee meetings where it's just a few members of Student Council and all of the Deans, up to the head dean. . . . They're definitely interested in what we say. I'm not saying that we can change the whole school. But, as an example, they raised the tuition a couple of months ago, and a group of students got together, mainly first years, went to [the administration] and explained how this was going to be really hard to pay this much money all at once, and they listened to them and they changed it. The tuition is still increasing, but they changed it and now it is a graduated increase over a few years . . . and that was all because of what the students said. . . . I think it really works.

—Stefani Wedl, M.D., '03

COMMUNITY

I never thought about "community." What I cared about was my interactions with the professors . . . with the house staff and with other students. Those interactions were great, and the School was very supportive in terms of helping people get housing.

—Ernst Schaefer, M.D., '72

I think that they did a lot to encourage interaction between medical students . . . both in terms of setting up scientific and medicine-related symposiums that we all could participate in openly, as well as public functions or recreational things. . . . There's a lot of professional and personal camaraderie that still exists.

—Michael Marin, M.D., '84

When I look back at my medical school experience, it was one of the best times of my life, particularly the first year. Mount Sinai, although it can be fragmented . . . into different levels of specialization . . . really is one big large happy family; one large community where ultimately, at the end of medical school, you really felt like you knew everybody, from the higher-ups in administration to the faculty to the student body, and even down to the workers and the janitors, and the people who worked in the labs and the people who served us lunch and coffee. It really was a very tightly knit community from that standpoint.

—Jeffrey Mechanick, M.D., '85

I came into an established class with established people who knew each other already for two years, and I have to say that once I got here, those people really did accept me with open arms and really made me feel like I was part of the team. We just had the dorm open across the street for the very first time, so there was a lot of excitement, a lot of camaraderie around all the activities that were going to be surrounding the new building. . . .

It wasn't just the learning experience of medicine, but it was also the learning experience of life. I developed some of my best friends during that time. It was a scary time. You're going through your third and fourth years of medical school, which is basically all hands-on and clinical, and it's not just whoever was able to excel on studying and answering questions any more; it was really what we were going to do, and touching people, and putting all of our knowledge together and then taking it the next step, feeling what it was like to become [a doctor].

It was both scary and exciting at the same time, and having all of those people going through it at the same time, and people you can talk to, made it much more helpful. A lot of times, the people who are outside of medicine, like your parents, every time you said you were on call or you were studying for an exam or you had a practice practical, they had no idea what you were talking about. These people understood where you were coming from, the fears that you were having, you're cramming, you're trying to still be a normal person and still become indoctrinated into what it was going to be like to be a doctor in the future.

—Marla Stern, M.D., '86

In any situation of intense work, you create community.

—Jeffrey Parvin, M.D., Ph.D., '89

I don't think [Sinai] fostered a sense of community. I went to Yeshiva University where most of the graduates go to Einstein. . . . I wasn't aware at that time that this place was very Jewish. . . . There was no Kosher cafeteria, there were very few people wearing yarmulkes walking around. . . . I also lived off campus, so that could be a factor in my feeling that it wasn't that close of a community.

—Ram Roth, M.D., '92

Obviously, there is a camaraderie about being at Mount Sinai. . . . I don't think that that's so common at other places. It's kind of a nice network of people who really feel for each other.

—Jennifer Trachtenberg, M.D., '93

While being a diverse group in cultural backgrounds and geographically where everyone was drawn from, to a certain extent, we sort of created a community atmosphere. I think the feeling was that generally anyone could approach anyone else if they had any type of need of any kind, whether academic or personal, and they were met pretty well.

—Alfio Carroccio, M.D., '96

I always felt like I knew a lot of people in the hospital. You get introduced early on. The people who give our lectures are the people who are attendings here. We met them early on, and then when we were doing our clinical years, we would work with them again. There was a nice sense of community.

It's easy to get isolated because there's so much work to do. . . . But then, on another part, I think it opened my eyes. . . . For me, I'd never been part of such a diverse group of people until I got to med school. Just seeing that, and people of so many different races and religions and backgrounds and then more so with the patients. Taking care of people from different countries and with different perspectives on things and having to think about how I interacted with people. That kind of opened my eyes more to things outside of my own little world. Then again, it's hard to look outside your little med student study group when you're just trying to get all the information down.

—Bethany Goldstein, M.D., '99

The fact that there's such a community attitude really helps to get through the experience because medical school, particularly the first year, is very hard. It's emotionally challenging, going through gross anatomy and dissecting a cadaver is initially scary, and you're coping with your own mortality dealing with this person who has donated their body and has now died for your benefit of learning from them, which is an amazing experience but very emotionally draining, and having to deal also with the rigors of lots and lots of new information. I think really what gets us through it is having each other for support.

—Talia Swartz, M.D., Ph.D. candidate

[The second-years] really take us under their wings. . . . There's a sense that you can go and ask someone who's been through it, what do you think? How does this work? . . . Even with the faculty; there are amazing faculty. You get to know people, and it's like they're friends. You know you can talk to them about personal things, a concern, or anything that's going on in your life. . . .

SEOM is Students of Equal Opportunity in Medicine. It's an umbrella organization for SNMA [Student National Medical Association], a historically black organization for students in medicine. . . . The only complaint I had coming to Sinai was the lack of community I felt amongst minorities. I think Sinai has come a long way. . . . I think it has a long way to go. Between minorities, there hasn't been that great of a community, and I think that there's very few of us. It's hard to form a huge community. There's not a huge support here for fostering a minority atmosphere, but I have to say it's changing. I think that the fact that myself with another guy are the only two Puerto Rican people in my class in New York City, that has the second largest Puerto Rican population in the world, says something. The second-year class only has one black male and four or five black women. [Sinai] has a really great percentage nationally wise. . . . Most schools have like 12 or 11 percent. . . . We're trying to make this organization [SEOM] so much more prevalent on campus. One of the things that we're doing for next year is to try to form a cohesiveness between the minority community at Mount Sinai, within the community at Sinai, between doctors and residents and fellows and attendings and everybody who's around.

—Melissa Alvarez, '05

RECREATION

Every Sunday morning we would play football in front of the Klingenstein Pavilion. We sort of had an agreement: from 9 to 12 we'd play
touch football almost year round, even in the snow. . . . Then, on Thurs-
days, we'd play basketball. . . . I would rebound, pass it out to this guy
. . . and he would hit from anywhere . . . and he was an attending. I was
pretty impressed that an attending would come and play basketball
with us guys.

—Ernst Schaefer, M.D., '72

I remember hanging out in the basic science laboratory, which was a
converted garage, just hanging out there in our laboratory, which was
our home base. Behind some corrugated iron was where we kept the
bodies for dissection, so there was this faint formaldehyde aroma there
the entire time.

—Peter Lang, M.D., '72

We had a couple of bridge games going. People would go in and out of
the lectures because the lectures were so small and the recreation area
was right outside.

—Marlene Marko, M.D., '72

During medical school, they did have a ping-pong table and a pool
table outside the lecture hall there, so there were those of us who
would rotate playing ping-pong and whoever lost had to go into the
lecture.

—Loren Skeist, M.D., '72

I think it would be fair to say that there was a difference between what
occupied a lot of our mental time and what occupied a lot of our actual
hours of the week. It might be possible to be in Washington for a march
or to be working for a candidate for just a few hours, but that would
have a big impact and that would be something that one might have
been working up towards for quite some time, whereas most of the
week's free time was spent in the laboratory. So it wasn't like all of that
free time was spent going door to door handing out pamphlets and
saying, you know, "have you burned your draft card?" But there was a
feeling that, for example, after Kent State, we didn't have classes here,

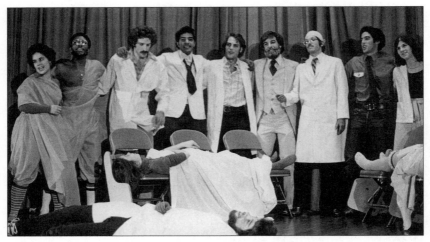

The class show of the Class of 1980, performed in Spring 1977. Each year, members of the first-year class put on a show that explores and spoofs the experiences in medical school.

either. There were meetings of the whole faculty to discuss what was the best thing for this community to do, and students were very involved.

—Kenneth Davis, M.D., '73

It was our class that started the tradition of the first-year play or the first-year skits. We're kind of proud of that because it's twenty-two-odd years later or more, actually twenty-five, almost from when we were first year students, more than twenty-five, and there's still a first-year play. . . . We had a bunch of cut-ups. I was pretty good at imitations. We obviously took no mercy on the faculty . . . but I think it actually was very well received, it wasn't mean spirited and it was a blast. I remember when we did it in those days, it was in the same lecture room we sat in every day, and the place was packed to the gills, and we just had a lot of fun. We just loved the place, we loved the faculty, we really cared about each other, and we had a good time.

—Scott Friedman, M.D., '79

I had far more of a personal life during medical school than I did before or any time afterwards. Even though it was rigorous, it was actually the most relaxed time, and there was a lot of camaraderie. I did a lot of things that were personally enriching. . . . Every Friday night, we would go down to Chinatown, then to Little Italy, then come back and watch *Dallas* on Friday nights. On weekends, we might take excursions to Fire Island. We went out a lot for dinner and movies. I went to Florida a couple times with friends and to Israel and Greece with classmates—those were two different summers—and then went to Indonesia with a classmate on the last break right before graduation. That was a very high-impact four years for me in terms of personal growth. I look back on it more for the personal experiences and the research experiences than the formal academic experiences. . . .

The first-year play . . . was a highlight of first-year of medical school. I was in it. I danced in my underwear, on stage.

—Jeffrey Mechanick, M.D., '85

We went out a lot. We experienced a lot of theater and eating and just enjoying Central Park, which was right across the street. I just remember it as a couple of the greatest years of my life.

—Marla Stern, M.D., '86

There was a lot of nightclub activity, there were a lot of parties that were organized by different medical schools, and there was a lot of clubs to go to. . . . I don't remember thinking that much about social life during medical school, though. . . . We didn't have a lot of time. We were also pre-Bell Commission, and so the medical students . . . for example, when I was doing OB-GYN, we would want to be there like the house doctors, the residents, the interns were doing, so we would do thirty-six-hour shifts and then go home and fall in a heap. And when you're in your third and fourth year of medical school and you're trying to prove your mettle as a future doctor, you know, you pretty much emphasize the work you're doing.

The social life was kind of a blur. If there was a special night to go out to a club or an evening, you would do that, but you didn't have a lot of hobbies or a lot of outside activities.

—Anthony Weiss, M.D., '88

I spent all of my free time in the lab. . . . I caught the research bug . . . and for anybody who has caught the research bug, they know exactly what that means.

—Jeffrey Parvin, M.D., Ph.D., '89

The infamous "orange couches" that were original equipment in the Levy Library and a favorite study and sleeping spot for medical students and others. These couches were replaced with reclining chairs in the late 1990s.

We had a softball league during the year. We had a lot of free time to be playing sports with. During the summer, I took trips.

—Alfio Carroccio, M.D., '96

A lot of people exercise. You can get free Y passes to the 92nd Street Y . . . learning about how good it is to exercise and people are stressed out.

—Stefani Wedl, M.D., '03

I think the most fun is being the anatomy TA (Teaching Assistant), because you go into it thinking that you're going to be the enlightenment for these students and that you have all the answers and you think that you're going to be like the tutor who teaches. But, really, what you find out is that you're a bunch of A students in anatomy teaching a bunch of A students that are just taking anatomy and it's not like tutoring someone who can't read or tutoring someone in calculus who can't do the integral. This is like, you're telling people information that they almost already know, so it becomes more of just a friendly and fun atmosphere and minutia and minutia games and mnemonics and stuff like that are what you end up teasing people about and quizzing people on, and it's a lot of fun.

—Robert Greenberg, M.D., '03

The fact that medical students were taking their time to put on and produce a legitimate one-hour-and-forty-five-minute full-length play, with full costumes, is great, is really terrific. And the guy who directed it was a Wesleyan grad who was also in the Humanities and Medicine program who was a theater and dance major.

—Peter Klatsky, M.D., '03

We set up a color war and divided the class into two teams and we had ice skating, a Pictionary thing, we had a trivia questions thing, a baking contest . . . weight lifting.

—Bethany Slater, M.D., '04

Everybody always goes to the park, Broadway shows. We have people who are training for the AIDS Ride New York, where they're going to bike like a thousand miles or something like that. Amazing. I know people who are training for the marathon, who sing in a choir . . . they perform in Carnegie Hall. We've had two plays. . . . We have Sinai coffee houses, where people come and sing and dance, and a band. Everybody has at least one thing that they're totally into.

—Melissa Alvarez, '05

COMMUNITY SERVICE

We did do community service in many ways. The first year we were in medical school was the famous, or some might say infamous, Woodstock Music Festival, and several of my classmates went up to volunteer in the medical facilities there. Some people went out and did some work in the community. . . . They worked very hard at getting us out into the Hispanic community that we served in East Harlem. We went to the East Harlem Health Department.

—Arthur Frank, M.D., Ph.D., '72

I remember very early on doing home visits . . . in East Harlem . . . a very exciting, special opportunity.

—Loren Skeist, M.D., '72

The class was too small for a lot of groups. . . . People did things, but they did them in larger groups. There were Physicians against the Vietnam War, Physicians for Social Responsibility. People would join that, but there wasn't a whole lot of generating our own chapters here. It didn't seem necessary. In other words, a lot of politics was not local. It was felt that you had to do something on a national level . . . but it was joining larger groups. It wasn't the notion that in order to be effective, one had to do something right here, although that was a message that Community Medicine was giving us, which was a message that resonated with a lot of students.

—Kenneth Davis, M.D., '73

Certainly Dr. George James, one of the founding fathers along with Dr. Hans Popper, both were very much involved with community service, George James having recently been the Commissioner of Health in New York City. So the fundamental basis of developing a school, it was clear that they were interested in community service and humanistic care, but I was still very young and was still just interested in learning my anatomy and physiology and internal medicine, pediatrics, etc. So I was not that involved and not that impressed with any tremendous effort to direct the students toward that.

—Mark H. Swartz, M.D., '73

The principle was clearly there: that we live in a community, that we serve the community, and that we need to understand how our background impacts on interactions with patients and how important it is for our institution to serve the community we work in.

—Scott Friedman, M.D., '79

It was known that there was this triad, this triple mission, for the medical school, where one prong of it was community service and community medicine. Although we had one rotation formally in Community Medicine during our third year, which was sort of one of those easy rotations to do because of the patient contact, there wasn't any perceived pressure or impetus to go out in the community, at least from my stand point. I know that a lot of my classmates took advantage of that and did that. . . . I know that some of the students were active. . . . I'm not sure how much of an impact they really had.

—Jeffrey Mechanick, M.D., '85

To be perfectly honest, we didn't really feel a lot of that by the time my generation of Sinai students got there. We were aware that it was part of the history and we were told, and we had the community medicine rotation, and the opportunity to do electives in the Department of Community Medicine . . . so that was available, but, really, by the time, I guess in the mid- to late '80s, when I went to medical school, you really felt that the focus of the medical school had changed. Rather than having community medicine as a focus, there was more of a sense that you were aspiring to be a top-tier medical school and more in the traditional mode . . . more basic science research and more of a traditional medical school program. . . .

I think [community service] was not something that was emphasized. . . . There were not any great social issues of the day; there were just a few people doing things. I remember once I got involved in a project where we were trying to raise medical supplies for South Africa. . . . Really, by the time you got to your third and fourth years, you were really very unlikely to be involved in any socially conscious activity.

—Anthony Weiss, M.D., '88

One of the programs that I did was . . . a rape crisis intervention program, that was sort of an outreach type program, and we were on call different days or nights for . . . you were sort of a counselor for the person, someone for them to talk to if they didn't have a family member, or even if they did have a family member—someone they could speak to so that they would know what was happening. . . . We were well versed on the rape kit and stuff that the physicians would have to perform. Someone to let them know their rights and what to do and what's this information going to be for.

—Jennifer Trachtenberg, M.D., '93

The school itself doesn't put out a flyer saying, "Do community service" . . . but I think in the way they've structured the school, they encourage it: the pass/fail system, the way the classes are arranged there's time in the afternoon . . . to do things. By setting up the school that way, they automatically encourage it. They encourage the Student Council, which [financially] supports the groups that organize the community service events.

—Stefani Wedl, M.D., '03

Student Council has a variety of different committees and . . . community service is probably the most active among those and the most demanding. . . . Formerly, . . . what community service representatives on that committee did was do the typical food drives, toy drives, clothing drives. They'd occasionally get things together to go out to soup kitchens . . . up until maybe until I got here, and maybe a little after, but since, it's taken on more responsibility. So now, instead of just organizing these sorts of events, we do a lot more of actually meeting with the students and sort of coordinating community service on a larger scale. Not to say that we organize more events, but what we do is, myself and a few other individuals started up something that we call the CSAC, Community Social Action Council, and that's a fancy title for the community service reps of Student Council having a meeting with people and actually facilitating community service events . . . mainly students, but really anyone can come. Faculty and deans have come. People in the community who work with things like the East Harlem Community Health Committee and a couple other local organizations have come, just to sort of sit in and see what we do and plug their activities. But, really, that organization is really at its birth.

Community service in the past has sort of been a student fed operation. . . . But now, it's taking on sort of a new face, and that is the organizational route. We don't necessarily organize the activities, but we bring people together. . . . That way we don't reinvent the wheel all over again every time we start an activity or do a group. We come together and talk about the problems with community service at Mount Sinai, and we talk about things that have been successful for each of the individual groups, and then, as the CSAC, we seek solutions to those problems and share those with each other.

—Robert Greenberg, M.D., '03

We're crazy busy medical students, but if we divide up the work and just do it, these are simple projects. So, now the one we are doing this summer is we're starting an AIDS education project. We're educating young girls, aged 13–17 who are at a boarding school, who are girls who have been orphaned by AIDS, so their parents have died of AIDS and they're at a boarding school. So we're going to go there and we're going to educate and teach them to be new AIDS educators, and they'll go to other high schools and junior high schools and teach their peers, people in their own age group, about HIV and AIDS. And they can be very real and say, "this is what happened to my family. This happens and here's how it happens, and here's how you can protect yourself," and it's an incredibly great preventative strategy.

—Peter Klatsky, M.D., '03

You could go and do something once a month, or one time and have fun with it. . . . I was very involved with homeless health. . . . We went to homeless help shelters and gave talks about health issues that were relevant to them, and I got like a group of four students to come with me. . . . We'd give half an hour or so talks about issues—we did pneumonia, the flu. I asked them every time what they wanted to speak about—what they wanted me to speak about, so I take their suggestions. . . . We go and we talk, and they ask questions, and they are like so into it. They ask all these questions and stuff, and then we actually have dinner with them afterwards. . . . Sinai definitely encourages people to do community service, and I feel like it's actually gotten more and more; people have gotten more involved with it . . . and they really try to get people active and into it.

—Bethany Slater, M.D., '04

STATS stands for Students Teaching AIDS to Students. Our mission is to go out into the community to high schools and junior high schools and just teach to kids and talk to them about AIDS. One thing that we're realizing is that there has been so much push in the media and in education to teach kids about AIDS and HIV and so a lot of students know. . . . And yet, then we say, "what's gonorrhea?" Blank. Nobody knows. Gonorrhea is the second largest STD, I believe, behind chlamydia. So our goal is definitely to get the education out about AIDS, because it's definitely so prevalent, but also to educate kids about everything.

—Melissa Alvarez, '05

Once we were trained, we would go out into East Harlem into high schools and present, as opposed to necessarily lecture, education around HIV-AIDS and that was a great experience also because often-times I was the only white person in the entire class, including the other copresenter. . . . It was definitely an interesting position to be in. To see that a lot of kids, at least underprivileged, socioeconomically

A graphic used in the 1984 yearbook, describing the condition of a graduating medical student.

disadvantaged . . . to be still not in a position where they have the full education that they need or the full knowledge that they need to make appropriate decisions about their sexual health or the process of growing up, which can be very stressful, and how to deal with that . . . but it's amazing, because you go there and it's a completely different world, and it's only two blocks away from Sinai. . . .

I've [also] been working with a colorectal surgeon down in Chelsea. Most of his patients are, I guess, gay men. . . . Even though I didn't consider that at first as community service, it's been incredibly enlightening for me to see the type of patients that come into his office, the majority of them who are HIV-positive, and the sort of health issues they deal with on a regular basis. . . . It's interesting to see a physician, in this case a surgeon, sort of keep people's bodies in check and put Band-Aids over things, but not really help them. . . . I'm learning about a group of people with a specific disease and a specific part of culture that's often stigmatized but receives, at least in his care, a great standard of care, but probably don't receive a great standard of care in other situations, and just to be able to see that is really enlightening for me.

—Adam Kawalek, '05

HUMANISM AND ETHICS IN MEDICINE

It was just sort of assumed. It was part of how we were taught, that the humanistic side was part of our education, where a lot of schools now have to put in courses on ethics and the humanism of medicine, and nobody thought of this as separate. It was just suffused throughout our education. . . .

It was wonderfully conceived. . . . The theme for the first five weeks was diabetes, and I still remember those first five weeks. . . . I remember also in those first months, we had sessions on do you or do you not tell a dying patient that they have a terminal condition. . . . It was a very open and nurturing environment for some of the issues in medicine that transcend just knowing your medicine and doing science. . . .

One thing we did on literally a weekly basis, every Monday at noon, the class, all thirty-six of us, got together over lunch and had a town hall community meeting. . . . We'd have ethical discussions. There was a lot of discussion about using animals in our training. Should we do a dog lab and euthanize the dogs when we were done? There were discussions about taking gifts from the drug companies. . . . I remember having a discussion about that with a very sage older physician at the time who had been at Sinai for his career. And his comment to me was: "If you think you can be bought by a book or a black bag and an examining hammer, you're going to be bought very cheap. And the rest of your career you're going to find all kinds of trinkets or all kinds of opportunities to be bought by the drug companies and those connected with the medical industry." It's amazing how much I remember from those years being in medical school: the lessons and the sage advice people would give.

—Arthur Frank, M.D., Ph.D., '72

I picketed with my Dalmatian. . . . It was a dog experiment that I felt was a completely useless waste of an animal's life. It wasn't a significant learning thing. It wasn't in preparation for doing surgery. It was some really dumb blood-pressure experiment and I refused. A couple of people picketed with me. I think they later . . . were very careful about which animal sacrifice experiments they would do.

—Marlene Marko, M.D., '72

Richard Gorlin, who was then chairman of Cardiology at Peter Bent Brigham, came down to Sinai when he became chairman of Medicine, and he was very, very interested in humanistic patient care, and looking at patients as people suffering from disease and not just people from whom you could learn about medicine or just the vehicle of learning about disease. And it was he who really got me on my track and my direction toward humanistic patient care . . . which was kind of becoming the old-fashioned doctor again and learning to listen to your patients and try to put yourself in the patient's shoes to see why they are experiencing the symptoms that they are experiencing in the manner that they are experiencing them, etc. Much of my feelings about improving patient care . . . stem from Richard Gorlin.

—Mark H. Swartz, M.D., '73

In retrospect, looking back, I now know what humanism is. I knew what humanism was as a resident looking back, but I really didn't know what humanism was as a student. . . . There was this Psych/Medicine liaison conference looking at "humanism," but being very naive as a first-, second-, or third-, or fourth-year medical student, you really don't have the context to understand what humanism is. I ended up actually writing a paper for the *Mount Sinai Journal of Medicine* about creativity in medicine, trying to bring in the way in which we cognitively approach diagnosis in terms of considering individual patients instead of being purely algorithmic or scientific. But that was more of an intellectual exercise. I think that any efforts to try to teach humanism at an undergraduate medical school level—it was just very difficult and, personally, I didn't really appreciate it. The push may have been there, but I just didn't feel it.

—Jeffrey Mechanick, M.D., '85

I never felt like I was dehumanized when I was either training as a medical student or as a resident, that I had to just cut off all of my emotions. But I did learn that there was a time and a place that you should have some distance with patients; even though you felt for them and you wanted to empathize with them, you learned how to separate all of your feelings from your management. In the beginning it was hard. . . .   I think that Mount Sinai really gave me the opportunity to really connect as a person but also gave you the opportunity to go into many different, varied fields and subspecialties and still always remember that you should treat the whole person, their identity, and not just their illness.

—Marla Stern, M.D., '86

Since I think that the approach to education was humanistic, I think you certainly didn't walk out of here feeling resentful, feeling like you'd been beaten, feeling like you weren't treated well. And I always emulated the people who were around, and that was the physicians at Mount Sinai.

—Adam Levine, M.D., '89

I'm not sure that that is something that can be taught. . . . The number one thing is to be a mensch, a human being, and that makes you considerate to others. And know your limitations and take care of patients as people. And I do give Mount Sinai credit for trying to teach that, but I don't think that it is something that anyone can teach. When I was in med school, we had the psychiatry and psychology part of the course, tried to have some rap sessions about things, and toward the end of school, the Morchand Center was being developed. That was probably one of the better attempts at teaching us to be more responsive to patients.

—Ram Roth, M.D., '92

I think that during the four years they have a lot of classes that focus on the psychosocial issues and the humanistic side of medicine. I think that the school does actually a very good job of bringing that to the forefront and making that a priority.

—Jason Brett, M.D., '94

I think that being a humanistic doctor came in part of our "Intro to Medicine," how to interact with patients. We would do role-playing: how to take a history and how to tell someone bad news. And we got to work with the actors. That, other than seeing real patients and getting feedback from them, was kind of the best way to pretend it was a real situation, but then also to get feedback from the patients themselves and from other people watching. I thought that was helpful.

—Bethany Goldstein, M.D., '99

We had ethics this year that really focused on [the humanistic aspect of learning], and we take "The Art and Science of Medicine." . . . It's a class that allows us to really spend a lot of time talking about the patient, practicing patient interviews in a group setting, and then being able to stop and say, "Well, how do you think the patient probably feels right now?" you know, using the standardized patients that come in, that are hired. That allows us to be able to role-play with the patient and say, "Gosh, she must be in a lot of pain," or "She's probably really upset. We'd like to talk more about it." It allowed us a lot of time to reflect in a practice setting, without actually being in real life with people who are actually going through these things.

—Melissa Alvarez, '05

BEING A WOMAN AT SINAI

I was just at the edge of the change in terms of medical school classes and women. When my husband started going to medical school in '68, 5 percent of all medical students were women, and the year that I started, nationally, 13 percent were women. So there was this interesting view that women were going into medicine, and it was a potentially good thing to do. . . . There was no sense among the women in my class, there were thirteen women there, that anybody gave me at least . . . that we didn't belong there. . . . They seemed very, very happy to have us.

—Terry Maratos-Flier, M.D., '76

I had no experience of a glass ceiling or prejudice or harassment, ever in my training . . . in terms of my perspective, I've been given every opportunity to grow.

—Deborah Marin, M.D., '84

We were just concerned about studying and passing, and I don't really think there was that much controversy at the time. . . . There was a clique of women at the time who were concerned that the lectures weren't as fair to women's rights. . . . There was one instance during sex ed when they took offense that it was termed the female response— you had the male sexual something and the female response—and they took offense at that terminology. At the time, there was a lot of controversy about that.

—Jeffrey Mechanick, M.D., '85

As I was going to med school, I think there were 40 percent women, and then there were 50 percent women and now I think we're over the 50 percent mark. I think that's wonderful for women. I wonder, I'm sure a lot of people wonder, if this is partially an economic thing. People don't run to medicine to make oodles of money and that leaves a lot more room open. Whether that's true or not, whether it's good or not, I'm glad women are in med school and enjoying it. When I was entering med school, there were few women who were teaching us. . . . It was rare. Once we got out to the wards, a lot of our residents were women, and that was fantastic. It's just interesting how that number has changed and has grown over the years.

—Ram Roth, M.D., '92

I think it's great. I've really only had . . . positive experiences. Maybe it's because I'm in pediatrics. . . . It's probably still far easier to do than to go into surgery, although there are a lot of great surgeons that I know that are female. . . . For me it's been a plus. I've had so many patients transfer to me because, one, even the parents, the moms, are really happy for role models for their children and that I was easy to speak to, issues that it would be easier to talk to a female about.

—Jennifer Trachtenberg, M.D., '93

Being a woman at Sinai, I never felt that that was any kind of weakness on Sinai's part. There were a lot of people in high positions who were women who were great. Clinical deans, you know, they were just great role models. . . . At least half of my class was female.

I felt that Sinai was very pro-women, just in terms of always having a lot of women lecturers, that we got a chance to meet a lot of strong women role models. . . . We'd always say, "Oh, they must be the top in the field," and it was a woman. It was nice to get that. I never felt different, that I had to prove myself just because I was female. . . . It just wasn't an issue at Sinai.

—Bethany Goldstein, M.D., '99

I think that being a woman in medicine is definitely interesting from the point of view of families. . . . I'm also president for a Life in Medicine society. . . . We have speakers come in once a month. We discuss things outside of medicine that people do with their lives. . . . We did one about being a female, having a family in medicine. I think that that's one of the biggest obstacles I will come across. When do you have kids? . . . Coming from a family where you would never hire a nanny—Puerto Ricans would never hire a nanny. Yet with the times, here I am. What will I do, being a good mom and being a good physician and balancing everything?

—Melissa Alvarez, '05

CHANGES

It lost a lot of that initial fervor that the faculty had for having students around. It went from, "Oh, my gosh, the students are here!" to, "Oh, yeah, the students are coming back, school's starting again." . . . That move and that change in environment [to the Annenberg building] probably modified things a little bit. I think as it grew clinically and as the school's philosophy changed, not the basic philosophy, but as the founders of the school were not as prominent anymore, the tripod wasn't as important. I think the school had lost a fair bit of its sense of the community piece. Probably still extremely strong, I'm sure, clinically, but even the basic sciences may be a world unto themselves a little more.

—Arthur Frank, M.D., Ph.D., '72

I left in 1973 and didn't come back until 1979. I think that those were critical years in which the school began to determine what the curriculum really was and the best way to teach it. So, when I came back in 1979, this was no longer the great experiment it was in 1969. . . . There was much more concern that the clerkship experiences would not just be a kind of loose confederation of preceptorships, with students tracking around after interns, residents, and occasionally attendings, but that there was a quality of information that needed to be communicated and that they knew what that was and that roles and objectives were being communicated and clarified. So, it was a much more formalized education.

—Kenneth Davis, M.D., '73

Medicine has changed tremendously since the 1970s, when I first started my medical training. [In the] 1970s, patients were able to stay in the hospital longer; they could do some convalescence of surgery or heart attack or whatever; they could be in the hospital for much longer periods of time. Then, in the 1980s, the recognition of the AIDS epidemic became obvious to all of us, and the patients in the hospital were very often patients who were HIV-positive or had frank AIDS.

And then with healthcare insurance and the big push toward length of stay and reducing length of stay, what has happened is that patients are in the hospital for a much shorter period of time because of insurance purposes, and they're here in the hospital during the sicker part of their illness, so that patients are being sent home from surgery much earlier. So patients are not being seen by students when they're convalescing in the same way that they were twenty or twenty-five years ago. . . . Therefore, students are seeing sicker patients. If patients are sicker, they're less likely to be able to spend time with a student who is really there primarily for his or her own education and not for really helping the patient.

I don't mean to say that medical students don't help patients. They do, and they do so by providing an ear and an opportunity for patients to communicate with students about a variety of issues, but the students don't have very much time to spend with the patients as they did in the days when I was a student. And that's unfortunate, and that's another reason why the use of standardized patients has become particularly important.

—Mark H. Swartz, M.D., '73

It almost seems like the first twenty years were to establish a credible and respected medical school. It seems to me like we are entering a new phase, which is one where there is no doubt that we are credible, we're widely respected, how do we become one of the best? How do we make that jump to one of the elite medical schools? And being in the top twenty or twenty-two medical schools in the U.S. after twenty or twenty-five years is a really impressive accomplishment. And so I think that this is a very challenging kind of task to set the bar higher. Okay, how do we become great? How do we become a world leader? Where is our strength, and how do we play to that strength? And how do we do it in an environment where it is harder and harder to keep hospitals fiscally solvent, where reimbursement for patient care is dropping, where grants are very demanding to get? And I guess the short answer is you need good leadership. . . .

It's a very interesting time in the life history of this institution, and I think that conflict between the old and the new, between the hospital-based world and the more academic medical school, is one thing that is playing out in many places around here, and I think it will change, for the better, the medical school. . . . I think we're seeing part of the natural evolution of the history of this institution, which is that there is a push on to create a top-flight academic school. . . . I think we are still trying to build critical mass in terms of the ability to do highly competitive science.

—Scott Friedman, M.D., '79

Managed care has not penetrated New York and is not a big percentage of what we do. We deal mainly with Medicare and Medicaid patients. Nonetheless, I think that length-of-stay issues, I don't think that that leads to us discharging people who are sicker, I think, rather, it has led to us being more aggressive and quickly managing people and becoming much more a seven-day-a-week hospital. The students aren't affected by it. . . . If you can get someone out of the hospital faster, I think that that's a good thing.

—Deborah Marin, M.D., '84

I came back to a place [Mount Sinai] that was undergoing reorganization . . . in terms of rapid changes in the medical center administration, in care delivery systems. I found a very different medical system, not having been here for approximately eight years. The way I can characterize it is that I saw that people were walking through the halls faster; that everyone seemed a little more focused; that there was a higher bar than there had been, even in the past. But I felt that the medical school had grown enormously, that the medical students who were being trained turned out even of a higher caliber, that the research that was sprouting in the '90s had been of a higher quality. Clinically . . . I saw a medical center that was clearly on a new trajectory and on track in many ways. I also found a medical center that, despite that, was still trying to resist some changes, frankly. I recognize that this was not just Mount Sinai.

—George Raptis, M.D., '87

I think that students now are less cynical than they were in the '80s. I think that a couple things have happened. Number one is people realized that they weren't going to make the same kind of money that doctors in the '70s and early '80s made. . . . So, once people saw that wasn't happening and they weren't going to be fabulously wealthy as doctors, two things happened. I think it attracted more people who were generally in it, more as caring individuals that weren't in it thinking about making a lot of money, and secondly, I think that people began to value lifestyle, and it wasn't such a dirty word to talk about having a home life.

Especially one thing that I think affected that was an increase in proportion of women in medical school. I think my class was about 40 percent women. Now, I think it was about four or five years ago, we had our first class that was a majority of women. And when you have people who are going to be raising families, whether the man's a doctor, nowadays with different expectations or if the woman's a doctor and expects to have children and be a physician, or people like myself, I married a physician, there's more of a sense of needing to have emphasized, defined time for home life and for raising kids. . . . Family life is important, and I think that those are changes for the better.

—Anthony Weiss, M.D., '88

I think we started seeing some of the changes . . . that were going on in medicine, how we would have to start treating patients with less testing as the reimbursement for testing would be less, which wasn't necessarily a bad thing and we were always given the training. I was taught as a student that you would only obtain tests to confirm what you could determine from both a history and physical exam; they were only there to confirm what you had if there was any doubt and just a barrage of tests basically to find something that you're not certain what you're looking for. . . . Legally, I think they've started putting a lot of emphasis on documentation and ordering tests that may not necessarily have been required but seem to be the wise thing to do in terms of legally trying to do what was the best thing for the patient.

—Alfio Carroccio, M.D., '96

It's unfortunate that a lot of the patients come in and, for us, it's the day of surgery. . . . They had all the workup and everything done as an outpatient. We're meeting them for the first time, and it's like, "Hi, I'm so and so, and this is my diagnosis." There's less of getting to go down in the emergency room and getting to see somebody—trying to figure out what's going on and putting the story together and getting the studies that you think you need and coming up with the diagnosis on your own. There is still some of that. We're presented with all of the information, and we still can formulate it in our heads and examine them and look at the studies that they've had, but it's different than what's kind of the joy of being a doctor, which is to figure things out from scratch.

—Bethany Goldstein, M.D., '99

I think that Sinai has had a big challenge in trying to be an academic institution without an undergraduate body close by to kind of integrate the academic atmosphere. It's always been an outstanding teaching hospital and research facility, but it's always been very clinically oriented, and they've really taken on a major challenge in creating a graduate school where the science becomes basic and in many regards may not even be biomedical. . . .

—Philip Mulieri, M.D./Ph.D., '03

MENTORS

There was a nephrologist by the name of Marvin Levitt, wonderful man, who would spend hours talking to you about things. . . . It was some of those older professors who really had a tremendous amount of personality, and their teaching was really very good. . . . It was Dr. Berson who encouraged people to ask deeper questions about things.

—Ernst Schaefer, M.D., '72

The people I remember most were the Baders, Richard Bader and Mortimer Bader. They were identical twins. They actually ran a journal called the *Green Journal*, I think. It's still the *American Journal of Medicine*. They were always ready to teach. You know, if you ran into them at 11:00 at night because they had come in to admit a patient, they would sit there and give you a lecture, like, you know, what's the differential diagnosis of a fever of unknown origin. . . . Dorothy Krieger . . . was a really amazing person. She was actually one of the few physician role models . . . that if you were a woman in 1972 you would want to follow. She was impeccably dressed, looked great, wore makeup . . . and she was really, really, really intelligent.

—Terry Maratos-Flier, M.D., '76

Probably the best thing that happened to me in terms of physical diagnosis was I became very interested in liver disease, even as a second-year student, based on the lectures of the great Hans Popper and Fenton Schaffner. And I sought out Fenton as an adviser, and I expressed an interest in going overseas and learning about liver disease. And he actually made an entrée to have me spend three months at the Royal Free Hospital, which is in Hampstead. . . . In those days, the liver unit was absolutely the most famous in the world. It was run by a woman named Sheila Sherlock, who was unforgettable, who was a dynamo and created a magnet for liver disease from physicians and patients throughout the world. And she was absolutely in her prime then, and I spent three months there . . . working closely with British physicians and house staff working on physical diagnosis. . . . And I had a real comfort level, which was fortunate because my first rotation was medicine, which is the toughest, and I was ready. I obviously lacked a lot of experience, but when it came to physical diagnosis and organizing my thoughts, I was much more comfortable than probably most of the other students who were starting medicine.

Hans Popper was just an extraordinary dynamo. He was full of energy. He was a creative thinker. Even then we recognized, and certainly now that I'm in the field, emerging as, I guess one of the leaders in the field now, his legacy in actually creating the specialty of liver diseases is still unforgettable. He is certainly not forgotten in the field, certainly by anybody over the age of forty or forty-five who was alive when he was in his prime, and I, having been a student here—how much better does it get than that? . . . These types of experiences absolutely seal your passion for something.

—Scott Friedman, M.D., '79

When I was a medical student, I had good relationships with the faculty, and I felt they were open. There were some I was very close with. There were people I worked with in the pathology department I'm still very friendly with today, and I've collaborated with on investigations of a variety of sorts over the past ten to fifteen years on multiple occasions. When I was a medical student, I was very interested in medical history, and there was a faculty member also had an interest in that. So we set up a little club, and we used to go out and have, you know, on a Friday, and maybe have a beer, with a group of medical students who shared this interest. And this sort of mentorlike faculty member would take us out and, fortunately, pick up the tab.

—Michael Marin, M.D., '84

In the best of all possible worlds, the mission is great. Students should be humanistic, and students should treat patients with respect and students should not be in classes so much, not be in lectures so much, and they should do a lot of self-teaching and use computer-aided educational devices, and all of these things are great. The problem is now, with the economic climate and cutbacks and the general prevailing pessimism in medicine, I am not sure that there are so many role models for medical students to look toward. There is a shortage of role models in medicine; the students are very confused.

—Jeffrey Mechanick, M.D., '85

I thought I had a great relationship with the faculty, particularly in the graduate school. . . . Dean Krulwich . . . was terrific. You could talk to her about anything, and you knew she was always on your side.

—Jeffrey Parvin, M.D./Ph.D., '89

I was impressed with people like Dr. Barry Salky, who came to our gross anatomy class to volunteer to help us during our dissection in first year. We were told that he was a surgeon who was here to brush up on his skills and was volunteering his time to be part of our education, and that was impressive. Later, I found out that he was one of the finer surgeons, and I appreciated that he worked hard for every step in his career, and he deserves his success.

—Ram Roth, M.D., '92

CLOSING THOUGHTS

There's certainly no greater loyalty, no greater emotional feeling that I have for any institution other than Mount Sinai. I still think back with great fondness, and probably there is hardly a day that goes by that something happens during my professional day that doesn't have its roots in terms of how I respond, how I was trained, how I think, how I approach a problem, that wasn't pretty much formed by the fifteen years I spent at Mount Sinai as a student, house officer, and faculty member.

—Arthur Frank, M.D., Ph.D., '72

This has always been a great place and will continue to be a great place because it has always attracted great people who are deeply committed to the science of medicine. . . . And this experiment, this medical school that we put in place with the notion that it would be necessary to continue to make this hospital great, will in fact prove true. . . . As we look at the next fifty years of Mount Sinai, as the medical school becomes a truly great medical school, it will again reflect the days when Mount Sinai was truly one of the leading places in the world for medical science.

—Kenneth Davis, M.D., '73

Students in the Class of 2003 celebrate Match Day, March 20, 2003. Stephanie Wedl, now M.D., is on the right.

The other thing is that they're New Yorkers, so they're really dynamic and outgoing and they say what they think. . . . It's a different culture. . . . New Yorkers are loud and they are emotional and they speak their mind and they're pushy and they really are all those things. But it was a really dynamic culture, because you had people arguing about things like diagnosis and how to treat, and they were kind of a little less polite and restrained . . . but they were also really good. They made medicine really exciting.

—Terry Maratos-Flier, M.D., '76

You're grounded in this place when you work here. You don't just pass through and punch your time card, figuratively or literally, and I think that that kind of spirit is what can make an institution great.

—Scott Friedman, M.D., '79

Sinai is a very inbred place, where a lot of people are medical students and then they stay on for residency and maybe fellowships, or maybe they go away for fellowships and then they come back. Because I think all of us have a sense of wanting to give back everything that we have been given. For me, that was something that was very important . . . and that's why I've been here so long.

My roots are Sinai. I've never left, even though now I'm not full-time faculty. It's still a part of me, and it's just an enormous place to be, even through all the ups and downs. That's just life. Life is ups and downs. But anyone who is lucky enough to train here in any capacity, it's a great place to be.

—Marla Stern, M.D., '86

APPENDIXES

Appendix A

Saul Horowitz, Jr. Memorial Award Recipients

Year Awarded		MSSM Class
1978	Kenneth Davis	1973
1979	Jeffrey S. Flier	1972
1980	Heidi Weissman nee Seitelbaum	1974
1981	Eric Prystowsky	1973
1982	Jeffrey P. Koplan	1970
1983	Arthur Frank	1972
1984	Mark Swartz	1973
1985	Steven Grant	1973
1986	Lloyd Mayer	1976
1987	Steven B. Heymsfield	1971
1988	John Morihisa	1976
1989	Ernst Schaefer	1972
	Janice L. Gabrilove	1977
1990	David S. Baskin	1978
	Steven S. Thacker	1973
1991	Mark Sobel	1975
	Idelle Weisman	1974
1992	Steven Itzkowitz	1979
1993	Charles Miller	1978
	Scott Friedman	1979
1994	Robert Friedland	1973
	Robert A. Phillips	1980

Year Awarded		MSSM Class
1995	Paul S. Teirstein	1980
1996	Steven P. Roose	1974
1997	David M. Nathan	1975
1998	Stephen A. Kaplan	1982
1999	Maria L. Padilla	1975
2000	Richard G. Stock	1988
2001	Michael L. Marin	1984
2002	Jonathan S. Stamler	1985
2003	Robert S. Shapiro	1984
2004	Lawrence Goldstein	1981

Appendix B

*Mount Sinai School of Medicine
Honorary Degree Recipients*

1972*   Annenberg, Walter
         Castle, William B., M.D.
         Moses, Mrs. Henry L.
         Rees, Mina S., Ph.D.

1975   Klingenstein, Joseph
         Jacob, Francois, M.D., D.Sc.

1976   Bergstrom, Sune, M.D., D. MSc.
         Lehman, Edith A.
         Levy, Gustave L.

1977   Rose, Alfred L.
         Baehr, George, M.D.
         Sherlock, Dame Sheila

1978   Black, James W., M.D.
         Fredrickson, Donald S., M.D.
         Hodgkin, Dorothy Crowfoot, Ph.D.

1979   Coons, Sheldon
         Fishberg, Arthur M., M.D.
         Ingelfinger, Franz J., M.D.
         Lederberg, Joshua, Ph.D.
         Petrie, Milton
         Popper, Hans, Ph.D., M.D.

1980   Ham, Thomas Hale, M.D.
         Hamburg, David A., M.D.
         Vane, Sir John Robert, F.R.S., M.D.
         Hexter, Maurice B., Ph.D.
         Rosengarten, Al
         Javits, The Hon. Jacob K.

1981   Crohn, Burrill, M.D.
        Ginzburg, Leon, M.D.
        Udenfriend, Sidney, M.D.

1982   Kety, Seymour, M.D.
        Kolff, Wilhelm, M.D.
        Hodes, Horace, M.D.
        Pepper, The Hon. Claude

1983   Gross, Ludwik, M.D.
        Hitchings, George, Ph.D., D.Sc.
        Stratton, Henry, Dr. Med. (Hon.)
        Wallen, Arthur (animal laboratory supervisor)

1984   Glaser, Robert, M.D.
        Kirschstein, Ruth, M.D.
        Gross, Alfred

1985   Cuatrecasas, Pedro, M.D.
        Leder, Philip, M.D.
        Turner-Warwick, Dame Margaret
        Turner-Warwick, Richard, D.M., M. Ch.

1986   Fisher, Bernard, M.D.
        Rogers, David E., M.D.
        Burke, James E.
        Carey, William D.
        Sackler, Arthur M., M.D.

1987   Hood, Leroy E., M.D., Ph.D.
        Kandel, Eric R., M.D.
        Rangel, The Hon. Charles B.
        Salk, Jonas E., M.D.

1988   Singer, Maxine Frank, Ph.D.
        Starzl, Thomas E., M.D., Ph.D.
        Califano, The Hon. Joseph A., Jr., LLB
        Laubach, Gerald D., Ph.D.

1989   Shumway, Norman E., M.D., Ph.D.
        Steitz, Joan Argetsinger, Ph.D.
        Thier, Samuel O., M.D.
        Weinberg, Robert A., Ph.D.
        Mahler, Halfdan, M.D.

1990	Baltimore, David, Ph.D. Fauci, Anthony S., M.D. Thomas, Lewis, M.D. Tisch, Wilma S.
1991	Braunwald, Eugene, M.D. Healy, Bernadine, M.D. Relman, Arnold S., M.D. Koshland, Daniel E., Ph.D. Walter, Henry G., Jr.
1992	McKusick, Victor A., M.D. Rosenberg, Leon, M.D. Lederman, Leon M., Ph.D. Mahoney, Margaret E. Perpich, Joseph G., M.D., J.D.
1993	Collins, Francis S., M.D., Ph.D. Hinshaw, Ada Sue, Ph.D. Petersdorf, Robert G., M.D. Tilghman, Shirley M., Ph.D. Bromley, D. Allan, Ph.D. Watkins, Admiral James D.
1994	Blobel, Gunter, M.D., Ph.D. Reinhart, Uwe, Ph.D. Massey, Walter E., Ph.D. Nicholson, Richard S., Ph.D. Wiesel, Elie
1995	Furchgott, Robert F., Ph.D. Moncada, Salvador, M.D., Ph.D., D.Sc. Gibbons, John H., Ph.D. McCarty, Maclyn, M.D. Novello, Antonia, M.D., M.P.H. Rubin, The Hon. Robert E.
1996	Bloom, Floyd, M.D. Folkman, Judah, M.D. Choppin, Purnell W., M.D. Klingenstein, Frederick A.

1997	Levine, Arnold J., Ph.D.
	Lewin, Benjamin, Ph.D.
	Vagelos, P. Roy, M.D.
	Stern, Alfred R.
1998	Alberts, Bruce, Ph.D.
	Kessler, David A., M.D., Ph.D.
	Klausner, Richard D., M.D.
	Steiner, Donald F., M.D.
1999	Satcher, David, M.D., Ph.D.
	Botstein, David, Ph.D.
	Browner, The Hon. Carol M.
	Lane, Neal F., Ph.D.
2000	Golden, William T.
	Peck, Stephen M.
	Porter, The Hon. John Edward
	Hendrickson, Wayne A. , Ph.D.
	Koplan, Jeffrey P., M.D., Ph.D.
2001	Kaplan, Helene L.
	Federman, Daniel D., M.D.
	Lander, Eric S., Ph.D.
	Varmus, Harold, M.D.
2002	Ruttenberg, Derald H.
	Evans, Martin, Ph.D.
	Coller, Barry, M.D.
2003	Aufses, Arthur H., Jr., M.D.
	Kase, Nathan G., M.D.
	Fuchs, Elaine, Ph.D.
	Graybiel, Ann M., Ph.D.

*No honorary degrees given in 1973 and 1974.

Appendix C

The Mount Sinai Leadership

CHAIRMEN* OF THE MOUNT SINAI HOSPITAL BOARD OF TRUSTEES, 1852–

Sampson Simson, 1852–1855

John I. Hart, 1855–1856

Benjamin Nathan, 1856–1870+

Emanuel B. Hart, 1870–1876

Adolph Hallgarten, 1876–1879

Harris Aronson, 1879+

Hyman Blum, 1879–1896

Isaac Wallach, 1896–1907

Isaac Stern, 1907–1910

George Blumenthal, 1911–1938

Leo Arnstein, 1938–1944

Waldemar Kops, 1944–1945+

George B. Bernheim, 1945–1948

Alfred L. Rose, 1948–1956

Joseph Klingenstein, 1956–1962

Gustave L. Levy, 1962–1976+

Alfred R. Stern, 1977–1985

Frederick A. Klingenstein, 1985–1995

Stephen M. Peck, 1995–2002

Peter W. May, 2002–

PRESIDENTS OF THE MOUNT SINAI MEDICAL CENTER, 1965–

George James, M.D., 1965–1972+

Hans Popper, M.D., Ph.D., 1972–1973

Thomas C. Chalmers, M.D., 1973–1983

James F. Glenn, M.D., 1983–1987

John W. Rowe, M.D., 1987–2000

Nathan G. Kase, M.D., 2001–2002 [Interim]

Kenneth I. Berns, M.D., Ph.D., 2002–2003

Kenneth Davis, M.D., 2003–

DEANS OF THE MOUNT SINAI SCHOOL OF MEDICINE, 1965–

George James, M.D., 1965–1972+

Hans Popper, M.D., 1972–1973

Thomas C. Chalmers, M.D., 1973–1983

James F. Glenn, M.D., 1983–1984 [Acting]

Lester B. Salans, M.D., July 1, 1984–September 1, 1984

James F. Glenn, M.D., September 1984–January 1985 [Acting]

Nathan Kase, M.D., 1985–1997

Arthur Rubenstein, M.B.B.Ch., 1997–2001

Nathan Kase, M.D., 2001–2002 [Interim]

Kenneth Davis, M.D., 2003–

PRESIDENTS OF THE MOUNT SINAI HOSPITAL, 1981–

Samuel Davis, 1981–1984

James F. Glenn, M.D., 1984–1987

Nathan Kase, M.D., 1987–1988

John W. Rowe, M.D., 1988–November 1998

Barry Freedman, December 1998–September 2002

Larry H. Hollier, M.D., September 2002–November 2003

Burton P. Drayer, M.D., November 2003–

DIRECTORS** OF THE MOUNT SINAI HOSPITAL, 1855–

Julius Raymond, 1855–1865

M. J. Bergman, 1865

Gabriel Schwarzbaum, 1866–1872

Dr. Treusch, 1872–1875

Leopold B. Simon, 1875–1878+

Theodore Hadel, 1878–1892

Leopold Minzesheimer, 1892–1898+

S. L. Fatman, 1898–1903

S. S. Goldwater, M.D., 1903–1929

Joseph Turner, M.D., 1929–1948

Martin Steinberg, M.D., 1948–1969

S. David Pomrinse, M.D., 1969–1975

Samuel Davis, 1975–1981

Barry Freedman, 1982–1995

Wendy Z. Goldstein, 1995–1998

Jeffrey Menkes, 2000–2002

Richard Celiberti, April 2003–December 2003 (Chief Operating Officer)

Wayne Keathley, December 2003– (Chief Operating Officer)

* Until 1969, this title was President of the Board of Trustees.
+ Died in office
** Until 1917, this position was called Superintendent.

Appendix D

The Mount Sinai Boards of Trustees, 2003

John S. Winkleman

Notes

NOTES TO THE PREFACE

1. A. H. Aufses, Jr. and B. J. Niss, *This House of Noble Deeds: The Mount Sinai Hospital, 1852–2002* (New York: New York University Press, 2002).
2. M. G. Lewis and S. M. Barker, *The Sinai Nurse: A History of Nursing at The Mount Sinai Hospital, New York, New York, 1852–2000* (West Kennebunk, ME: Phoenix Publishing, 2001).

NOTES TO CHAPTER 1

1. K. M. Ludmerer, "The Origins of Mount Sinai School of Medicine," *J Hist Med* 45 (1990): 469–89. This article is based on extensive research in the primary sources.
2. Martin Steinberg oral history interview, INT6, p. 44, The Mount Sinai Archives.
3. Minutes, Joint Committee on Medical Education, March 1958.
4. For more on the development of full-time faculty at Mount Sinai, see chapter 10, on the Faculty Practice Plan.
5. Department of Health, Education, and Welfare, *Physicians for a Growing America* (Bane Report), October 1959.
6. Negotiations with Columbia ended in 1965 at Columbia's behest. Minutes of the Board of Trustees, Mount Sinai School of Medicine, May 17, 1965, The Mount Sinai Archives.
7. L. Block, *Let us not sponsor a medical school*, 1961.
8. Ibid, p. 8.
9. Ibid., p. 10.
10. Ibid., p. 16.
11. Ibid., p. 16.
12. *Dedication Program, The Esther and Joseph Klingenstein Clinical Center of The Mount Sinai Hospital of New York, November 14, 1962*, The Mount Sinai Archives.
13. Aufses and Niss, *This House of Noble Deeds*, pp. 294–96; P. D. Berk, F. Schaffner, and R. Schmid, *Hans Popper: A Tribute* (New York: Raven Press, 1992), pp. 2, 7–8.

14. H. Popper, "Gustave L. Levy and the development of the Medical School," *Mt Sinai J Med* 44 (1977): 585–93.

15. December 14, 1961, report on research by W. B. Castle, M.D., and W. B. Wood, Jr., M.D., found in Mount Sinai School of Medicine, Early Papers, "Advisory Committee" file, b. 2, f. 2, The Mount Sinai Archives.

16. See Aufses and Niss, *This House of Noble Deeds*, pp. 26–32.

17. G. James, "Mount Sinai School of Medicine of The City University of New York," in *Case Histories of Ten New Medical Schools*, ed. V. W. Lippard and E. F. Purcell (New York: The Josiah Macy Jr. Foundation, 1972), p. 221.

18. "Taking the Education Challenge . . . ," *Medical Tribune*, October 10–11, 1964, p. 21.

19. G. James in Lippard and Purcell, *Case Histories*, p. 220.

20. "Report to the Liaison Committee," September 9, 1964, p. 3, Mount Sinai School of Medicine, Early Papers, "LCME" file, b.6, f.1, The Mount Sinai Archives.

21. Ibid., p. 2.

22. For a full discussion of the development and evolution of the curriculum, see chapter 2, on the curriculum.

23. H. Popper, "The Mount Sinai Concept," *Clinical Research* 13 (1965): 503–4.

24. In an interview in December 1987, Hans Popper noted that the community medicine arm of the tripod did surprisingly well in reaching out to the community and in instilling the concept of humanities and community service into the student body. At that point, he actually believed that this factor had succeeded beyond hopes, while the other two legs of research and patient care, the ones he considered most important, had lagged. INT33, Interview with Hans Popper, December 22, 1987, The Mount Sinai Archives.

25. Minutes of the Advisory Committee meeting, March 22, 1963. See also subsequent minutes for discussion of the Deanship. Mount Sinai School of Medicine, Early Papers, "Dean Selection" file, b.3, f.5, The Mount Sinai Archives.

26. R. K. H. Kinne and P. Eggena, "Basic Sciences and Medicine in the Career of Irving L. Schwartz," in *Biology and Medicine into the 21st Century: Issues in Biomedicine*, ed. M. A. Hardy and R. K. H. Kinne (Basel: Karger, 1991), vol. 15, p. 4. See also chapter 4, on the basic sciences, for more on Schwartz's scientific career.

27. Interview with Irving L. Schwartz, M.D., April 10, 2003, The Mount Sinai Archives. See also INT91, Interview with Reba Nosoff, January 23, 2003, The Mount Sinai Archives.

28. Minutes of the Board of Trustees, Mount Sinai School of Medicine, January 18, 1965, The Mount Sinai Archives. See also chapter 3, on the Graduate

School of Biological Sciences, and chapter 4, on the basic sciences, for more on Schwartz's role in these positions.

29. Nosoff interview.

30. Kinne and Eggena, "Basic Sciences," p. 5.

31. Mount Sinai School of Medicine. Early Papers, "Dean Selection" file, b.3, f.5, The Mount Sinai Archives.

32. See chapter 10, on the *Faculty Practice Plan*, for more on Sisselman.

33. Letter from George James to Gustave Levy, September 10, 1965. Appendix IV to "Material for discussion of the Medical School Board building program by the Board of Trustees at their meeting on Tuesday, October 12, 1965." Mount Sinai School of Medicine. Board of Trustees. Minutes, 1965 folder, The Mount Sinai Archives.

34. Mount Sinai School of Medicine Board of Trustees. Minutes, March 27, 1967, The Mount Sinai Archives.

35. *The Mount Sinai HospitalAnnual Report, 1967*, p. 78, The Mount Sinai Archives.

36. Ibid., p. 67.

37. See chapter 2, on the Curriculum, for more on the content.

38. Interview with Arthur Frank, M.D., Ph.D., Class of 1972, 2002, The Mount Sinai Archives.

39. Interview with Kenneth Davis, M.D., Class of 1973, 2002, The Mount Sinai Archives.

40. Interview with Marlene Marko, M.D., Class of 1972, 2002, The Mount Sinai Archives.

41. For student reminiscences about these meetings and achieving representation on the committees, see chapter 12.

42. For more information on these individuals and their departments, see chapter 4, on the Basic Sciences.

43. For more on Berson and his pioneering work on the development of radioimmunoassay, see Aufses and Niss, *This House of Noble Deeds,* pp. 36–39.

44. Interview with Peter Lang, M.D., Class of 1972, 2002, The Mount Sinai Archives.

45. Interview with Scott Friedman, M.D., Class of 1979, 2002, The Mount Sinai Archives.

46. See chapter 10, on the *Faculty Practice Plan*.

47. "Meta-analysis in the breech," *Science* 249 (1990): 478.

48. Obituary of Thomas C. Chalmers, M.D., *New York Times*, Dec. 28, 1995, p. A31.

49. Interview with Arthur Frank, M.D., Ph.D., 2002, The Mount Sinai Archives.

50. Edra Spilman memo to Thomas C. Chalmers, "Take-Over of the Last Floor in Annenberg 10 to 26," February 14, 1975, from President Office Files (Chalmers), "BLDG4: Annenberg, 1976," file, The Mount Sinai Archives.

51. For more on Stimmel's role, see chapter 2, on the curriculum, and chapter 9, on Graduate and Postgraduate Education.

52. See Aufses and Niss, *This House of Noble Deeds*, chapter 19, pp. 205–17.

53. The Mount Sinai Hospital, *Annual Report, 1989*, p. 49.

54. P. J. Anderson and C. J. Stapleton, "Planning for Change at Mount Sinai," June 1, 1977, p. 1, The Mount Sinai Archives.

55. Ibid.

56. C. J. Stapleton and T. C. Chalmers, "The Most Distinguished and Least Preferred Private Academic Medical Centers: A Comparative Report" (New York: Mount Sinai Medical Center, 1980). The Mount Sinai Archives.

57. See INT79, interview with S. David Pomrinse, M.D., December 11, 1998, The Mount Sinai Archives, for information on this gift.

58. "Overview and Recommendations for Management/Organization," prepared by the Subcommittee of the Planning Committee of the Board of Trustees on Management/Organization, November 16, 1981, p.1, in Office of the President, MSMC (Glenn) Records, b. 11, f.3, The Mount Sinai Archives.

59. Ibid, p. 6.

60. For more on Kase, see Aufses and Niss, *This House of Noble Deeds*, pp. 248–50.

61. B. Stimmel, "The educational process at the Mount Sinai School of Medicine: The first 25 years," *Mt Sinai J Med* 56 (1989): 381–82.

62. Memo from Lynn Kasner Morgan to Dean Kase, September 12, 1988, Dean's files, The Mount Sinai Archives.

63. See special theme issue of *Academic Medicine* 78 (2003): 429–30, devoted to efforts to recruit underrepresented minorities into medicine.

64. M. R. Rifkin, K. D. Smith, B. D. Stimmel, et al., "The Mount Sinai Humanities and Medicine Program: an alternative pathway to medical school," *Academic Medicine* 75 (October Supplement) (2000): S124–26.

65. B. Stimmel and M. Serber, "The role of curriculum in influencing students to select generalist training: a 21-year longitudinal study," *J Urban Health* 76 (1999): 117–26.

66. See chapter 9, on Graduate and Postgraduate Education, for more on this.

67. See chapter 2, on the Curriculum, and chapter 8, on the Department of Health Policy, for more on the Department of Nursing's role in education and research.

68. E-mail from Nathan Kase to Arthur Aufses, June 10, 2002. Personal communication.

69. "NIH Support to U.S. Medical Schools, FYs 1970–2000—The Mount Sinai School of Medicine of NYU, New York, New York," March 14, 2001,

available at http://grants1.nih.gov/grants/award/trends/medsup7000.txt, accessed April 25, 2003.

70. "Mount Sinai-New York University" clippings file, The Mount Sinai Archives. See especially *Crain's Health Pulse*, January 9, 1998.

71. See also chapter 2, on the Curriculum.

NOTES TO CHAPTER 2

1. H. Popper, "A new curriculum," *Ann NY Acad Sci* 128 (1965): 552–60.

2. H. Popper, "The Mount Sinai concept," *Clin Res* 13 (1965): 500–4.

3. Popper, "A new curriculum."

4. G. James, "Mount Sinai School of Medicine of the City University of New York," in *Case Histories of Ten New Medical Schools*, ed. V. W. Lippard and E. F. Purcell (New York: Josiah Macy Jr. Foundation, 1972), pp. 209–39.

5. T. Barka, G. Christakis, H. L. Gadboys, et al., "Proposed Curriculum for the Mount Sinai School of Medicine," *J Mt Sinai Hosp* 34 (1967): 366–77.

6. Curriculum proposal, January 25, 1967, p. 5. Files of the Curriculum Committee, The Mount Sinai Archives.

7. Minutes of Curriculum Committee, April 14, 1967, The Mount Sinai Archives.

8. A. H. Aufses, Jr., and B. Dana, "Introduction to Medicine: A Developmental History and Analysis of an Interdepartmental Course," in *The Changing Medical Curriculum*, ed. V. W. Lippard and E. F. Purcell (New York: Josiah Macy Jr. Foundation, 1972), pp. 117–29.

9. See chapter 6, on the Department of Community and Preventive Medicine, for more on Dana.

10. Aufses returned to Mount Sinai in 1974 as Chairman of the Department of Surgery, a position he then held for twenty-two years. See A. H. Aufses, Jr., and B. Niss, *This House of Noble Deeds: The Mount Sinai Hospital, 1852–2002* (New York: New York University Press, 2002), for more on Aufses.

11. Minutes of Curriculum Committee, April 14, 1967, The Mount Sinai Archives.

12. Ibid.

13. Mount Sinai School of Medicine *Student Bulletin*, 1968–1969, p. 43, The Mount Sinai Archives.

14. A. Frank, "MSSM: the first year," *Spectrum* (a publication of the Mount Sinai Alumni), no. 8 (Spring 1978): 15–16.

15. B. Stimmel, "To: Curriculum Retreat Participants," *Curriculum Retreat Workbook*, November 7, 1978, The Mount Sinai Archives.

16. B. Stimmel, "The Curriculum," *Data Book*, Curriculum Retreat, November 25, 1980, p. 63, The Mount Sinai Archives.

17. Ibid.

18. See also chapter 19, on the Department of Geriatrics and Adult Development, in Aufses and Niss, *This House of Noble Deeds*, pp. 205–17.

19. B. D. Stimmel, "The educational process at the Mount Sinai School of Medicine: the first 25 years," *Mt Sinai J Med* 56 (1989): 375–82.

20. Ibid., p. 380.

21. Mount Sinai School of Medicine *Student Bulletin*, 1995–1996, p. 17, The Mount Sinai Archives.

22. M. H. Swartz and J. A. Colliver, "Using standardized patients for assessing clinical performance: an overview," *Mt Sinai J Med* 63 (1996): 241–49.

23. See also chapter 9, on Graduate Medical Education.

24. LCME committee II, *Final Report on the Educational Program for the MD Degree,* Mount Sinai School of Medicine, 2003, Office of the Dean of Medical Education.

25. A. Stagnaro-Green and L. Smith, "Mount Sinai School of Medicine," *Acad Med* 75 (2000): S239–40.

NOTES TO CHAPTER 3

1. S. Kupfer and T. A. Krulwich, "The Graduate School of Biological Sciences: The first twenty-five years," *Mt Sinai J Med* 56 (1989): 383–91.

2. See chapter 1, on the History of the School, for more on Schwartz and his administrative roles. See chapter 4, on the Basic Sciences, for more on Schwartz's work in science.

3. "Dr. Irving L. Schwartz: Dean for Graduate Studies," *The Mount Sinai Hospital News*, Science insert, July 1965, pp. 2–3, The Mount Sinai Archives.

4. I. L. Schwartz, "Graduate education and medical education: a synergism," *J Mt Sinai Hosp* 34 (1967): 242–48.

5. Available at http://www.mssm.edu/gradschool.

6. T. A. Krulwich and P. J. Friedman, "Integrity in the education of researchers," *Acad Med* 68 (1993): S14–18.

7. The Careers and Professional Activities of Graduates of the NIGMS Medical Scientist Training Program, September 1998. (September 9, 1998), NIGMS, NIH, Bethesda, Maryland. Available at www.nigms.nih.gov/news/reports/mstpstudy/mstpstudy.html.

8. L. E. Rosenberg, "Physician-scientists—endangered and essential," *Science* 283 (1999): 331–32.

9. T. R. Zemlo, H. H. Garrison, N. C. Partridge, and T. J. Ley, "The physician-scientist: career issues and challenges at the year 2000," *FASEB J* 14 (2000): 221–30.

NOTES TO CHAPTER 4

1. Minutes of the Medical Board, The Mount Sinai Hospital, January, 1934, The Mount Sinai Archives.

2. See chapter 7, on the Department of Human Genetics.

3. See chapter 1, on the History of the School.

4. It is impossible in this book to list all the publications of the faculty. We have cited those that we believe are the most important papers published since the School opened its doors in 1968. We have also cited those papers published prior to a faculty member's coming to Mount Sinai when we believed that they were of major import. A listing of every faculty member's peer-reviewed publications since 1966 can be found in PubMed on the website of the National Library of Medicine, a branch of the National Center for Biotechnology Information. The address is http://www.ncbi.nlm.nih.gov/entrez/query.fcgi.

5. A. H. Aufses, Jr. and B. Niss, *This House of Noble Deeds: The Mount Sinai Hospital, 1852–2002* (New York: New York University Press, 2002), p. 13.

6. T. Barka and P. J. Anderson, *Histochemistry: Theory, Practice and Bibliography* (New York: P. B. Hoeber, 1963).

7. T. Barka and P. J. Anderson, "Histochemical methods for acid phosphatase using hexazonium pararosanilin as coupler," *J Histochem Cytochem* 10 (1962): 741–53. Citation Classics. *Current Contents*, "The articles most cited in 1961–1982," no. 8 (February 2, 1978): 7.

8. T. Barka, "Induced cell proliferation: the effect of isoproterenol," *Exp Cell Res* 37 (1965): 662–79.

9. M. Levitan and A. Montagu, *Textbook of Human Genetics* (New York: Oxford University Press, 1971).

10. M. Levitan, "Spontaneous chromosome aberrations in *Drosophila robusta*," *Proc Natl Acad Sci U S A* 48 (1962): 930–37.

11. M. Levitan, "Chromosomal breaks with attachments to the nucleolus," *Chromosoma* 31 (1970): 452–67.

12. M. Levitan and M. Verdonck, "Twenty-five years of a unique chromosome-breakage system. I. Principal features and comparison to other sytems," *Mutat Res* 161 (1986): 135–42.

13. M. Levitan and S. J. Scheffer, "Studies of linkage in populations. X. Altitude and autosomal gene arrangements in *Drosophila robusta*," *Genet Res* 61 (1993): 9–20.

14. M. Levitan, "Suppressor genes with gender differences in activity in natural populations of *Drosophila robusta*: another approach to wild-type," *Genet Res* 69 (1997): 81–88.

15. M. Levitan, "Studies of linkage in populations. XIV. Historical changes in frequencies of gene arrangements and arrangement combinations in natural populations of *Drosophila robusta*," *Evolution Int J Org Evolution* 55 (2001): 2359–62.

16. See chapter 2, on the curriculum, for more on Shriver.

17. T. Barka, "Anatomy," The Mount Sinai School of Medicine, *Annual Report for 1969*, p. 105.

18. T. Barka, R. M. Gubits, and H. M. van der Noen, "B-adrenergic stimulation of c-fos gene expression in the mouse submandibular gland," *Mol Cell Biol* 6 (1986): 2984–89.

19. T. Barka and H. M. van der Noen, "Retrovirus-mediated gene transfer into salivary glands in vivo," *Human Gene Ther* 7 (1996): 613–18.

20. P. A. Shaw, J. L. Cox, T. Barka, and Y. Naito, "Cloning and sequencing of cDNA encoding a rat salivary cysteine proteinase inhibitor inducible by beta-adrenergic agonists," *J Biol Chem* 263 (1988): 18133–37.

21. P. A. Shaw and T. Barka, "Beta-adrenergic induction of a cysteine-protease-inhibitor messenger RNA in rat salivary glands," *Biochem J* 257 (1989): 685–89.

22. P. A. Shaw, T. Barka, A. Woodin, et al., "Expression and induction of beta-adrenergic agonists of the cystatin S gene in submandibular glands of developing rats," *Biochem J* 265 (1990): 115–20.

23. P. A. Shaw and W. H. Yu, "Sympathetic and parasympathetic regulation of cystatin S gene expression," *Life Sci* 70 (2001): 301–13.

24. P. A. Shaw, X. Zhang, A. F. Russo, et al., "Homeobox protein, Hmx3, in postnatally developing rat submandibular glands," *J Histochem Cytochem* 51 (2003): 385–96.

25. The Mount Sinai Medical Center, *Annual Report for 1989,* p. 50.

26. G. M. Small, M. J. Santos, T. Imanaka, et al., "Peroxisomal integral membrane proteins in livers of patients with Zellweger syndrome, infantile Refsum's disease and X-linked adrenoleukodystrophy," *J Inherit Metab Dis* 11 (1988): 358–71.

27. P. E. Purdue, J. W. Zhang, M. Skoneczny, and P. B. Lazarow, "Rhizomelic chondrodysplasia punctata is caused by deficiency of human PEX7, a homologue of the yeast PTS2 receptor," *Nat Genet* 15 (1997): 381–84.

28. For a complete review of Lazarow's work on peroxisomes, see P. E. Purdue and P. B. Lazarow, "Peroxisome biogenesis," *Annu Rev Cell Dev Biol* 17 (2001): 701–52, and P. B. Lazarow, "Peroxisome biogenesis: advances and conundrums," *Curr Opin Cell Biol* 15 (2003): 489–97.

29. M. LeDizet and G. Piperno, "The light chain p28 associates with a subset of inner dynein arm heavy chains in *Chlamydomonas axonemes*," *Mol Biol Cell* 6 (1995): 697–711.

30. G. Piperno and K. Mead, "Transport of a novel complex in the cytoplasmic matrix of *Chlamydomonas flagella*," *Proc Natl Acad Sci U S A* 94 (1997): 4457–62.

31. G. Piperno, E. Siuda, S. Henderson, et al., "Distinct mutants of retro-

grade intraflagellar transport (IFT) share similar morphological and molecular defects," *J Cell Biol* 143 (1998): 1591–601.

32. C. Iomini, V. Babaev-Khaimov, M. Sassaroli, and G. Piperno, "Protein particles in *Chlamydomonas flagella* undergo a transport cycle consisting of four phases," *J Cell Biol* 153 (2001): 13–24.

33. L. Ossowski, "Invasion of connective tissue by human carcinoma cell lines: requirement for urokinase, urokinase receptor, and interstitial collagenase," *Cancer Res* 52 (1992): 6754–60.

34. N. Busso, S. K. Masur, D. Lazega, et al., "Induction of cell migration by pro-urokinase binding to its receptor: possible mechanism for signal transduction in human epithelial cells," *J Cell Biol* 126 (1994): 259–70.

35. W. Yu, J. Kim, and L. Ossowski, "Reduction in surface urokinase receptor forces malignant cells into a protracted state of dormancy," *J Cell Biol* 137 (1997): 767–77.

36. L. Ossowski and J. A. Aguirre-Ghiso, "Urokinase receptor and integrin partnership: coordination of signaling for cell adhesion, migration and growth," *Curr Opin Cell Biol* 12 (2000): 613–20.

37. J. T. Laitman and J. S. Reidenberg, "Specializations of the human upper respiratory and upper digestive systems as seen through comparative and developmental anatomy," *Dysphagia* 8 (1993): 318–25.

38. J. S. Reidenberg and J. T. Laitman, "Position of the larynx in odontoceti (toothed whales)," *Anat Rec* 218 (1987): 98–106.

39. J. S. Reidenberg and J. T. Laitman, "Existence of vocal folds in the larynx of odontoceti (toothed whales)," *Anat Rec* 221 (1988): 884–91.

40. J. S. Reidenberg and J. T. Laitman, "Anatomy of the hyoid apparatus in Odontoceti (toothed whales): specializations of their skeleton and musculature compared with those of terrestrial mammals," *Anat Rec* 240 (1994): 598–624.

41. J. T. Laitman and J. S. Reidenberg, "The human aerodigestive tract and gastroesophageal reflux: an evolutionary perspective," *Am J Med* 103 (5A) (1997): 2S–8S.

42. J. T. Laitman, J. S. Reidenberg, S. Marquez, and P. J. Gannon, "What the nose knows: new understandings of Neanderthal upper respiratory tract specializations," *Proc Natl Acad Sci U S A* 93 (1996): 10543–45.

43. J. T. Laitman and I. Tattersall, "*Homo erectus newyorkensis*: An Indonesian fossil rediscovered in Manhattan sheds light on the middle phase of human evolution," *Anat Rec* 262 (2001): 341–43.

44. J. S. Reidenberg and J. T. Laitman, "The new face of gross anatomy," *Anat Rec* 269 (2002): 81–88.

45. American Association of Anatomists Web page, available at http://www.anatomy.org. Accessed April 2003.

46. M. Y. McGinnis and D. F. Kahn. "Inhibition of male sexual behavior by intracranial implants of the protein synthesis inhibitor anisomycin into the medial preoptic area of the rat," *Horm Behav* 31 (1997): 15–23.

47. W. H. Yu and M. Y. McGinnis, "Androgen receptors in cranial nerve motor nuclei of male and female rats," *J Neurobiol* 46 (2001): 1–10.

48. M. Y. McGinnis and R. M. Dreifuss, "Evidence for a role of testosterone-androgen receptor interactions in mediating masculine sexual behavior in male rats," *Endocrinology* 124 (1989): 618–26.

49. M. E. Vagell and M. Y. McGinnis, "The role of gonadal steroid receptor activation in the restoration of sociosexual behavior in adult male rats," *Horm Behav* 33 (1998): 163–79.

50. M. E. Breuer, M. Y. McGinnis, A. R. Lumia, and B. P. Possidente, "Aggression in male rats receiving anabolic androgenic steroids: effects of social and environmental provocation," *Horm Behav* 40 (2001): 409–18.

51. A. M. Fannon and D. R. Colman, "A model for central synaptic junctional complex formation based on the differential adhesive specificities of the cadherins," *Neuron* 17 (1996): 423–34.

52. D. R. Colman, "Neurites, synapses, and cadherins reconciled," *Mol Cell Neurosci* 10 (1997): 1–6.

53. J. P. Doyle, J. G. Stempak, P. Cowin, et al., "Protein zero, a nervous system adhesion molecule, triggers epithelial reversion in host carcinoma cells," *J Cell Biol* 131 (1995): 465–82.

54. See chapter 5, on the Centers and Institutes, section on Neurobiology, for more of Colman's publications.

55. K. M. Mak, M. A. Leo, and C. S. Lieber, "Alcoholic liver injury in baboons: transformation of lipocytes to transitional cells," *Gastroenterology* 87 (1984): 188–200.

56. K. M. Mak and C. S. Lieber, "Portal fibroblasts and myofibroblasts in baboons after long-term alcohol consumption," *Arch Path Lab Med* 110 (1986): 513–16.

57. P. S. Haber, R. T. Gentry, K. M. Mak, et al., "Metabolism of alcohol by human gastric cells: relation to first-pass metabolism," *Gastroenterology* 111 (1996): 863–70.

58. L. J. Mi, K. M. Mak, and C. S. Lieber, "Attenuation of alcohol-induced apoptosis of hepatocytes in rat livers by polyenylphosphatidylcholine (PPC)," *Alcohol Clin Exp Res* 24 (2000): 207–12.

59. See chapter 1, on the History of the School, and chapter 3, on the Graduate School of Biological Sciences, for more on Schwartz.

60. The Mount Sinai Medical Center, *Annual Report for 1968*, p. 121, The Mount Sinai Archives.

61. Ibid.

62. I. L. Schwartz, E. P. Cronkite, H. A. Johnson, et al., "Radioautographic localization of the 'satiety center,'" *Trans Asso Am Phys* 74 (1961): 300–17.

63. A listing of all of the pertinent papers by this group relating to the neurohypophyseal hormones is beyond the scope of this work. For a review and a listing of the most important papers see H. R. Wyssbrod, W. R. Laws, and R. B. Merrifield, "Peptide Biology: Progress and Prospects," in *Biology and Medicine into the 21st Century. Issues in Biomedicine*, ed. M. A. Hardy and R. K. H. Kinne (Basel: Karger, 1991), vol. 15, pp. 107–16. Wyssbrod was a member of the Department; Laws was a member of the Department of Biochemistry.

64. L. F. Johnson, I. L. Schwartz, and R. Walter, "Oxytocin and neuro-hypophyseal peptides: spectral assignment and conformational analysis by 220 MHz nuclear magnetic resonance," *Proc Natl Acad Sci U S A* 64 (1969): 1269–75.

65. See R. K. H. Kinne and P. Eggena, "Basic sciences and medicine in the career of Irving L. Schwartz," in Hardy and Kinne, *Biology and Medicine*, vol. 15, pp. 1–9, for more information on this work and additional citations.

66. H. Burlington, E. P. Cronkite, U. Reincke, and E. D. Zanjani, "Erythro-poietin production in cultures of goat renal glomeruli," *Proc Natl Acad Sci U S A* 69 (1972): 3547–50.

67. H. Burlington and E. P. Cronkite, "Characteristics of cell cultures derived from renal glomeruli," *Proc Soc Exp Biol Med* 142 (1973): 143–49.

68. E. P. Cronkite, H. Burlington, A. D. Chanana, and D. D. Joel, "Regula-tion of granulopoiesis," *Prog Clin Biol Res* 184 (1985): 129–44.

69. T. P. Schilb and W. A. Brodsky, "Transient acceleration of transmural water flow by inhibition of sodium transport in turtle bladders," *Am J Physiol* 219 (1970): 590–96.

70. W. A. Brodsky and G. Ehrenspeck, "The localization of ion-selective pumps and paths in the plasma membranes of turtle bladders," *Adv Exp Med Biol* 84 (1977): 41–66.

71. W. A. Brodsky, J. H. Durham, and G. Ehrenspeck, "Bicarbonate and chloride transport in relation to the acidification or alkalinization of the urine," *Ann N Y Acad Sci* 341 (1980): 210–24.

72. W. A. Brodsky, "Acid and alkali secretion by the turtle urinary bladder. A model system for neurohormonal regulation of acid-base homeostasis," *Ann N Y Acad Sci* 574 (1989): 463–79.

73. J. S. Eisenman, H. M. Edinger, J. L. Barker, and D. O. Carpenter, "Neu-ronal thermosensitivity," *Science* 172 (1971): 1360–62.

74. J. S. Eisenman, "Depression of preoptic thermosensitivity by bacterial pyrogen in rabbits," *Am J Physiol* 227 (1974): 1067–73.

75. J. S. Eisenman, "Sensory organs and thermogenesis," *Isr J Med Sci* 12 (1976): 916–23.

76. P. Eggena, I. L. Schwartz, and R. Walter, "Effects of neurohypophyseal hormones, theophylline and nucleotides on the smooth muscle of the toad bladder," *Life Sci* 7 (1968): 979–88.

77. R. Walter, B. M. Dubois, P. Eggena, and I. L. Schwartz, "Comparison of the mode of action of oxytocin and lysine-vasopressin on the isolated rat uterus," *Experientia* 25 (1969): 33–34.

78. P. Eggena, "Vasopressin resistance of toad urinary bladder: *in vivo* and *in vitro* studies," *Endocrinology* 108 (1981): 1125–31.

79. P. Eggena, "Disk method for measuring effects of neurohypohyseal hormones on urea permeability of toad bladder," *Am J Physiol* 250 (1986): E31–34.

80. P. Eggena, *The Physiological Basis of Primary Care,* 2nd ed. (Carmel, N.Y.: Novateur Medmedia, LLC, 2002).

81. See the section on Pharmacology for publications by Weinstein while a member of that department.

82. Interview with Harel Weinstein, Ph.D., by Arthur Aufses, M.D., April 17, 2003, The Mount Sinai Archives.

83. D. Zhang and H. Weinstein, "Signal transduction by a 5-HT2 receptor: a mechanistic hypothesis from molecular dynamics simulations of the three-dimensional model of the receptor complexed to ligands," *J Med Chem* 36 (1993): 934–38.

84. N. Almaula, B. J. Ebersole, D. Zhang, et al., "Mapping the binding site pocket of the serotonin 5-Hydroxytryptamine2A receptor. Ser3.36(159) provides a second interaction site for the protonated amine of serotonin but not of lysergic acid diethylamide or bufotenin," *J Biol Chem* 271 (1996): 14672–75.

85. B. J. Ebersole, I. Visiers, H. Weinstein, and S. C. Sealfon, "Molecular basis of partial agonism: orientation of indoleamine ligands in the binding pocket of the human serotonin 5-HT2A receptor determines relative efficacy," *Mol Pharmacol* 63 (2003): 36–43.

86. M. Filizola, O. Olmea, and H. Weinstein, "Prediction of heterodimerization interfaces of G-protein coupled receptors with a new subtractive correlated mutation method," *Protein Eng* 15 (2002): 881–85.

87. J. Sun, H. Viadiu, A. K. Aggarwal, and H. Weinstein, "Energetic and structural considerations for the mechanism of protein sliding along DNA in the nonspecific BamHI-DNA complex," *Biophys J* 84 (2003): 3317–25.

88. I. Visiers, J. A. Ballesteros, and H. Weinstein, "Three-dimensional representations of G protein-coupled receptor structures and mechanisms," *Methods Enzymol* 343 (2002): 329–71.

89. From the website of the Department of Physiology and Biophysics, http://www.mssm.edu/physbio/research/shtml. Accessed May 1, 2003.

90. S. K. McLaughlin, P. J. McKinnon, and R. F. Margolskee, "Gustducin is

a taste-cell-specific G protein closely related to the transducins," *Nature* 357 (1992): 563–69.

91. G. T. Wong, K. S. Gannon, and R. F. Margolskee, "Transduction of bitter and sweet taste by gustducin," *Nature* 381 (1996): 796–800.

92. L. Huang, Y. G. Shanker, J Dubauskaite, et al., "Ggamma 13 colocalizes with gustducin in taste receptor cells and mediates IP3 responses to bitter denatonium," *Nat Neurosci* 2 (1999): 1055–62.

93. C. A. Perez, L. Huang, M. Rong, et al., "A transient receptor potential channel expressed in taste receptor cells," *Nat Neurosci* 5 (2002): 1169–76.

94. S. Damak, M. Rong, K. Yasumatsu, et al., "Detection of sweet and umami taste in the absence of taste receptor T1r3," *Science* 301 (2003): 850–53.

95. M. Max, Y. G. Shanker, L. Huang, et al., "Tas1r3, encoding a new candidate taste receptor, is allelic to the sweet responsiveness locus Sac," *Nat Genet* 28 (2001): 58–63.

96. D. V. Smith and R. F. Margolskee, "Making sense of taste," *Sci Am* 284 (2001): 32–39.

97. M. H. Milekic and C. M. Alberini, "Temporally graded requirement for protein synthesis following memory reactivation," *Neuron* 36 (2002): 521–25.

98. C. L. Taylor Clelland, L. Craciun, C. Bancroft, and T. Lufkin, "Mapping and developmental expression analysis of the WD-repeat gene *Preb*," *Genomics* 63 (2000): 391–99.

99. M. A. Kirchberger, M. Tada, D. I. Repke, and A. M. Katz, "Cyclic adenosine 3′,5′-monophosphate-dependent protein kinase stimulation of calcium uptake by canine cardiac microsomes," *J Mol Cell Cardiol* 4 (1972): 673–80.

100. M. A. Kirchberger, M. Tada, and A. M. Katz, "Adenosine 3′,5′-monophosphate-dependent protein kinase-catalyzed phosphorylation reaction and its relationship to calcium transport in cardiac sarcoplasmic reticulum," *J Biol Chem* 249 (1974): 6166–73.

101. M. Tada, M. A. Kirchberger, D. I. Repke, and A. M. Katz, "The stimulation of calcium transport in cardiac sarcoplasmic reticulum by adenosine 3′:5′-monophosphate-dependent protein kinase," *J Biol Chem* 249 (1974): 6174–80.

102. K. R. Weiss, V. Brezina, E. C. Cropper, et al., "Physiology and biochemistry of peptidergic cotransmission in Aplysia," *J Physiol Paris* 87 (1993): 141–51.

103. J. Heierhorst, B. Kobe, S. C. Feil, et al., "Ca2+/S100 regulation of giant protein kinases," *Nature* 380 (1996): 636–39.

104. V. Brezina, I. V. Orekhova, and K. R. Weiss, "Functional uncoupling of linked neurotransmitter effects by combinatorial convergence," *Science* 273 (1996): 806–10.

105. F. S. Vilim, E. C. Cropper, D. A. Price, et al., "Release of peptide cotransmitters in Aplysia: regulation and functional implications," *J Neurosci* 16 (1996): 8105–14.

106. V. Brezina, I. V. Orekhova,and K. R. Weiss, "Control of time-dependent biological processes by temporally patterned input," *Proc Natl Acad Sci U S A* 94 (1997): 10444–49.

107. J. Jing and K. R. Weiss, "Interneuronal basis of the generation of related but distinct motor programs in Aplysia: implications for current neuronal models of vertebrate intralimb coordination," *J Neurosci* 22 (2002): 6228–38.

108. H. Y. Koh, F. S. Vilim, J. Jing, and K. R. Weiss, "Two neuropeptides colocalized in a command-like neuron use distinct mechanisms to enhance its fast synaptic connection," *J Neurophysiol* 90 (2003): 2074–79.

109. A. Proekt and K. R. Weiss, "Convergent mechanisms mediate preparatory states and repetition priming in the feeding network of Aplysia," *J Neurosci* 23 (2003): 4029–33.

110. A. K. Aggarwal and D. A. Wah, "Novel site-specific DNA endonucleases," *Curr Opin Struct Biol* 8 (1998): 19–25.

111. C. Dhalluin, J. E. Carlson, L. Zeng, et al., "Structure and ligand of a histone acetyltransferase bromodomain," *Nature* 399 (1999): 491–96.

112. C. M. Lukacs, R. Kucera, I. Schildkraut, and A. K. Aggarwal, "Understanding the immutability of restriction enzymes: crystal structure of BglII and its DNA substrate at 1.5 A resolution," *Nat Struct Biol* 7 (2000): 134–40.

113. Z. Topcu, D. L. Mack, R. A. Hromas, and K. L. Borden, "The promyelocytic leukemia protein PML interacts with the proline-rich homeodomain protein PRH: a RING may link hematopoiesis and growth control," *Oncogene* 18 (1999): 7091–100.

114. A. Kentsis, R. E. Gordon, and K. L. Borden, "Control of biochemical reactions through supramolecular RING domain self-assembly," *Proc Natl Acad Sci U S A* 99 (2002): 15404–49.

115. L. Zeng and M. M. Zhou, "Bromodomain: an acetyl-lysine binding domain," *FEBS Lett* 513 (2002): 124–28.

116. L. Zeng, C. H. Chen, M. Muller, and M. M. Zhou, "Structure-based rational design of chemical ligands for AMPA-subtype glutamate receptors," *J Mol Neurosci* 20 (2003): 345–48.

117. L. Laakkonen, W. Li, J. H. Perlman, et al., "Restricted analogues provide evidence of a biologically active conformation of thyrotropin-releasing hormone," *Mol Pharmacol* 49 (1996): 1092–96.

118. A. Rosenhouse-Dantsker and R. Osman, "Application of the primary hydration shell approach to locally enhanced sampling simulated annealing: computer simulation of thyrotropin-releasing hormone in water," *Biophys J* 79 (2000): 66–79.

119. R. Osman, M. Fuxreiter, and N. Luo, "Specificity of damage recognition and catalysis of DNA repair," *Comput Chem* 24 (2000): 331–39.

120. E. L. Rachofsky, J. B. Ross, and R. Osman, "Conformation and dynam-

ics of normal and damaged DNA," *Comb Chem High Throughput Screen* 4 (2001): 675–706.

121. D. E. Logothetis, Y. Kurachi, J. Galper, et al., "The beta gamma subunits of GTP-binding proteins activate the muscarinic K+ channel in heart," *Nature* 325 (1987): 321–26.

122. T. Jin, L. Peng, T. Mirshahi, et al., "The (beta)gamma subunits of G proteins gate a K(+) channel by pivoted bending of a transmembrane segment," *Mol Cell* 10 (2002): 469–81.

123. C. M. Lopes, H. Zhang, T. Rohacs, et al., "Alterations in conserved Kir channel-PIP2 interactions underlie channelopathies," *Neuron* 34 (2002): 933–44.

124. E. Kobrinsky, T. Mirshahi, H. Zhang, et al., "Receptor-mediated hydrolysis of plasma membrane messenger PIP2 leads to K+-current desensitization," *Nat Cell Biol* 2 (2000): 507–14.

125. H. Zhang, L. C. Craciun, T. Mirshahi, et al., "PIP(2) activates KCNQ channels, and its hydrolysis underlies receptor-mediated inhibiton of M currents," *Neuron* 37 (2003): 963–75.

126. See also chapter 3, on the Graduate School.

127. F. C. Bancroft, G.-J. Wu, and G. Zubay, "Cell-free synthesis of rat growth hormone," *Proc Nat Acad Sci USA* 70 (1973): 3646–49.

128. Interview of F. Carter Bancroft, Ph.D., by Arthur Aufses, M.D., June 4, 2003, The Mount Sinai Archives.

129. F. Guarnieri, M. Fliss, and C. Bancroft, "Making DNA add," *Science* 273 (1996): 220–23.

130. C. T. Clelland, V. Risca, and C. Bancroft, "Hiding messages in DNA microdots," *Nature* 399 (1999): 533–34.

131. See Aufses and Niss, *This House of Noble Deeds*, pp. 92 and 379–81, for more on Schneierson.

132. *Atlas of Diagnostic Microbiology*, photographs by S. Stanley Schneierson, text by Alan F. Sewell, 1st ed. (North Chicago, Ill: Abbott Laboratories, 1971).

133. The Mount Sinai Medical Center, *Annual Report for 1969*, p. 119, The Mount Sinai Archives.

134. Ibid.

135. E. D. Kilbourne, "Future influenza vaccines and the use of genetic recombinants," *Bull World Health Organ* 41 (1969): 643–45.

136. Letter from Edwin D. Kilbourne, M.D., to Barbara Niss, Archivist, the Mount Sinai Medical Center, dated May 15, 1991, The Mount Sinai Archives.

137. Ibid.

138. E. D. Kilbourne, J. L. Schulman, G. C. Schild, et al., "Related studies of a recombinant influenza-virus vaccine. I. Derivation and characterization of virus and vaccine," *J Infect Dis* 124 (1971): 449–62.

139. J. L. Schulman and E. D. Kilbourne, "Correlated studies of a recombinant influenza-virus vaccine. II. Definition of antigenicity in experimental animals," *J Infect Dis* 124 (1971): 463–72.

140. R. B. Couch, R. G. Douglas Jr., D. S. Fedson, and J. A. Kasel, "Correlated studies of a recombinant influenza-virus vaccine. III. Protection against experimental influenza in man," *J Infect Dis* 124 (1971): 473–80.

141. A. Leibovitz, R. L. Coultrip, E. D. Kilbourne, et al., "Correlated studies of a recombinant influenza-virus vaccine. IV. Protection against naturally occurring influenza in military trainees," *J Infect Dis* 124 (1971): 481–87.

142. J. L. Schulman, "Experimental transmission of influenza virus infection in mice. 3. Differing effects of immunity induced by infection and by inactivated influenza virus vaccine on transmission of infection," *J Exp Med* 125 (1967): 467–78.

143. J. L. Schulman, "Experimental transmission of influenza virus infection in mice. IV. Relationship of transmissibility of differential strains of virus and recovery of airborne virus in the environment of infector mice," *J Exp Med* 125 (1967): 479–88.

144. J. L. Schulman and E. D. Kilbourne, "The antigenic relationship of the neuraminidase of Hong Kong virus to that of other human strains of influenza A virus," *Bull World Health Organ* 41 (1969): 425–28.

145. J. L. Schulman, "The role of antineuraminidase antibody in immunity to influenza virus infection," *Bull World Health Organ* 41 (1969): 647–50.

146. J. L. Schulman and E. D. Kilbourne, "Independent variation in nature of hemagglutinin and neuraminidase antigens of influenza virus: distinctiveness of hemagglutinin antigen of Hong Kong-68 virus," *Proc Natl Acad Sci U S A* 63 (1969): 326–33.

147. Interview with Peter Palese, Ph.D., by Arthur Aufses, M.D., February 24, 2003, The Mount Sinai Archives.

148. P. Palese, J. L. Schulman, G. Bodo, and P. Meindl, "Inhibition of influenza and parainfluenza virus replication in tissue culture by 2-deoxy-2,3-dehydro-N-trifluoroacetylneuraminic acid (FANA)," *Virology* 59 (1974): 490–98.

149. P. Palese, "The genes of influenza virus," *Cell* 10 (1977): 1–10.

150. W. Luytjes, M. Krystal, M. Enami, et al., "Amplification, expression, and packaging of foreign gene by influenza virus," *Cell* 59 (1989): 1107–13.

151. M. Enami, W. Luytjes, M. Krystal, and P. Palese, "Introduction of site-specific mutations into the genome of influenza virus," *Proc Natl Acad Sci U S A* 87 (1990): 3802–5.

152. A. Garcia-Sastre, T. Muster, W. S. Barclay, et al., "Use of a mammalian internal ribosomal entry site element for expression of a foreign protein by a transfectant influenza virus," *J Virol* 68 (1994): 6254–61.

153. E. Fodor, L. Devenish, O. G. Engelhardt, et al., "Rescue of influenza A virus from recombinant DNA," *J Virol* 73 (1999): 9679–82.

154. J. Talon, M. Salvatore, R. E. O'Neill, et al., "Influenza A and B viruses expressing altered NS1 proteins: a vaccine approach," *Proc Natl Acad Sci U S A* 97 (2000): 4309–14.

155. G. K. Geiss, M. Salvatore, T. M. Trumpey, et al., "Cellular transcriptional profiling in influenza A virus-infected lung epithelial cells: the role of the nonstructural NS1 protein in the evasion of the host innate defense and its potential contribution to pandemic influenza," *Proc Natl Acad Sci U S A* 99 (2002): 10736–41.

156. C. F. Basler, X. Wang, E. Muhlberger, et al., "The Ebola virus VP35 protein functions as a type I IFN antagonist," *Proc Natl Acad Sci U S A* 97 (2000): 12289–94.

157. M. Aubert and J. A. Blaho, "The herpes simplex virus type I regulatory protein ICP27 is required for the prevention of apoptosis in infected human cells," *J Virol* 73 (1999): 2803–13.

158. L. E. Pomeranz and J. A. Blaho, "Modified VP22 localizes to the cell nucleus during synchronized herpes simplex virus type I infection," *J Virol* 73 (1999): 6769–81.

159. L. E. Pomeranz and J. A. Blaho, "Assembly of infectious Herpes simplex virus type I virions in the absence of full-length VP22," *J Virol* 74 (2000): 10041–54.

160. A. Kotsakis, L. E. Pomeranz, A. Blouin, and J. A. Blaho, "Microtubule reorganization during herpes simplex virus type I infection facilitates the nuclear localization of VP22, a major virion tegument protein," *J Virol* 75 (2001): 8697–711.

161. M. Aubert, J. O'Toole, and J. A. Blaho, "Induction and prevention of apoptosis in human HEp-2 cells by herpes simplex virus type I," *J Virol* 73 (1999): 10359–70.

162. M. Aubert, S. A. Rice, and J. A. Blaho, "Accumulation of herpes simplex virus type I early and leaky-late proteins correlates with apoptosis prevention in infected human HEp-2 cells," *J Virol* 75 (2001): 1013–30.

163. M. L. Goodkin, A. T. Ting, and J. A. Blaho, "NF-kappaB is required for apoptosis prevention during herpes simplex virus type I infection," *J Virol* 77 (2002): 7261–80.

164. M. Aubert and J. A. Blaho, "Viral oncoapoptosis of human tumor cells," *Gene Ther* 10 (2003): 1437–45.

165. J. G. Wetmur and N. Davidson, "Kinetics of renaturation of DNA," *J Mol Biol* 31 (1968): 349–70.

166. *Current Contents, Life Sciences*, no. 3 (1983): 17.

167. T. W. Durso, "Eight researchers accept the National Medal of Science for 1996," *The Scientist* 10(16):3 (August 19, 1996), available at http://www.the-scientist.com/yr1996/august/medal_960819.html.

168. Ibid.

169. Ibid.

170. J. G. Wetmur, "Acceleration of DNA renaturation rates," *Biopolymers* 14 (1975): 2517–24.

171. J. G. Wetmur, "Hybridization and renaturation kinetics of nucleic acids," *Ann Rev Biophys Bioeng* 5 (1976): 337–61.

172. R. Wieder and J. G. Wetmur, "One hundredfold acceleration of DNA renaturation rates in solution," *Biopolymers* 20 (1981): 1537–47.

173. R. Wieder and J. G. Wetmur, "Factors affecting the kinetics of DNA reassociation in phenol-water emulsion at high DNA concentrations," *Biopolymers* 21 (1982): 665–77.

174. R. S. Quartin, M. Plewinska, and J. G. Wetmur, "Branch migration mediated DNA labeling and cloning," *Biochemistry* 28 (1989): 8676–82.

175. J. G. Wetmur, "DNA probes: applications of the principles of nucleic acid hybridization," *Crit Rev Biochem Mol Biol* 26 (1991): 227–59.

176. R. J. Schneider and J. G. Wetmur, "Kinetics of transfer of Escherichia coli single strand deoxyribonucleic acid binding protein between single-stranded deoxyribonucleic acid molecules," *Biochemistry* 21 (1982): 608–15.

177. J. G. Wetmur, D. M. Wong, B. Ortiz, et al., "Cloning, sequencing, and expression of RecA proteins from three distantly related thermophilic eubacteria," *J Biol Chem* 269 (1994): 25928–35.

178. J. Tong and J. G. Wetmur, "Cloning, sequencing, and expression of ruvB and characterization of RuvB proteins from two distantly related thermophilic eubacteria," *J Bacteriol* 178 (1996): 2695–700.

179. H. G. Rao, A. Rosenfeld, and J. G. Wetmur, "Methanococcus jannaschii flap endonuclease: expression, purification, and substrate requirements," *J Bacteriol* 180 (1998): 5406–12.

180. S. Gonzalez, A. Rosenfeld, D. Szeto, and J. G. Wetmur, "The ruv proteins of Thermotoga maritima: branch migration and resolution of Holliday junctions," *Biochem Biophys Acta* 1494 (2000): 217–25.

181. C. D. Putnam, S. B. Clancy, H. Tsuruta, et al., "Structure and mechanism of the RuvB Holliday junction branch migration motor," *J Mol Biol* 311 (2001): 297–310.

182. J. G. Wetmur, D. F. Bishop, L. Ostasiewicz, and R. J. Desnick, "Molecular cloning of a cDNA for human delta-aminolevulinate dehydratase," *Gene* 43 (1986): 123–30.

183. J. G. Wetmur, D. F. Bishop, C. Cantelmo, and R. J. Desnick, "Human delta-aminolevulinate dehydratase: nucleotide sequence of a full-length cDNA clone," *Proc Natl Acad Sci U S A* 83 (1986): 7703–7.

184. J. G. Wetmur, A. H. Kaya, M. Plewinska, and R. J. Desnick, "Molecular characterization of the human delta-aminolevulinate dehydratase 2 (ALAD2) allele: implications for molecular screening of individuals for genetic susceptibility to lead poisoning," *Am J Hum Genet* 49 (1991): 757–63.

185. L. Claudio, T. Lee, M. S. Wolff, and J. G. Wetmur, "A murine model of genetic susceptibility to lead bioaccumulation," *Fundam Appl Toxicol* 35 (1997): 84–90.

186. J. Chen, S. Germer, R. Higuchi, et al., "Kinetic polymerase chain reaction on pooled DNA: a high-throughput, high-efficiency alternative in genetic epidemiological studies," *Cancer Epidemiol Biomarkers Prev* 11 (2002): 131–36.

187. J. Chen, M. Kumar, W. Chan, et al., "Increased influence of genetic variation on PON1 activity in neonates," *Environ Health Perspect* 111 (2003): 1403–9.

188. J. D. Capra and J. M. Kehoe, "Variable region sequences of five human immunoglobulin heavy chains of the VH3 subgroup: definitive identification of four heavy chain hypervariable regions," *Proc Natl Acad Sci U S A* 71 (1974): 845–48.

189. J. D. Capra and J. M. Kehoe, "Structure of antibodies with shared idiotypy: the complete sequence of the heavy chain variable regions of two immunoglobulin M anti-gamma globulins," *Proc Natl Acad Sci U S A* 71 (1974): 4032–36.

190. J. D. Capra and J. M. Kehoe, "Hypervariable regions, idiotypy, and the antibody-combining site," *Adv Immunol* 20 (1975): 1–40.

191. R. L. Wasserman and J. D. Capra, "Amino acid sequence of the Fc region of a canine immunoglobulin M: interspecies homology for the IgM class," *Science* 200 (1978): 1159–61.

192. M. B. Prystowsky, J. M. Kehoe, and B. W. Erickson, "Inhibition of the classical complement pathway by synthetic peptides from the second constant domain of the heavy chain of human immunoglobulin G," *Biochemistry* 20 (1981): 6349–56.

193. C. Bona, A. Anteunis, R. Robineaux, and A. Astesano, "Transfer of antigenic macromolecules from macrophages to lymphocytes. I. Autoradiographic and quantitative study of (14C)endotoxin and (125I)haemocyanin transfer," *Immunology* 23 (1972): 799–816.

194. C. Bona, C. Damais, and L. Chedid, "Blastic transformation of mouse spleen lymphocytes by a water-soluble mitogen extracted from Nocardia," *Proc Natl Acad Sci U S A* 71 (1974): 1602–6.

195. W. E. Paul and C. Bona, "Regulatory idiotypes and immune network: a hypothesis," *Immunol Today* 3 (1982): 230–4.

196. C. Bona, "V genes encoding antibodies: molecular and phenotypic characteristics," *Ann Rev Immunol* 6 (1988): 327–58.

197. C. Bona, "Postulates defining pathogenic antibodies and T cells," *Autoimmunity* 10 (1991): 169–72.

198. H. Zaghouani, M. Krystal, H. Kuzu, et al., "Cells expressing an H chain Ig gene carrying a viral T cell epitope are lysed by specific cytolytic T cells," *J Immunol* 148 (1992): 3604–9.

199. H. Zaghouani, R. Steinman, R. Nonacs, et al., "Presentation of a viral T cell epitope expressed in the CDR3 region of a self immunoglobulin molecule," *Science* 259 (1993): 224–27.

200. C. A. Bona, C. Murai, S. Casares, et al., "Structure of the mutant fibrillin-1 gene in the tight skin (TSK) mouse," *DNA Res* 4 (1997): 267–71.

201. F. K. Tan, F. C. Arnett, S. Antohi, et al., "Autoantibodies to the extracellular matrix microfibrillar protein, fibrillin-1 gene, in patients with scleroderma and other connective tissue diseases," *J Immunol* 163 (1999): 1066–72.

202. T. Kodera, T. L. McGaha, R. Phelps, et al., "Disrupting the IL-4 gene rescues mice homozygous for the tight-skin mutation from embryonic death and diminishes TGF-beta production by fibroblasts," *Proc Natl Acad Sci U S A* 99 (2002): 3800–5.

203. T. L. McGaha, R. G. Phelps, H. Spiera, and C. Bona, "Halofuginone, an inhibitor of type-I collagen synthesis and skin sclerosis, blocks transforming-growth-factor-beta-mediated Smad3 activation in fibroblasts," *J Invest Dermatol* 118 (2002): 461–70.

204. T. M. Moran, H. Isobe, A. Fernandez-Sesma, and J. L. Schulman, "Interleukin-4 causes delayed virus clearance in influenza virus-infected mice," *J Virol* 70 (1996): 5230–35.

205. C. B. Lopez, A. Garcia-Sastre, B. R. Williams, and T. M. Moran, "Type I interferon induction pathway, but not released interferon, participates in the maturation of dendritic cells induced by negative-strand RNA viruses," *J Infect Dis* 187 (2003): 1126–36.

206. M. K. Brimnes, L. Bonifaz, R. M. Steinman, and T. M. Moran, "Influenza virus-induced dendritic cell maturation is associated with the induction of strong T cell immunity in a coadministered, normally nonimmunogenic protein," *J Exp Med* 198 (2003): 133–44.

207. S. Casares, C. A. Bona, and T.-D. Brumeanu, "Engineering and characterization of a murine MHC class II-imunoglobulin chimera expressing an immunodominant CD4 T viral epitope," *Protein Eng* 10 (1997): 1295–1301.

208. S. Casares, A. C. Stan, C. A. Bona, and T.-D. Brumeanu, "Antigen-specific downregulation of T cells by doxorubicin delivered through a recombinant MHC II-peptide chimera," *Nat Biotechnol* 19 (2001): 142–47.

209. S. Casares, A. Hurtado, R. C. McEvoy, et al., "Down-regulation of diabetogenic CD4+ T cells by a soluble dimeric peptide-MHC class II chimera," *Nat Immunol* 3 (2002): 383–91.

210. B. Poon, D. Dixon, L. Ellis, et al., "Molecular basis of the activation of the tumorigenic potential of Gag-insulin receptor chimeras," *Proc Natl Acad Sci U S A* 88 (1991): 877–81.

211. C. S. Zong and L. H. Wang, "Modulatory effect of the transmembrane domain of the protein-tyrosine kinase encoded by oncogene ros: biological function and substrate interaction," *Proc Natl Acad Sci U S A* 91 (1994): 10982–86.

212. J. Chen, H. B. Sadowski, R. A. Kohanski, and L. H. Wang, "Stat5 is a physiological substrate of the human insulin receptor," *Proc Natl Acad Sci U S A* 94 (1997): 2295–300.

213. C. S. Zong, J. Chan, D. E. Levy, et al., "Mechanism of STAT3 activation by insulin-like growth factor I receptor," *J Biol Chem* 275 (2000): 15099–105.

214. L. Zeng, P. Sachdev, L. Yan, et al., "Vav3 mediates receptor protein tyrosine kinase signaling, regulates GTPase activity, modulates cell morphology, and induces cell transformation," *Mol Cell Biol* 20 (2000): 9212–24.

215. U. Hermanto, C. S. Zong, and L. H. Wang, "Inhibition of mitogen-activated protein kinase kinase selectively inhibits cell proliferation in human breast cancer cells displaying enhanced insulin-like growth factor I-mediated mitogen-activated protein kinase activation," *Cell Growth Differ* 11 (2000): 655–64.

216. U. Hermanto, C. S. Zong, and L H. Wang, "ErbB2-overexpressing human mammary carcinoma cells display an increased requirement for phosphatidylinositol 3-kinase signaling pathway in anchorage-independent growth," *Oncogene* 20 (2001): 7551–62.

217. P. Sachdev, Y. X. Jiang, W. Li, et al., "Differential requirement for Rho family GTPases in an oncogene insulin-like growth factor-I receptor-induced cell transformation," *J Biol Chem* 276 (2001): 26461–71.

218. *Inside Mount Sinai*, July 14–20, 2003, p. 1, The Mount Sinai Archives.

219. P. G. Katsoyannis, "Insulin generation by recombination of synthetic and natural A- and B-chains," *Science* 150 (1965): 376.

220. P. G. Katsoyannis, "The chemical synthesis of human and sheep insulin," *Am J Med* 40 (1966): 652–61.

221. Interview with Panayotis Katsoyannis, Ph.D., by Arthur Aufses, M.D., February 25, 2003, The Mount Sinai Archives.

222. The Mount Sinai Medical Center, *Annual Report for 1968*, p. 96, The Mount Sinai Archives.

223. Ibid.

224. A. Horvat, E. Li, and P. G. Katsoyannis, "Cellular binding sites for insulin in rat liver," *Biochim Biophys Acta* 382 (1975): 609–20.

225. G. T. Burke, G. Schwartz, and P. G. Katsoyannis, "Nature of the B10 amino acid residue. Requirements for high biological activity of insulin," *Int J Pept Protein Res* 23 (1984): 394–401.

226. G. P. Schwartz, G. T. Burke, and P. G. Katsoyannis, "A superactive insulin: [B10-aspartic acid]insulin(human)," *Proc Natl Acad Sci U S A* 84 (1987): 6408–11.

227. G. P. Schwartz, G. T. Burke, and P. G. Katsoyannis, "A highly potent insulin: des-(B26-B30)-[AspB10,TyrB25-NH2]insulin(human)," *Proc Natl Acad Sci U S A* 86 (1989): 458–61.

228. G. T. Burke, S. Q. Hu, N. Ohta, et al., "Superactive insulins," *Biochem Biophys Res Commun* 173 (1990): 982–87.

229. S. Q. Hu, G. T. Burke, G. P. Schwartz, et al., "Steric requirements at position B12 for high biological activity in insulin," *Biochemistry* 32 (1993): 2631–35.

230. S. Q. Hu, G. T. Burke, and P. G. Katsoyannis, "Contribution of the B16 and B26 tyrosine residues to the biological activity of insulin," *J Protein Chem* 12 (1993): 741–47.

231. W. R. Laws, G. P. Schwartz, E. Rusinova, et al., "5-Hydroxytryptophan: an absorption and fluorescence probe which is a conservative replacement for [A14 tyrosine] in insulin," *J Protein Chem* 14 (1995): 225–32.

232. A. C. Trakatellis, "Effect of sparsomycin on protein synthesis in the mouse liver," *Proc Natl Acad Sci U S A* 59 (1968): 854–60.

233. A. C. Trakatellis and G. Schwartz, "The biosynthesis of ferredoxin in a cell-free system," *Proc Natl Acad Sci U S A* 63 (1969): 436–41.

234. S. B. Koritz and A. M. Kumar, "On the mechanism of action of the adrenocorticotrophic hormone. The stimulation of the activity of enzymes involved in pregnenolone synthesis," *J Biol Chem* 245 (1970): 152–59.

235. J. M. Fry and S. B. Koritz, "The intracellular localization in the rat adrenal of enzymes which degrade 3'Phosphoadenosine-5'-phosphosulfate," *Proc Soc Exp Biol Med* 140 (1972): 1275–78.

236. S. B. Koritz, R. Wiesner, and I. L. Schwartz, "Concerning the relationship between protein synthesis and adenosine-3',5'-cyclic phosphate-stimulated steroidogenesis in isolated rat adrenal cells," *Proc Soc Exp Biol Med* 149 (1975): 779–81.

237. S. B. Koritz and A. M. Moustafa, "Some characteristics of adrenal steroidogenesis and their possible relationships to the action of the adrenocorticotrophic hormone," *Arch Biochem Biophys* 174 (1976): 20–26.

238. S. B. Koritz, G. Bhargava, and E. Schwartz, "ACTH action on adrenal steroidogenesis," *Ann N Y Acad Sci* 297 (1977): 329–35.

239. R. B. Iyer, A. Chauhan, and S. B. Koritz, "The stimulation by adrenocorticotropin of the phosphorylation of adrenal inhibitor-1: a possible role in steroidogenesis," *Mol Cell Endocrinol* 60 (1988): 61–69.

240. S. Rosenblatt, W. P. Leighton, and J. D. Chanley, "A novel approach to the investigation of norepinephrine metabolism in human sympathetic nerve," *J Neurochem* 17 (1970): 1105–8.

241. S. Rosenblatt, W. P. Leighton, and J. D. Chanley, "Dopamine-beta-hydroxylase: evidence for increased activity in sympathetic neurons during psychotic states," *Science* 182 (1973): 923–24.

242. W. P. Leighton, S. Rosenblatt, and J. D. Chanley, "Lithium-induced changes in plasma amino acid levels during treatment of affective disorders," *Psychiatry Res* 8 (1983): 33–40.

243. S. Rosenblatt, J. D. Chanley, and R. L. Segal, "The effect of lithium on vitamin D metabolism," *Biol Psychiatry* 26 (1989): 206–8.

244. D. S. Beattie, "Studies on the biogenesis of mitochondrial protein components in rat liver slices," *J Biol Chem* 243 (1968): 4027–33.

245. D. S. Beattie, "The biosynthesis of the protein and lipid components of the inner and outer membranes of rat liver mitochondria," *Biochem Biophys Res Commun* 35 (1969): 67–74.

246. D. S. Beattie, "The possible relationship between heme synthesis and mitochondrial biogenesis," *Arch Biochem Biophys* 147 (1971): 136–42.

247. J. R. Paterniti, Jr., C. I. Lin, and D. S. Beattie, "Metabolic interrelationships during aging: the heme metabolic pathways," *Mt Sinai J Med* 47 (1980): 111–16.

248. Y. S. Chen and D. S. Beattie, "Two forms of cytochrome b in yeast mitochondria: purification, characterization, and localization in the inner mitochondrial membrane," *Biochemistry* 20 (1981): 7557–65.

249. E. Finzi, L. Clejan, and D. S. Beattie, "Effect of temperature on protein synthesis and leucine transport by yeast mitochondria," *Membr Biochem* 5 (1985) 291–307.

250. H. G. Zachau, G. Acs, and F. Lipmann, "Isolation of adenosine amino acid esters from a ribonuclease digest of soluble, liver ribonucleic acid," *Proc Natl Acad Sci U S A* 44 (1958): 885–89.

251. O. Greengard, M. A. Smith, and G. Acs, "Relation of cortisone and synthesis of ribonucleic acid to induced and developmental enzyme formation," *J Biol Chem* 238 (1963): 1548–51.

252. M. Schonberg, S. C. Silverstein, D. H. Levin, and G. Acs, "Asynchronous synthesis of the complementary strands of the reovirus genome," *Proc Natl Acad Sci U S A* 68 (1971): 505–8.

253. M. A. Sells, M. L. Chen, and G. Acs, "Production of hepatitis B virus particles in Hep G2 cells transfected with cloned hepatitis B virus DNA," *Proc Natl Acad Sci U S A* 84 (1987): 1005–9.

254. G. Acs, M. A. Sells, R. H. Purcell, et al., "Hepatitis B virus produced by transfected Hep G2 cells causes hepatitis in chimpanzees," *Proc Natl Acad Sci U S A* 84 (1987): 4641–44.

255. See chapter 3, on the Graduate School of Biological Sciences, for more on Krulwich.

256. T. A. Krulwich, J. G. Federbush, and A. A. Guffanti, "Presence of a nonmetabolizable solute that is translocated with Na+ enhances Na+ -dependent pH homeostasis in an alkalophilic Bacillus," *J Biol Chem* 260 (1985): 4055–58.

257. D. B. Hicks and T. A. Krulwich, "Purification and reconstitution of the F1FO-ATP synthase from alkaliphilic Bacillus firmus OF4. Evidence that the enzyme translocates H+ but not Na+," *J Biol Chem* 265 (1990): 20547–54.

258. J. Cheng, D. B. Hicks, and T. A. Krulwich, "The purified Bacillus subtilis tetracycline efflux protein TetA(L) reconstitutes both tetracycline-cobalt/H+ and Na+(K+)/H+ exchange," *Proc Natl Acad Sci U S A* 93 (1996): 14446–51.

259. M. Ito, A. A. Guffanti, and T. A. Krulwich, "Mrp-dependent Na(+)/H(+) antiporters of Bacillus exhibit characteristics that are unanticipated for completely secondary active transporters," *FEBS Lett* 496 (2001): 117–20.

260. J. Cheng, A. A. Guffanti, and T. A. Krulwich, "The chromosomal tetracycline resistance locus of Bacillus subtilis encodes a Na+/H+ antiporter that is physiologically important at elevated pH," *J Biol Chem* 269 (1994): 27365–71.

261. A. A. Guffanti, J. Cheng, and T. A. Krulwich, "Electrogenic antiport activities of the Gram-positive Tet protein include a Na+(K+)/K+ mode that mediates net K+ uptake," *J Biol Chem* 273 (1998): 26447–54.

262. J. Jin, A. A. Guffanti, C. Beck, and T. A. Krulwich, "Twelve-transmembrane-segment (TMS) version (DeltaTMS VII-VIII) of the 14-TMS Tet(L) antibiotic resistance protein retains monovalent cation transport modes but lacks tetracycline efflux capacity," *J Bacteriol* 183 (2001): 2667–71.

263. A. I. Cederbaum, C. S. Lieber, D. S. Beattie, and E. Rubin, "Effect of chronic ethanol ingestion on mitochondrial permeability and the transport of reducing equivalents," *Biochem Biophys Res Commun* 49 (1972): 649–55.

264. A. I. Cederbaum, and E. Rubin, "Molecular injury to mitochondria produced by ethanol and acetaldehyde," *Fed Proc* 34 (1975): 2045–51.

265. A. I. Cederbaum, E. Dicker, and G. Cohen, "Role of hydroxyl radicals in microsomal oxidation of alcohols," *Adv Exp Med Biol* 132 (1980): 1–10.

266. A. I. Cederbaum, "Regulation of pathways of alcohol metabolism by the liver," *Mt Sinai J Med* 47 (1980): 317–28.

267. A. I. Cederbaum, "Oxygen radical generation by microsomes: role of iron and implications for alcohol metabolism and toxicity," *Free Radic Biol Med* 7 (1989): 559–67.

268. Y. Dai and A. I. Cederbaum, "Cytotoxicity of acetaminophen in human cytochrome P4502E1-transfected HepG2 cells," *J Pharmacol Exp Ther* 273 (1995): 1497–505.

269. Q. Chen and A. I. Cederbaum, "Cytotoxicity and apoptosis produced by cytochrome P450 2E1 in Hep G2 cells," *Mol Pharmacol* 53 (1998): 638–48.

270. N. Nieto, S. I. Friedman, P. Greenwel, and A. I. Cederbaum, "CYP2E1-mediated oxidative stress induces collagen type I expression in rat hepatic stellate cells," *Hepatology* 30 (1999): 987–96.

271. N. Nieto, S. I. Friedman, and A. I. Cederbaum, "Cytochrome P450 2E1-derived reactive oxygen species mediate paracrine stimulation of collagen I protein synthesis by hepatic stellate cells," *J Biol Chem* 277 (2002): 9853–64.

272. D. H. Bechhofer and K. H. Zen, "Mechanism of erythromycin-induced ermC mRNA stability in Bacillus subtilis," *J Bacteriol* 171 (1989): 5803–11.

273. J. F. DiMari and D. H. Bechhofer, "Initiation of mRNA decay in Bacillus subtilis," *Mol Microbiol* 7 (1993): 705–17.

274. Y. Wei and D. H. Bechhofer, "Tetractycline induces stabilization of mRNA in Bacillus subtilis," *J Bacteriol* 184 (2002): 889–94.

275. J. S. Sharp and D. H. Bechhofer, "Effect of translational signals on mRNA decay in Bacillus subtilis," *J Bacteriol* 185 (2003): 5372–79.

276. See chapter 5, on The Centers and Institutes, for more on Lazzarini, Ramirez, and Rifkin.

277. The Mount Sinai Medical Center, *Annual Report for 1968*, p. 119, The Mount Sinai Archives.

278. J. P. Green, "Uptake, storage and release of histamine. Uptake and binding of histamine," *Fed Proc* 26 (1967): 211-18.

279. J. P. Green and S. Maayani, "Tricyclic antidepressant drugs block H2 receptor in brain," *Nature* 269 (1977): 163-65.

280. L. B. Hough and J. P. Green, "Possible functions of brain histamine," *Psychopharmacol Bull* 16 (1980): 42-44.

281. P. I. Szilagyi, D. E. Schmidt, and J. P. Green, "Microanalytical determination of acetylcholine, other choline esters, and choline by pyrolysis-gas chromatography," *Anal Chem* 40 (1968): 2009–13.

282. D. E. Schmidt, P. I. Szilagyi, D. L. Alkon, and J. P. Green, "A method for measuring nanogram quantities of acetylcholine by pyrolysis-gas chromatography: the demonstration of acetylcholine in effluents from the rat phrenic nerve-diaphragm preparation," *J Pharm Exp Ther* 174 (1970): 337–45.

283. S. Kang and J. P. Green, "Steric and electronic relationships among some hallucinogenic compounds," *Proc Natl Acad Sci U S A* 67 (1970): 62–67.

284. S. Kang, C. L. Johnson, and J. P. Green, "Theoretical studies on the conformations of psilocin and mescaline," *Mol Pharmacol* 9 (1973): 640–48.

285. J. P. Green, C. L. Johnson, H. Weinstein, and S. Maayani, "Antagonism of histamine-activated adenylate cyclase in brain by D-lysergic acid diethylamide," *Proc Natl Acad Sci U S A* 74 (1977): 5697–701.

286. J. P. Green, H. Weinstein, and S. Maayani, "Defining the histamine H2-receptor in brain: the interaction with LSD," *NIDA Res Monogr* 22 (1978): 38–59.

287. H. Weinstein, J. P. Green, R. Osman, and W. D. Edwards, "Recognition and activation mechanisms on the LSD/serotonin receptor: the molecular basis of structure activity relationships," *NIDA Res Monogr* 22 (1978): 333–58.

288. J. P. Green, S. Maayani, H. Weinstein, and L. B. Hough, "Histamine and psychotropic drugs," *Psychopharmacol Bull* 16 (1980): 36–38.

289. J. P. Green, "Psychopharmacologic activity: quantum chemical study," *Psychopharmacol Bull* 11 (1975): 44–45.

290. The Mount Sinai Medical Center, *Annual Report for 1969*, p. 129, The Mount Sinai Archives.

291. H. Weinstein and J. P. Green (eds.), "Quantum Chemistry in Biomedical Sciences," *Ann N Y Acad Sci* 367 (1981): 1–552.

292. S. Topiol, R. Osman, and H. Weinstein, "Effective potential methods for use in electronic structure calculations of large molecules," *Ann N Y Acad Sci* 367 (1981): 17–34.

293. R. Osman, H. Weinstein, and S. Topiol, "Models for active sites of metalloenzymes. II. Interactions with a model substrate," *Ann N Y Acad Sci* 367 (1981): 356–69.

294. H. Weinstein, R. Osman, S. Topiol, and J. P. Green, "Quantum chemical studies on molecular determinants for drug action," *Ann N Y Acad Sci* 367 (1981): 434–51.

295. H. Weinstein, D. Chou, C. L. Johnson, et al., "Tautomerism and the receptor action of histamine: a mechanistic model," *Mol Pharmacol* 12 (1976): 738–45.

296. S. Wilk and M. Orlowski, "Cation-sensitive neutral endopeptidase: isolation and purification of the bovine pituitary enzyme," *J Neurochem* 35 (1980): 1172–82.

297. M. Orlowski and S. Wilk, "A multicatalytic protease complex from pituitary that forms enkephalin and enkephalin containing peptides," *Biochem Biophys Res Comm* 101 (1981): 814–22.

298. S. Wilk and M. Orlowski, "Evidence that pituitary cation-sensitive endopeptidase is a multicatalytic protease complex," *J Neurochem* 40 (1983): 842–49.

299. M. Orlowski, Letter to Arthur Aufses, October 2, 2003, The Mount Sinai Archives.

300. M. Orlowski, "The multicatalytic proteinase complex, a major extra-lysosomal proteolytic system," *Biochemistry* 29 (1990): 10289–97.

301. M. Orlowski and A. Szewczuk, "A note on the occurrence of gamma-glutamyl transpeptidase in human serum," *Clin Chim Acta* 6 (1961): 430–32.

302. M. Orlowski, "The role of gamma-glutamyl transpeptidase in the internal diseases clinic," *Arch Immunol Ther Exp (Warsz)* 11 (1963): 1–61.

303. M. Orlowski and A. Meister, "Gamma-glutamyl-p-nitroanilide: a new convenient substrate for determination and study of l- and d-gamma-glutamyltranspeptidase activities," *Biochim Biophys Acta* 73 (1963): 679–81.

304. M. Orlowski and A. Meister, "The gamma-glutamyl cycle: a possible transport system for amino acids," *Proc Natl Acad Sci U S A* 67 (1970): 1248–55.

305. M. Orlowski, D. F. Reingold, and M. E. Stanley, "D- and L-stereoisomers of allylglycine: convulsive action and inhibition of brain L-glutamate decarboxylase," *J Neurochem* 28 (1977): 349–53.

306. J. Almenoff and M. Orlowski, "Membrane-bound kidney neutral metalloendopeptidase: interaction with synthetic substrates, natural peptides, and inhibitors," *Biochemistry* 22 (1983): 590-99.

307. M. Orlowski, C. Michaud, and T. G. Chu, "A soluble metalloendopeptidase from rat brain. Purification of the enzyme and determination of

specificity with synthetic and natural peptides," *Eur J Biochem* 135 (1983): 81–88.

308. R. Z. Orlowski, J. R. Eswara, A. Lafond-Walker, et al., "Tumor growth inhibition induced in a murine model of human Burkitt's lymphoma by a proteasome inhibitor," *Cancer Res* 58 (1998): 4342–48.

309. R. Z. Orlowski, T. E. Stinchcombe, B. S. Mitchell, et al., "Phase I trial of the proteasome inhibitor PS-341 in patients with refractory hematologic malignancies," *J Clin Oncol* 20 (2002): 4420–27.

310. P. G. Richardson, B. Barlogic, J. Berenson, et al., "A phase 2 study of bortezomid in relapsed, refractory myeloma," *New Engl J Med* 348 (2003): 2609–17.

311. S. Wilk, S. E. Gitlow, M. Mendlowitz, et al., "A quantitative assay for vanillylmandelic acid (VMA) by gas-liquid chromatography," *Anal Biochem* 13 (1965): 544–51.

312. J. M. Griscavage, S. Wilk, and L. J. Ignarro, "Serine and cysteine proteinase inhibitors prevent nitric oxide production by interfering with transcription of the inducible NO synthase gene," *Biochem Biophys Res Comm* 215 (1995): 721–29.

313. S. Wilk, E. Wilk, and R. P. Magnusson, "Purification, characterization, and cloning of a cytosolic aspartyl aminopeptidase," *J Biol Chem* 273 (1998): 15961–70.

314. J. Saranak and J. Goldfarb, "Effects of electrolytic and 6-hydroxy-dopamine lesions of the lateral hypothalamus on rotation evoked by electrical stimulation of the substantia nigra in rats," *Brain Res* 208 (1981): 81–95.

315. P. Blandina, J. Goldfarb, and J. P. Green, "Activation of the 5-HT3 receptor releases dopamine from rat striatal slice," *Europ J Pharmacol* 155 (1988): 349–50; P. Blandina, J. Goldfarb, B. Royal-Craddock, and J. P. Green, "Release of endogenous dopamine by stimulation of 5-hydroxytryptamine3 receptors in rat striatum," *J Pharmecol Exp Ther* 251 (1989): 803–9.

316. K. A. Berg, S. Maayani, J. Goldfarb, et al., "Effector pathway-dependent relative efficacy at serotonin type 2A and 2C receptors: evidence for agonist-directed trafficking of receptor stimulus," *Mol Pharmacol* 54 (1998): 94–104.

317. S. R. Neves, P. T. Ram, and R. Iyengar, "G protein pathways," *Science* 296 (2002): 1636–39.

318. J. D. Jordan and R. Iyengar, "Modes of interaction between signaling pathways," *Biochem Pharmacol* 55 (1998): 1347–52.

319. U. S. Bhalla and R. Iyengar, "Emergent properties of networks of biological signaling pathways," *Science* 283 (1999): 381–87.

320. S. R. Neves and R. Iyengar, "Modeling of signaling networks," *Bioessays* 24 (2002): 1110–17.

321. R. Iyengar and L. Birnbaumer, *G Proteins* (San Diego: Academic Press, 1990).

322. R. Iyengar, *Heterotrimeric G proteins* (San Diego: Academic Press, 1994).

323. R. Iyengar, *Heterotrimeric G-protein Effectors* (San Diego: Academic Press, 1994).

324. R. Iyengar and J. D. Hildebrandt, *G Protein Pathways. Part A: Receptors* (San Diego: Academic Press, 2002).

325. R. Iyengar and J. D. Hildebrandt, *G Protein Pathways. Part B: G Proteins and Their Regulators* (San Diego: Academic Press, 2002).

326. R. Iyengar and J. D. Hildebrandt, *G Protein Pathways. Part C: Effector Mechanisms* (San Diego: Academic Press, 2002).

327. J. P. Green, "Physicians practicing other occupations, especially literature," *Mt Sinai J Med* 60 (1993): 132–55.

328. A. Anantharam and M. A. Diverse-Pierluissi, "Biochemical approaches to study interaction of calcium channels with RGS12 in primary neuronal cultures," *Methods Enzymol* 345 (2002): 60–70.

329. M. L. Schiff, D. P. Siderovski, J. D. Jordan, et al., "Tyrosine-kinase-dependent recruitment of RGS12 to the N-type calcium channel," *Nature* 408 (2000): 723–27.

330. G. P. Brown, R. D. Blitzer, J. H. Connor, et al., "Long-term potentiation induced by theta frequency stimulation is regulated by a protein phosphatase-1-operated gate," *J Neurosci* 20 (2000): 7880–87.

331. R. D. Blitzer, T. Wong, M. G. Giovanni, et al., "Amyloid beta peptides activate the phosphoinositide signaling pathway in oocytes expressing rat brain RNA," *Brain Res Mol Brain Res* 76 (2000): 115–20.

332. J. Hama, H. Xu, M. Goldfarb, and D. C. Weinstein, "SNT-1/FRS2alpha physically interacts with Laloo and mediates mesoderm induction by fibroblast growth factor," *Mech Dev* 109 (2001): 195–204.

333. J. Hama, C. Suri, T. Haremaki, and D. C. Weinstein, "The molecular basis of Src kinase specificity during vertebrate mesoderm formation," *J Biol Chem* 277 (2002): 19806–10.

334. D. P. Healy, "Radioimmunocytochemical localization of corticotropin-releasing factor and adrenocorticotropin in the hypothalamo-hypophyseal system of the rat: effects of adrenalectomy," *J Histochem Cytochem* 40 (1992): 969–78.

335. M. Troyanovskaya, L. Song, G. Jayaraman, and D. P. Healy, "Expression of aminopeptidase A, an angiotensinase, in glomerular mesangial cells," *Hypertension* 27 (1996): 518–22.

336. M. Troyanovskaya, G. Jayaraman, L. Song, and D. P. Healy, "Aminopeptidase-A. I. CDNA cloning and expression and localization in rat tissues," *Am J Physiol Regul Integr Comp Physiol* 278 (2000): R413–24.

337. C. P. Cardozo, C. Michaud, M. C. Ost, et al., "C-terminal Hsp-interacting protein slows androgen receptor synthesis and reduces its rate of degradation," *Arch Biochem Biophys* 410 (2003): 134–40.

338. O. D. Gil, T. Sakurai, A. E. Bradley, et al., "Ankyrin binding mediates LICAM interactions with static components of the cytoskeleton and inhibits retrograde movement of LICAM on the cell surface," *J Cell Biol* 162 (2003): 719–30.

339. I. Gomes, J. Filipovska, and L. A. Devi, "Opioid receptor oligomerization. Detection and functional characterization of interacting receptors," *Methods Mol Med* 84 (2003): 157–83.

340. See also chapter 2, on the Curriculum.

NOTES TO CHAPTER 5

1. See chapter 4, on the basic sciences.

2. Interview with Nathan Kase, M.D., by Arthur Aufses, M.D., July 16, 2003, The Mount Sinai Archives.

3. Ibid.

4. Interview with Robert Lazzarini, Ph.D., by Arthur Aufses, M.D., April 30, 2003, The Mount Sinai Archives.

5. G. Stamminger and R. A. Lazzarini, "Analysis of the RNA of defective VSV particles," *Cell* 3 (1974): 85–93.

6. R. A. Lazzarini, G. H. Weber, L. D. Johnson, and G. M. Stamminger, "Covalently linked message and anti-message (genomic) RNA from a defective vesicular stomatitis virus particle," *J Mol Biol* 97 (1975): 289–307.

7. J. D. Keene, M. H. Schubert, R. A. Lazzarini, and M. Rosenberg, "The Terminal Nucleotide Sequences of RNA from VSV and Its Defective Interfering Particles." In *Persistent Viruses, ICN-UCLA. Symposia on Molecular and Cellular Biology*, ed. J. Stevens, G. Todaro, and C. F. Fox (New York: Academic Press, 1978), vol. 11, pp. 285–97.

8. R. A. Lazzarini, J. D. Keene, and M. Schubert, "The origins of defective interfering particles of the negative-strand RNA viruses," *Cell* 26 (1981): 145–54.

9. F. de Ferra, H. Engh, L. Hudson, et al., "Alternative splicing accounts for the four forms of myelin basic protein," *Cell* 43 (1985): 721–27.

10. L. D. Hudson, J. A. Berndt, C. Puckett, et al., "Aberrant splicing of proteolipid protein mRNA in the dysmyelinating jimpy mutant mouse," *Proc Natl Acad Sci U S A* 84 (1987): 1454–58.

11. A. Gow, V. L. Friedrich, Jr., and R. A. Lazzarini, "Myelin basic protein gene contains separate enhancers for oligodendrocyte and Schwann cell expression," *J Cell Biol* 119 (1992): 605–16.

12. A. Gow, V. L. Friedrich, Jr., and R. A. Lazzarini, "Intracellular transport and sorting of the oligodendrocyte transmembrane proteolipid protein," *J Neurosci Res* 37 (1994): 563–73.

13. B. Lee, M. Godfrey, E. Vitale, et al., "Linkage of Marfan syndrome and a phenotypically related disorder to two different fibrillin genes," *Nature* 352 (1991): 330–34.

14. L. Pereira, K. Andrikopoulos, J. Tian, et al., "Targetting of the gene encoding fibrillin-1 recapitulates the vascular aspect of Marfan syndrome," *Nat Genet* 17 (1997): 218–22.

15. L. Pereira, S. Y. Lee, B. Gayraud, et al., "Pathogenetic sequence for aneurysm revealed in mice underexpressing fibrillin-1," *Proc Natl Acad Sci USA* 96 (1999): 3819–23.

16. W. Han, Y. Yu, N. Altan, and L. Pick, "Multiple proteins interact with the *fushi tarazu* proximal enhancer," *Mol Cell Biol* 13 (1993): 5549–59.

17. Y. Yu and L. Pick, "Non-periodic cues generate seven *ftz* stripes in the *Drosophila* embryo," *Mech Dev* 50 (1995): 163–75.

18. L. Pick, "Segmentation: painting stripes from flies to vertebrates," *Dev Genet* 23 (1998): 1–10.

19. J. Song, K. Wu, Z. Chew, et al., "Axons guided by insulin receptor in *Drosophila* visual system," *Science* 300 (2003): 502–5.

20. N. Azpiazu and M. Frasch, "*tinman* and *bagpipe*: two homeo box genes that determine cell fate in the dorsal mesoderm of *Drosophila*," *Genes Dev* 7 (1993): 1325–40.

21. M. Frasch, "Induction of visceral and cardiac mesoderm by ectodermal Dpp in the early *Drosophila* embryo," *Nature* 374 (1995): 464–67.

22. W. Wang, P. Lo, M. Frasch, and T. Lufkin, "*Hmx*: an evolutionary conserved homeobox gene family expressed in the developing nervous system in mice and *Drosophila*," *Mech Dev* 99 (2000): 123–37.

23. Manfred Frasch web page. The Mount Sinai School of Medicine web site, at http://adsr13.mssm.edu/domains/dept/facultyInfo.epl?objname= biomo l&user=frascm01&sid=30403_23. Accessed August 5, 2003.

24. M. Boutros, J. Mihaly, T. Bouwmeester, and M. Mlodzik, "Signaling specificity by Frizzled receptors in *Drosophila*," *Science* 288 (2000): 1825–28.

25. J. Curtiss, G. Halder, and M. Mlodzik, "Selector and signalling molecules cooperate in organ patterning," *Nat Cell Biol* 4 (2002): E48–51.

26. I. J. Miller and J. J. Bieker, "A novel, erythroid cell-specific murine transcription factor that binds to the CACCC element and is related to the Krüppel family of nuclear proteins," *Molec Cell Biol* 13 (1993): 2776–86.

27. J. J. Bieker, "Krüppel-like factors: three fingers in many pies," *J Biol Chem* 276 (2001): 34355–58.

28. J. K. Wang, H. Xu, H. C. Li, and M. Goldfarb, "Broadly expressed SNT-like proteins link FGF receptor stimulation to activators of Ras," *Oncogene* 13 (1996): 721–29.

29. H. Xu, K. W. Lee, and M. Goldfarb, "Novel recognition motif on FGF

receptor mediates direct association and activation of SNT adaptor proteins," *J Biol Chem* 273 (1998): 17987–90.

30. H. Hartung, B. Feldman, H. Lovec, et al., "Murine FGF-12 and FGF-13: expression in embryonic nervous system, connective tissue and heart," *Mech Dev* 64 (1997): 31–39.

31. J. Schoorlemmer and M. Goldfarb, "Fibroblast growth factor homologous factors are intracellular signaling proteins," *Curr Biol* 11 (2001): 793–97.

32. W. Wang, X. Chen, H. Xu, and T. Lufkin, "Msx3: a novel murine homologue of the *Drosophila* msh homeobox gene restricted to the dorsal embryonic nervous system," *Mech Dev* 58 (1996): 203–15.

33. T. Lufkin, "Transcriptional regulation of vertebrate Hox genes during embryogenesis," *Crit Rev Eukaryot Gene Expr* 7 (1997): 195–213.

34. C. Tribioli and T. Lufkin, "Molecular cloning, chromosomal mapping and developmental expression of BAPX1, a novel homeobox-containing gene homologous to *Drosophila bagpipe*," *Gene* 203 (1997): 225–33.

35. See A. H. Aufses Jr. and B. J. Niss, *This House of Noble Deeds: The Mount Sinai Hospital, 1852–2002* (New York: New York University Press, 2002), pp. 39–41, for more on Gorlin as Chairman of the Department of Medicine.

36. Ibid., pp. 50–51, for more on Taubman.

37. J. D. Licht, C. Chomienne, A. Goy, et al., "Clinical and molecular characterization of a rare syndrome of acute promyleocytic leukemia associated with translocation (11;17).," *Blood* 85 (1995): 1083–94.

38. S. Hosono, X. L. Luo, D. P. Hyink, et al., "WT1 expression induces features of renal differentiation in mesenchymal fibroblasts," *Oncogene* 18 (1999): 417–27.

39. M. A. English and J. D. Licht, "Tumor-associated WT1 missense mutants indicate that transcriptional activation by WT1 is critical for growth control," *J Biol Chem* 274 (1999): 13258–63.

40. A. Melnick, K. F. Ahmad, S. Arai, et al., "In-depth mutational analysis of the promyelocytic leukemia zinc finger BTB/POZ domain reveals motifs and residues required for biological and transcriptional functions," *Mol Cell Biol* 20 (2000): 6550–67.

41. I. Gross, B. Bassit, M. Benezra, and J. D. Licht, "Mammalian sprouty proteins inhibit cell growth and differentiation by preventing ras activation," *J Biol Chem* 276 (2001): 46460–68.

42. M. Benezra, N. Chevallier, D. J. Morrison, et al., "BRCA1 augments transcription by the NF-KappaB transcription factor by binding to the Rel domain of the p65/RelA subunit," *J Biol Chem* 278 (2003): 26333–41.

43. See chapter 1, on the History of the School, for more on Rifkin and the Humanities and Medicine Program.

44. Y. Yu and J. P. Hirsch, "An essential gene pair in *Saccharomyces cerevisiae* with a potential role in mating," *DNA Cell Biol* 14 (1995): 411–18.

45. J. Kim and J. P. Hirsch, "A nucleolar protein that affects mating efficiency in *Saccharomyces cerevisiae* by altering the morphological response to pheromone," *Genetics* 149 (1998): 795–805.

46. J. D. Bleil and P. M. Wassarman, "Structure and function of the zona pellucida: identification and characterization of the proteins of the mouse oocyte's zona pellucida," *Dev Biol* 76 (1980): 185–202.

47. J. D. Bleil and P. M. Wassarman, "Synthesis of zona pellucida proteins by denuded and follicle-enclosed mouse oocytes during culture *in vitro*," *Proc Natl Acad Sci USA* 77 (1980): 1029–33.

48. J. D. Bleil and P. M. Wassarman, "Mammalian sperm-egg interaction: identification of a glycoprotein in mouse egg zonae pellucidae possessing receptor activity for sperm," *Cell* 20 (1980): 873–82.

49. P. M. Wassarman, "Mammalian fertilization: molecular aspects of gamete adhesion, exocytosis, and fusion," *Cell* 96 (1999): 175–83.

50. P. M. Wassarman, L. Jovine, and E. Litscher, "A profile of fertilization in mammals," *Nat Cell Biol* 3 (2001): E59–E64.

51. P. M. Wassarman, "Channels of communication in the ovary," *Nat Cell Biol* 4(S1) (2002): S7–S9.

52. F. Relaix, X. Wei, W. Li, et al., "Pw1/Peg3 is a potential cell death mediator and cooperates with Siah1a in p53-mediated apoptosis," *Proc Natl Acad Sci USA* 97 (2000): 2105–10.

53. C. Miller, K. Degenhardt, and D. Sassoon, "Fetal exposure to DES results in de-regulation of Wnt7A during uterine morphogenesis," *Nat Genet* 20 (1998): 228–30.

54. H. Sun, B. Lu, R. A. Flavell, et al., "Defective T-cell activation and autoimmune disorder in Stra13-deficient mice," *Nature Immunol* 2 (2001): 1040–47.

55. Interview with Paul Wassarman, Ph.D., by Arthur Aufses, M.D., April 28, 2003, The Mount Sinai Archives.

56. J. L. Roberts, J. R. Lundblad, J. H. Eberwine, et al., "Hormonal regulation of POMC expression in pituitary," *Ann NY Acad Sci* 512 (1987): 275–85.

57. J. R. Lundblad and J. L. Roberts, "Regulation of proopiomelanocortin gene expression in pituitary," *Endocr Rev* 9 (1988): 135–58.

58. D. Autelitano, J. R. Lundblad, M. Blum, and J. L. Roberts, "Hormonal repletion of POMS gene expression," *Ann Rev Physiol* 51 (1989): 715–26.

59. D. I. Lugo, J. L. Roberts, and J. E. Pintar, "Analysis of proopiomelanocortin gene expression during prenatal development of the rat pituitary gland," *Mol Endocrinol* 3 (1989): 1313–24.

60. N. Levin, M. Blum, and J. L. Roberts, "Modulation of basal and corticotropin-releasing factor-stimulated proopiomelanocortin gene expression by vasopressin in rat anterior pituitary," *Endocrinology* 125 (1989): 29957–66.

61. M. Blum, J. L. Roberts, and S. L. Wardlaw, "Androgen regulation of proopiomelanocortin gene expression and peptide content in the basal hypothalamus," *Endocrinology* 124 (1989): 2283–88.

62. M. Woloschak and J. L. Roberts, "Characterization and regulation of upstream transcripts of the murine proopiomelanocortin gene," *Biochem Biophys Res Commun* 210 (1995): 281–87.

63. A. C. Gore, J. L. Roberts, and M. J. Gibson, "Mechanisms for the regulation of gonadotropin-releasing hormone gene expression in the developing mouse," *Endocrinology* 140 (1999): 2280–87.

64. S. R. Salton, "Nucleotide sequence and regulatory studies of VGF, a nervous system-specific mRNA that is rapidly and relatively selectively induced by nerve growth factor," *J Neurochem* 57 (1991): 991–96.

65. S. R. Salton, D. J. Fischberg, and K. W. Dong, "Structure of the gene encoding VGF, a nervous system-specific mRNA that is rapidly and selectively induced by nerve growth factor in PC12 cells," *Mol Cell Biol* 11 (1991): 2335–49.

66. S. Halm, T. Mizuno, T. Wu, et al., "Targeted deletion of the Vgf gene indicates that the encoded secretory peptide precursor plays a novel role in the regulation of energy balance," *Neuron* 23 (1999): 537–48.

67. L. M. Lazar and M. Blum, "Regional distributon and developmental expression of epidermal growth factor and transforming growth factor-alpha mRNA in mouse brain by a quantitative nuclease protection assay," *J Neurosci* 12 (1992): 1688–97.

68. D. Casper, G. Roboz, and M. Blum, "Epidermal growth factor and basic fibroblast growth factor have independent actions on mesencephalic dopamine neurons in culture," *J Neurochem* 62 (1994): 2166–77.

69. A. Ho and M. Blum, "Induction of interleukin-1 associated with compensatory dopaminergic sprouting in the denervated striatum of young mice: model of aging and neurodegenerative disease," *J Neurosci* 18 (1998): 5614–29.

70. P. R. Hof, K. Cox, and J. H. Morrison, "Quantitative analysis of a vulnerable subset of pyramidal neurons in Alzheimer's disease: I. Superior frontal and inferior temporal cortex," *J Comp Neurol* 301 (1990): 44–54.

71. P. R. Hof and J. H. Morrison, "Quantitative analysis of a vulnerable subset of pyramidal neurons in Alzheimer's disease: II. Primary and secondary visual cortex," *J Comp Neurol* 301 (1990): 55–64.

72. J. H. Morrison and P. R. Hof, "Life and death of neurons in the aging brain," *Science* 278 (1997): 412–19.

73. A. H. Gazzaley, S. J. Siegel, J. H. Kordower, et al., "Circuit-specific alterations of N-methyl-D-aspartate receptor subunit 1 in the dentate gyrus of aged monkeys," *Proc Natl Acad Sci USA* 93 (1996): 3121–25.

74. M. M. Adams, R. A. Shah, W. G. Janssen, and J. H. Morrison, "Different modes of hippocampal plasticity in response to estrogen in young and aged female rats," *Proc Natl Acad Sci USA* 98 (2001): 8071–76.

368   NOTES TO CHAPTER 5

75. A. C. Gore, T. J. Wu, J. J. Rosenberg, and J. L. Roberts, "Gonadotropin-releasing hormone and NMDA receptor gene expression and colocalization change during puberty in female rats," *J Neurosci* 16 (1996): 5281–89.

76. A. C. Gore and J. L. Roberts, "Regulation of gonadotropin-releasing hormone gene expression *in vivo* and *in vitro*," *Front Neuroendocrinol* 18 (1997): 209–45.

77. A. C. Gore, "Gonadotropin-releasing hormone neurons, NMDA receptors, and their regulation by steroid hormones across the reproductive life cycle," *Brain Res Brain Res Rev* 37 (2001): 235–48.

78. B. H. Miller and A. C. Gore, "N-Methyl-D-Aspartate receptor subunit expression in GnRH hormone neurons changes during reproductive senescence in the female rat," *Endocrinology* 143 (2002): 3568–74.

79. A. C. Gore, *GnRH: The Master Molecule of Reproduction* (Norwell, Mass.: Kluwer Academic Publishers, 2002).

80. J. Salama, T. R. Chakraborty, L. Ng, and A. C. Gore, "Effects of polychlorinated biphenyls on estrogen receptor-beta expression in the anteroventral periventricular nucleus," *Environ Health Perspect* 111 (2003): 1278–82.

81. S. S. Daftary and A. C. Gore, "Developmental changes in hypothalamic insulin-like growth factor-1: relationship to gonadotropin-releasing hormone neurons." *Endocrinology* 144 (2003): 2034–45.

82. C. V. Mobbs, "Molecular hysteresis: residual effects of hormones and glucose on genes during aging," *Neurobiol Aging* 15 (1994): 523–34.

83. T. Mizuno, H. Bergen, S. Kleopoulos, et al., "Effects of nutritional status and aging on leptin gene expression in mice: importance of glucose," *Horm Metab Res* 28 (1996): 679–84.

84. T. Mizuno, H. Bergen, T. Funabashi, et al., "Obese gene expression: reduction by fasting and stimulation by insulin and glucose in lean mice, and persistent elevation in acquired (diet-induced) and genetic (yellow agouti) obesity," *Proc Natl Acad Sci U S A* 93 (1996): 3434–38.

85. T. Mizuno, S. P. Kleopoulos, H. T. Bergen, et al., "Hypothalamic proopiomelanocortin mRNA is reduced by fasting and [corrected] in ob/ob and db/db mice, but is stimulated by leptin," *Diabetes* 47 (1998): 294–97.

86. From the Charles Vernon Mobbs Web page, at http://adsr13.mssm.edu/domains/dept/facultyInfo.epl?objname=neuro bio&user=mobbsc01&sid=31196_19. Accessed August 25, 2003.

87. T. M. Mizuno, K. A. Kelley, G. M. Pasinetti, et al., "Transgenic neuronal expression of proopiomelanocortin attenuates hyperphagic response to fasting and reverses metabolic impairments in leptin-deficient obese mice," *Diabetes* 52 (2003): 2675–83.

88. D. L. Benson, D. R. Colman, and G. W. Huntley, "Molecules, maps and synapse specificity," *Nature Rev Neurosci* 2 (2001): 899–909.

89. G. W. Huntley, D. L. Benson, and D. R. Colman, "Structural remodeling of the synapse in response to physiological activity," *Cell* 108 (2002): 1–4.

90. D. L. Benson, L. M. Schnapp, L. Shapiro, and G. W. Huntley, "Making memories stick: cell-adhesion molecules in synaptic plasticity," *Trends Cell Biol* 10 (2000): 473–82.

91. G. W. Huntley and D. L. Benson, "Neural (N)-cadherin at developing thalamocortical synapses provides an adhesion mechanism for the formation of somatopically organized connections," *J Comp Neurol* 407 (1999): 453–71.

92. P. R. Rapp, J. H. Morrison, and J. A. Roberts, "Cyclic estrogen replacement improves cognitive function in aged ovariectomized rhesus monkeys," *J Neurosci* 23 (2003): 5708–14.

93. Interview with John Morrison, Ph.D., by Arthur Aufses, M.D., February 13, 2003, The Mount Sinai Archives.

94. Aufses and Niss, *This House of Noble Deeds*, p. 102.

95. A. G. Gerster, "On the surgical dissemination of cancer," *NY Med J* 41 (1885): 233–36.

96. C. Leuchtenberger, R. Lewisohn, D. Laszlo, and R. Leuchtenberger, "Folic acid, a tumor growth inhibitor," *Proc Soc Exp Biol* 55 (1944): 204–5.

97. R. Lewisohn, "Five years of cancer research," *J Mt Sinai Hosp* 8 (1942): 771–76.

98. Ibid., p. 776.

99. C. Friend, "Cell-free transmission in adult Swiss mice of a disease having the character of a leukemia," *J Exp Med* 105 (1957): 307–18.

100. C. Friend, W. Scher, J. G. Holland, and T. Sato, "Hemoglobin synthesis in murine virus-induced leukemic cells *in vitro*: stimulation of erythroid differentiation by dimethyl sulfoxide," *Proc Natl Acad Sci U S A* 68 (1971): 378–82.

101. See Aufses and Niss, *This House of Noble Deeds*, pp. 105–8, for more on Holland.

102. Interview with Stuart Aaronson, M.D., by Arthur Aufses, M.D., on July 15, 2003, The Mount Sinai Archives.

103. S. A. Aaronson and G. J. Todaro, "Development of 3T3-like lines from BALB/c mouse embryo cultures: transformation susceptibility to SV40," *J Cell Physiol* 72 (1968): 141–48.

104. J. L. Jainchill, S. A. Aaronson, and G. J. Todaro, "Murine sarcoma and leukemia viruses: assay using clonal lines of contact-inhibited mouse cells," *J Virol* 4 (1969): 549–53.

105. D. J. Giard, S. A. Aaronson, G. J. Todaro, et al., "*In vitro* cultivation of human tumors: establishment of cell lines derived from a series of solid tumors," *J Natl Cancer Inst* 51 (1973): 1417–23.

106. R. F. Doolittle, M. W. Hunkapiller, L. E. Hood, et al., "Simian sarcoma *onc* gene, *v-sis*, is derived from the gene (or genes) encoding a platelet-derived growth factor," *Science* 221 (1983): 275–77.

107. K. C. Robbins, H. N. Antoniades, S. G. Devare, et al., "Structural and immunological similarities between simian sarcoma virus gene product(s) and human platelet-derived growth factor," *Nature* 305 (1983): 605–8.

108. C. R. King, M. H. Kraus, and S. A. Aaronson, "Amplification of a novel *v-erbB*-related gene in a human mammary carcinoma," *Science* 229 (1985): 974–76.

109. J. S. Rubin, H. Osada, P. W. Finch, et al., "Purification and characterization of a newly identified growth factor specific for epithelial cells," *Proc Natl Acad Sci U S A* 86 (1989): 802–6.

110. P. W. Finch, J. S. Rubin, T. Miki, et al., "Human KGF is FGF-related with properties of a paracrine effector of epithelial cell growth," *Science* 245 (1989): 752–55.

111. V. Adler, Z. Yin, S. Y. Fuchs, et al., "Regulation of JNK signaling by GSTp," *EMBO J* 18 (1999): 1321–34.

112. T. Buschmann, O. Potapova, A. Bar-Shira, et al., "Jun NH2-terminal kinase phosphorylation of p53 on Thr-81 is important for p53 stabilization and transcriptional activities in response to stress," *Mol Cell Biol* 21 (2001): 2743–54.

113. H. Habelhah, K. Shah, L. Huang, et al., "ERK phosphorylation drives cytoplasmic accumulation of hnRNP-K and inhibition of mRNA translation," *Nat Cell Biol* 3 (2001): 325–30.

114. H. Habelhah, I. J. Frew, A. Laine, et al., "Stress-induced decrease in TRAF2 stability is mediated by Siah2," *EMBO J* 21 (2002): 5756–65.

115. V. N. Ivanov, A. Bhoumik, M. Krasilnikov, et al., "Cooperation between STAT3 and c-Jun suppresses Fas transcription," *Mol Cell* 7 (2001): 517–28.

116. A. Bhoumik, T. G. Huang, V. Ivanov, et al., "An ATF2-derived peptide sensitizes melanomas to apoptosis and inhibits their growth and metastasis," *J Clin Invest* 110 (2002): 643–50.

117. P. Tan, S. Y. Fuchs, A. Chen, et al., "Recruitment of a ROC1-CUL1 ubiquitin ligase by Skp1 and HOS to catalyze the ubiquitination of I kappa B alpha," *Mol Cell* 3 (1999): 527–33.

118. D. C. Dias, G. Dolios, R. Wang, and Z.-Q. Pan, "CUL7: a DOC domain-containing cullin selectively binds Skp1.Fbx29 to form an SCF-like complex," *Proc Natl Acad Sci U S A* 99 (2002): 16601–6

119. K. Wu, K. Yamoah, G. Dolios, et al., "DEN1 is a dual function protease capable of processing the C terminus of Nedd8 and deconjugating hyper-neddylated CUL1," *J Biol Chem* 278 (2003): 28882–91.

120. K. Wu, S. Y. Fuchs, A. Chen, et al., "The SCF(HOS/beta-TRCP)-ROC1 E3 ubiquitin ligase utilizes two distinct domains within CUL1 for substrate targeting and ubiquitin ligation," *Mol Cell Biol* 20 (2000): 1382–93.

121. K. Wu, A. Chen, and Z.-Q. Pan, "Conjugation of Nedd8 to CUL1 enhances the ability of the ROC1-CUL1 complex to promote ubiquitin polymerization," *J Biol Chem* 275 (2000): 32317–24.

122. S. Han, B. Zheng, D. G. Schatz, et al., "Neoteny in lymphocytes: Rag1 and Rag2 expression in germinal center B cells," *Science* 274 (1996): 2094–97.

123. M. J. Difilippantonio, C. J. McMahan, Q. M. Eastman, et al., "RAG1 mediates signal sequence recognition and recruitment of RAG2 in V(D)J recombination," *Cell* 87 (1996): 253–62.

124. A. Villa, S. Santagata, F. Bozzi, et al., "Partial V(D)J recombination activity leads to Omenn syndrome," *Cell* 93 (1998): 885–96.

125. V. Aidinis, T. Bonaldi, M. Beltrame, et al., "The RAG1 homeodomain recruits HMG1 and HMG2 to facilitate recombination signal sequence binding and to enhance the intrinsic DNA-bending activity of RAG1-RAG2," *Mol Cell Biol* 19 (1999): 6532–42.

126. C. A. Gomez, L. M. Ptaszek, A. Villa, et al., "Mutations in conserved regions of the predicted RAG2 kelch repeats block initiation of V(D)J recombination and result in primary immunodeficiencies," *Mol Cell Biol* 20 (2000): 5653–64.

127. S. Santagata, T. J. Boggon, C. L. Baird, et al., "G-protein signaling through tubby proteins," *Science* 292 (2001): 2041–50.

128. L. Resnick-Silverman, S. St. Clair, M. Maurer, et al., "Identification of a novel class of genomic DNA-binding sites suggests a mechanism for selectivity in target gene activation by the tumor suppressor protein p53," *Genes Dev* 12 (1998) 2102–7.

129. E. C. Thornborrow and J. J. Manfredi, "The tumor suppressor protein p53 requires a cofactor to activate transcriptionally the human BAX promoter," *J Biol Chem* 276 (2001): 15598–608.

130. E. C. Thornborrow, S. Patel, A. E. Mastropietro, et al., "A conserved intronic response element mediates direct p53-dependent transcriptional activation of both the human and murine bax genes," *Oncogene* 21 (2002): 990–99.

131. A. M. Chan, T. Miki, K. A. Meyers, and S. A. Aaronson, "A human oncogene of the RAS superfamily unmasked by expression cDNA Cloning," *Proc Natl Acad Sci U S A* 91 (1994)): 7558–62.

132. R. Saez, A. M. Chan, T. Miki, and S. A. Aaronson, "Oncogenic activation of human R-ras by point mutations analogous to those of prototype H-ras oncogenes," *Oncogene* 9 (1994): 2977–82.

133. A. Kimmelman, T. Tolkacheva, M. V. Lorenzi, et al., "Identification and characterization of R-ras3: a novel member of the RAS gene family with a non-ubiquitous pattern of tissue distribution," *Oncogene* 15 (1997): 2675–85.

134. M. Osada, T. Tolkacheva, W. Li, et al., "Differential roles of Akt, Rac, and Ral in R-Ras-mediated cellular transformation, adhesion, and survival," *Mol Cell Biol* 19 (1999): 6333–44.

135. A. C. Kimmelman, N. Nuñez Rodriguez, and A. M. Chan, "R-Ras3/M-Ras induces neuronal differentiation of PC12 cells through cell-type-specific activation of the mitogen-activated protein kinase cascade," *Mol Cell Biol* 22 (2002): 5946–61.

136. G. Narla, K. E. Heath, H. L. Reeves, et al., "KLF6, a candidate tumor suppressor gene mutated in prostate cancer," *Science* 294 (2001): 2563–66.

137. R. B. Hazan, G. R. Phillips, R. F. Qiao, et al., "Exogenous expression of N-cadherin in breast cancer cells induces cell migration, invasion, and metastasis," *J Cell Biol* 149 (2000): 779–90.

138. K. Suyama, I. Shapiro, M. Guttman, and R. B. Hazan, "A signaling pathway leading to metastasis is controlled by N-cadherin and the FGF receptor," *Cancer Cell* 2 (2002): 301–14.

139. S. Podgrabinska, P. Braun, P. Velasco, et al., "Molecular characterization of lymphatic endothelial cells," *Proc Natl Acad Sci USA* 99 (2002): 16069–74.

140. M. Sugrue, D. Y. Shin, S. W. Lee, and S. A. Aaronson, "Wild-type p53 triggers a rapid senescence program in human tumor cells lacking functional p53," *Proc Natl Acad Sci U S A* 94 (1997): 9648–53.

141. L. Y. Marmorstein, T. Ouchi, and S. A. Aaronson, "The BRCA2 gene product functionally interacts with p53 and RAD51," *Proc Natl Acad Sci U S A* 95 (1998): 13869–74.

142. S. W. Lee, L. Fang, M. Igarashi, et al., "Sustained activation of Ras/RaF/mitogen-activated protein kinase cascade by the tumor suppressor p53," *Proc Natl Acad Sci U S A* 97 (2000): 8302–5.

143. L. Fang, G. Li, G. Liu, et al., "p53 induction of heparin-binding EGF-like growth factor counteracts p53 suppression through activation of MAPK and PI3K/Akt signaling cascades," *EMBO J* 20 (2001): 1931–39.

144. A. Bafico, G. Liu, A. Yaniv, et al., "Novel mechanism of Wnt signalling inhibition mediated by Dickkopf-1 interaction with LRP6/Arrow," *Nat Cell Biol* 3 (2001): 683–86.

145. G. Liu, A. Bafico, V. K. Harris, and S. A. Aaronson, "A novel mechanism for Wnt activation of canonical signaling through the LRP6 receptor," *Mol Cell Biol* 23 (2003): 5825–35.

146. J. Dong, R. G. Phelps, R. Qiao, et al., "BRAF oncogenic mutations correlate with progression rather than initiation of human melanoma," *Cancer Res* 63 (2003): 3883–85.

147. T Ouchi, A. N. A. Monteiro, A. August, et al., "BRCA1 regulates p53-dependent gene expression," *Proc Natl Acad Sci U S A* 95 (1998): 2302–6.

148. T. Ouchi, S. W. Lee, M. Ouchi, et al., "Collaboration of signal transducer and activator of transcription 1 (STAT1) and BRCA1 in differential regulation of IFN-gamma target genes," *Proc Natl Acad Sci U S A* 97 (2000): 5208–13.

149. S. Okada and T. Ouchi, "Cell cycle differences in DNA damage-induced BRCA1 phosphorylation affect its subcellular localization," *J Biol Chem* 278 (2003): 2015–20.

150. J. C. Kwak, P. P. Ongusaha, T. Ouchi, and S. W. Lee, "IFI16 as a negative regulator in the regulation of p53 and p21 (Waf1)," *J Biol Chem* 278 (2003): 40899–904.

151. W. H. Redd, G. H. Montgomery, and K. N. DuHamel, "Behavioral intervention for cancer treatment side effects," *J Natl Cancer Inst* 93 (2001): 810–23.

152. S. Manne, N. Nereo, K. DuHamel, et al., "Anxiety and depression in mothers of children undergoing bone marrow transplant: symptom prevalence and use of the Beck depression and Beck anxiety inventories as screening instruments," *J Consult Clin Psychol* 69 (2001): 1037–47.

153. G. H. Montgomery, K. N. DuHamel, and W. H. Redd, "A meta-analysis of hypnotically induced analgesia: how effective is hypnosis?" *Int J Clin Exp Hypn* 48 (2000): 138–53.

154. S. Nowell, C. Sweeney, M. Winters, et al., "Association between sulfotransferase 1A1 genotype and survival of breast cancer patients receiving tamoxifen therapy," *J Natl Cancer Inst* 94 (2002): 1635–40.

155. C. B. Ambrosone, "Oxidants and antioxidants in breast cancer," *Antioxid Redox Signal* 2 (2000): 903–17.

156. C. B. Ambrosone, C. Sweeney, B. F. Coles, et al., "Polymorphisms in glutathione S-transferases (GSTM1 and GSTT1) and survival after treatment for breast cancer," *Cancer Res* 61 (2001): 7130–35.

157. P. A. Thompson, D. M. DeMarini, F. F. Kadlubar, et al., "Evidence for the presence of mutagenic arylamines in human breast milk and DNA adducts in exfoliated breast ductal epithelial cells," *Environ Mol Mutagen* 39 (2002): 134–42.

158. S. M. Gold, S. G. Zakowski, H. B. Valdimarsdottir, and D. H. Bovbjerg, "Stronger endocrine responses after brief psychological stress in women at familial risk of breast cancer," *Pychoneuroendocrinology* 28 (2003): 584–93.

159. J. Erblich, G. H. Montgomery, H. B. Valdimarsdottir, et al., "Biased cognitive processing of cancer-related information among women with family histories of breast cancer: evidence from a cancer stroop task," *Health Psych* 22 (2003): 235–44.

160. G. H. Montgomery and D. H. Bovbjerg, "Pre-infusion expectations predict post-treatment nausea during repeated adjuvant chemotherapy infusions for breast cancer," *Br J Health Psych* 5 (2000): 105–19.

161. H. B. Valdimarsdottir, S. G. Zakowski, W. Gerin, et al., "Heightened psychobiological reactivity to laboratory stressors in healthy women at familial risk for breast cancer," *J Behav Med* 25 (2002): 51–65.

162. J. Erblich, D. H. Bovbjerg, C. Norman, et al., "It won't happen to me: lower perception of heart disease risk among women with family histories of breast cancer," *Prev Med* 31 (2000): 714–21.

163. J. Erblich, Y. Boyarsky, B. Spring, et al., "A family history of smoking predicts heightened levels of stress-induced cigarette craving," *Addiction* 98 (2003): 657–64.

164. J. Erblich, D. H. Bovbjerg, and H. B Valdimarsdottir, "Looking forward and back: distress among women at familial risk for breast cancer," *Ann Behav Med* 22 (2000): 53–59.

165. H. S. Thompson, H. B. Valdimarsdottir, C. Duteau-Buck, et al., "Psychosocial predictors of BRCA counseling and testing decisions among African-American women," *Cancer Epidemiol Biomarkers Prev* 11 (2002): 1579–85.

166. G. H. Montgomery, D. David, G. Winkel, et al., "The effectiveness of adjunctive hypnosis with surgical patients: a meta-analysis," *Anesth Analg* 94 (2002): 1639–45.

167. See Aufses and Niss, *This House of Noble Deeds*, p. 108, for more on Gabrilove.

168. Interview with Savio Woo, Ph.D., by Arthur Aufses, M.D., July 9, 2003, The Mount Sinai Archives.

169. K. J. Robson, T. Chandra, R. T. MacGillivray, and S. L. Woo, "Polysome immunoprecipitation of phenylalanine hydroxylase mRNA from rat liver and cloning of its cDNA," *Proc Natl Acad Sci* 79 (1982): 4701–5.

170. S. L. Woo, A. Lidsky, F. Guttler, et al., "Cloned human phenylalanine hydroxylase gene allows prenatal diagnosis and carrier detection of classical phenylketonuria," *Nature* 306 (1983): 151–55.

171. J. Kidd, M. S. Golbus, R. B. Wallace, et al., "Prenatal diagnosis of alpha 1-antitrypsin deficiency by direct analysis of the mutation site in the gene," *New Engl J Med* 310 (1984): 639–42.

172. S. L. C. Woo, "Molecular basis and population genetics of phenylketonuria," *Biochemistry* 28 (1989): 1–7.

173. Y. Okano, R. C. Eisensmith, F. Guttler, et al., "Molecular basis of phenotypic heterogeneity in phenylketonuria," *N Engl J Med* 324 (1991): 1232–38.

174. Woo interview.

175. M. Caruso, K. Pham-Nguyen, Y. L. Kwong, et al., "Adenovirus-mediated interleukin-12 gene therapy for metastatic colon carcinoma," *Proc Natl Acad Sci* 93 (1996): 11302–6.

176. S. J. Hall, S. H. Chen, and S. L. Woo, "The promise and reality of cancer gene therapy," *Am J Hum Genet* 61 (1997): 785–89.

177 S. H. Chen, K. B. Pham-Nguyen, O. Martinet, et al., "Rejection of disseminated metastases of colon carcinoma by synergism of IL-12 gene therapy and 4-1BB costimulation," *Mol Ther* 2 (2000): 39–46.

178. B. V. Sauter, O. Martinet, W. J. Zhang, et al., "Adenovirus-mediated gene transfer of endostatin *in vivo* results in high level of transgene expression and inhibition of tumor growth and metastases," *Proc Natl Acad Sci U S A* 97 (2000): 4802–7.

179. M. Kennedy, M. Firpo, K. Choi, et al., "A common precursor for primitive erythropoiesis and definitive haematopoiesis," *Nature* 386 (1997): 488–93.

180. K. Choi, M. Kennedy, A. Kazarov, et al., "A common precursor for hematopoietic and endothelial cells," *Development* 125 (1998): 725–32.

181. S. Robertson, M. Kennedy, and G. Keller, "Hematopoietic commitment during embryogenesis," *Ann N Y Acad Sci* 872 (1999): 9–15.

182. G. Lacaud, S. Robertson, J. Palis, et al., "Regulation of hemangioblast development," *Ann N Y Acad Sci* 938 (2001): 96–107.

183. G. Lacaud, L. Gore, M. Kennedy, et al., "Runx1 is essential for hematopoietic commitment at the hemangioblast stage of development *in vitro*," *Blood* 100 (2002): 458–66.

184. R. Y. Lin, A. Kubo, G. M. Keller, and T. F. Davies, "Committing embryonic stem cells to differentiate into thyrocyte like cells *in vitro*," *Endocrinology* 144 (2003): 2644–49.

185. S. F. Weekx, J. Plum, P. Van Cauwelaert, et al., "Developmentally regulated responsiveness to transforming growth factor-beta is correlated with functional differences between human adult and fetal primitive hematopoietic progenitor cells," *Leukemia* 13 (1999): 1266–72.

186. E. Henckaerts, H. Geiger, J. C. Langer, et al., "Genetically determined variation in the number of phenotypically defined hematopoietic progenitor and stem cells and in their response to early-acting cytokines," *Blood* 99 (2002): 3947–54.

187. L. A. DeBruyne, K. Li, D. K. Bishop, and J. S. Bromberg, "Gene transfer of virally encoded chemokine antagonists vMIP-II and MC148 prolongs cardiac allograft survival and inhibits donor-specific immunity," *Gene Ther* 7 (2000): 575–82.

188. Y. Bai, J. Liu, Y. Wang, et al., "L-selectin-dependent lymphoid occupancy is required to induce alloantigen-specific tolerance," *J Immunol* 168 (2002): 1579–89.

189. Y. Ding, L. Qin, S. V. Kotenko, et al., "A single amino acid determines the immunostimulatory activity of interleukin 10," *J Exp Med* 191 (2000): 213–24.

190. L. Qin, Y. Ding, H. Tahara, and J. S. Bromberg, "Viral IL-10-induced immunosuppression requires Th2 cytokines and impairs APC function within the allograft," *J Immunol* 166 (2001): 2385–93.

191. D. Chen, R. Sung, and J. S. Bromberg, "Gene therapy in transplantation," *Transpl Immunol* 9 (2002): 301–14.

192. J. S. Bromberg, P. Boros, Y. Ding, et al., "Gene transfer methods for transplantation," *Methods Enzymol* 346 (2002): 199–224.

193. G. J. Randolph and M. B. Furie, "Mononuclear phagocytes egress from an *in vitro* model of the vascular wall by migrating across endothelium in the basal to apical direction: role of intercellular adhesion molecule 1 and the CD11/CD18 integrins," *J Exp Med* 183 (1996): 451–62.

194. G. J. Randolph, S. Beaulieu, S. Lebecque, et al., "Differentiation of monocytes into dendritic cells in a model of transendothelial trafficking," *Science* 282 (1998): 480–83.

195. G. J. Randolph, K. Inaba, D. F. Robbiani, et al., "Differentiation of phagocytic monocytes into lymph node dendritic cells *in vivo*," *Immunity* 11 (1999): 753–61.

196. G. J. Randolph, S. Beaulieu, M. Pope, et al., "A physiologic function for p-glycoprotein (MDR-1) during the migration of dendritic cells from skin via afferent lymphatic vessels," *Proc Natl Acad Sci U S A* 95 (1998): 6924–29.

197. D. F. Robbiani, R. A. Finch, D. Jager, et al., "The leukotriene C(4) transporter MRP1 regulates CCL19 (MIP-3beta, ELC)-dependent mobilization of dendritic cells to lymph nodes," *Cell* 103 (2000): 757–68.

198. G. J. Randolph, G. Sanchez-Schmitz, R. M. Liebman, and K. Schakel, "The CD16(+)(FegammaRIII(+)) subset of human monocytes preferentially becomes migratory dendritic cells in a model tissue setting," *J Exp Med* 196 (2002): 517–27.

199. N. Dutheil, F. Shi, T. Dupressoir, and R. M. Linden, "Adeno-associated virus site-specifically integrates into a muscle-specific DNA region," *Proc Natl Acad Sci U S A* 97 (2000): 4862–66.

200. R. M. Linden, P. Ward, C. Giraud, et al., "Site-specific integration by adeno-associated virus," *Proc Natl Acad Sci U S A* 93 (1996): 11288–94.

201. D. H. Smith, P. Ward, and R. M. Linden, "Comparative characterization of rep proteins from the helper-dependent adeno-associated virus type 2 and the autonomous goose parvovirus," *J Virol* 73 (1999): 2930–37.

202. P. Ward and R. M. Linden, "A role for single-stranded templates in cell-free adeno-associated virus DNA replication," *J Virol* 74 (2000): 744–54.

203. Woo interview.

204. See Aufses and Niss, *This House of Noble Deeds*, pp. 56–57, for more on Mayer and the Division of Clinical Immunology.

205. S. R. Targan, S. B. Hanauer, S. J. van Deventer, et al., "A short-term study of chimeric monoclonal antibody cA2 to tumor necrosis factor alpha for Crohn's disease," *New Engl J Med* 337 (1997): 1029–35.

206. P. Rutgeerts, G. D'Haens, S. Targan, et al., "Efficacy and safety of retreatment with anti-tumor necrosis factor antibody (infliximab) to maintain remission in Crohn's disease," *Gastroenterology* 117 (1999): 761–69.

207. D. H. Present, P. Rutgeerts, S. Targan, et al., "Infliximab for the treatment of fistulas in patients with Crohn's disease," *N Engl J Med* 340 (1999): 1398–405.

208. L. Mayer and D. Eisenhardt, "Lack of induction of suppressor T cells by intestinal epithelial cells from patients with inflammatory bowel disease," *J Clin Invest* 86 (1990): 1255–60.

209. L. Mayer, D. Eisenhardt, P. Salomon, et al., "Expression of class II molecules on intestinal epithelial cells in humans. Differences between normal and inflammatory bowel disease," *Gastroenterology* 100 (1991): 3–12.

210. P. Salomon, A. Pizzimenti, A. Panja, et al., "The expression and regulation of class II antigens in normal and inflammatory bowel disease peripheral blood monocytes and intestinal epithelium," *Autoimmunity* 9 (1991): 141–49.

211. L. S. Toy, X. Y. Yio, A. Lin, et al., "Defective expression of gp180, a novel CD8 ligand on intestinal epithelial cells, in inflammatory bowel disease," *J Clin Invest* 100 (1997): 2062–71.

212. N. A. Campbell, M. S. Park, L. S. Toy, et al., "A non-class I MHC intestinal epithelial surface glycoprotein, gp 180, binds to CD8," *Clin Immunol* 102 (2000): 267–74.

213. M. Allez, J. Brimmes, I. Dotan, and L. Mayer, "Expansion of CD8+ T cells with regulatory function after interaction with intestinal epithelial cells," *Gastroenterology* 123 (2002): 1516–26.

214. Y. Li, X. Y. Yio, and L. Mayer, "Human intestinal epithelial cell-induced CD8+ T cell activation is mediated through CD8 and the activation of CD8-associated p56lck," *J Exp Med* 182 (1995): 1079–88.

215. L. Mayer, D. Sherris, R. Huang, and M. Cidon, "Cytokines regulating human B cell growth and differentiation," *Clin Immunol Immunopathol* 61 (1991): S28–36.

216. Z. Qu, J. Odin, J. D. Glass, and J. C. Unkeless, "Expression and characterization of a truncated murine Fc gamma receptor," *J Exp Med* 167 (1988): 1195–1210.

217. J. A. Odin, J. C. Edberg, C. J. Painter, and J. C. Unkeless, "Regulation of phagocytosis and [Ca2+]; flux by distinct regions of an Fc receptor," *Science* 254 (1991): 1785–88.

218. C. T. Lin, Z. Chen, P. Boros, and J. C. Unkeless, "Fc receptor-mediated signal transduction," *J Clin Immunol* 14 (1994): 1–13.

219. J. C. Unkeless, Z. Chen, C. Lin, and E. DeBeus, "Function of human Fc gamma RIIA and Fc gamma RIIIB," *Semin Immunol* 7 (1995): 37–44.

220. F. Y. Chuang, M. Sassaroli, and J. C. Unkeless, "Convergence of Fc gamma receptor IIA and Fc gamma receptor IIIB signaling pathways in human neutrophils," *J Immunol* 164 (2000): 350–60.

221. H. Xiong, C. Zhu, H. Li, et al., "Complex formation of the interferon (IFN) consensus sequence-binding protein with IRF-1 is essential for murine macrophage IFN-gamma-induced iNOS gene expression," *J Biol Chem* 278 (2003): 2271–77.

222. H. Palosaari, J. P. Parisien, J. J. Rodriguez, et al., "STAT protein interference and suppression of cytokine signal transduction by measles virus V protein," *J Virol* 77 (2003): 7635–44.

223. J. F. Lau, I. Nusinzon, D. Burakov, et al., "Role of metazoan mediator proteins in interferon-responsive transcription," *Mol Cell Biol* 23 (2003): 620–28.

224. C. M. Ulane, J. J. Rodriguez, J. P. Parisien, and C. M. Horvath, "STAT3 ubiquitylation and degradation by mumps virus suppress cytokine and oncogene signaling," *J Virol* 77 (2003): 6385–93.

225. S. A. Lira, P. Zalamea, J. N. Heinrich, et al., "Expression of the chemokine N51/KC in the thymus and epidermis of transgenic mice results in marked infiltration of a single class of inflammatory cells," *J Exp Med* 180 (1994): 2039–48.

226. T. Y. Yang, S. C. Chen, M. W. Leach, et al., "Transgenic expression of the chemokine receptor encoded by human herpesvirus 8 induces an angioproliferative disease resembling Kaposi's sarcoma," *J Exp Med* 191 (2000): 445–54.

227. P. Cortes, F. Weis-Garcia, Z. Misulovin, et al., "*In vitro* V(D)J recombination: signal joint formation," *Proc Natl Acad Sci U S A* 93 (1996): 14008–13.

228. E. Spanopoulou, P. Cortes, C. Shih, et al., "Localization, interaction, and RNA binding properties of the V(D)J recombination-activating proteins RAG1 and RAG2," *Immunity* 3 (1995): 715–26.

229. S. Santagata, A. Villa, C. Sobacchi, et al., "The genetic and biochemical basis of Omenn syndrome," *Immunol Rev* 178 (2000): 64–74.

230. K. L. He and A. T. Ting, "A20 inhibits tumor necrosis factor (TNF) alpha-induced apoptosis by disrupting recruitment of TRADD and RIP to the TNF receptor 1 complex in Jurkat T cells," *Mol Cell Biol* 22 (2002): 6034–45.

231. M. L. Goodkin, A. T. Ting, and J. A. Blaho, "NF-kappaB is required for apoptosis prevention during herpes simplex virus type 1 infection," *J Virol* 77 (2003): 7261–80.

232. D. Serrano, K. Becker, C. Cunningham-Rundles, and L. Mayer, "Characterization of the T cell receptor repertoire in patients with common variable immunodeficiency: oligoclonal expression of CD8(+) T cells," *Clin Immunol* 97 (2000): 248–58.

233. S. Rakoff-Nahoum, H. Chen, T. Kraus, et al., "Regulation of class II expression in monocytic cells after HIV-1 infection," *J Immunol* 167 (2001): 2331–42.

234. K. Sperber, P. Beuria, N. Singha, et al., "Induction of apoptosis by HIV-1-infected monocytic cells," *J Immunol* 170 (2003): 1566–78.

235. K. Sperber, M. Louie, T. Kraus, et al., "Hydroxychloroquine treatment of patients with human immunodeficiency virus type 1," *Clin Ther* 17 (1995): 622–36.

236. A. Savarino, L. Gennero, K. Sperber, and J. R. Boelaert, "The anti-HIV-1 activity of chloroquine," *J Clin Virol* 20 (2001): 131–35.

237. H. A. Sampson, "The evaluation and management of food allergy in atopic dermatitis," *Clin Dermatol* 21 (2003): 183–92.

238. H. A. Sampson, "Clinical practice. Peanut allergy," *N Engl J Med* 346 (2002): 1294–99.

239. K. Beyer, L. Bardina, G. Grishina, and H. A. Sampson, "Identification of sesame seed allergens by 2-dimensional proteomics and Edman sequencing: seed storage proteins as common food allergens," *J Allergy Clin Immunol* 110 (2002): 154–59.

240. A. Nowak-Wegrzyn, H. A. Sampson, R. A. Wood, and S. H. Sicherer, "Food protein-induced enterocolitis syndrome caused by solid food proteins," *Pediatrics* 111 (2003): 829–35.

241. X. M. Li, K. Srivastava, J. W. Huleatt, et al., "Engineered recombinant peanut protein and heat-killed Listeria monocytogenes coadministration protects against peanut-induced anaphylaxis in a murine model," *J Immunol* 170 (2003): 3289–95.

242. D. Y. Leung, H. A. Sampson, J. W. Yunginger, et al., "Effect of anti-IgE therapy in patients with peanut allergy," *N Engl J Med* 348 (2003): 986–93.

243. K. S. Zier and B. Gansbacher, "IL-2 gene therapy of solid tumors: an approach for the prevention of signal transduction defects in T cells," *J Mol Med* 74 (1996): 127–34.

244. S. Salvadori and K. Zier, "Gene therapy with modified tumor cells enables T-cell activation by stimulating pathways required for signal transduction," *Cytokines Mol Ther* 2 (1996): 171–75.

245. K. Zier, K. Johnson, J. M. Maddux, et al., "IFNgamma secretion following stimulation with total tumor peptides from autologous human tumors," *J Immunol Methods* 241 (2000): 61–68.

NOTES TO CHAPTER 6

1. H. Popper, "The Mount Sinai concept," *Clin Res* 13 (1965): 500–4.

2. H. Popper and D. Koffler, "The goal," *J Mt Sinai Hosp* 34 (1967): 401–7.

3. A. H. Aufses, Jr. and B. Dana, "Introduction to Medicine: A Developmental History and Analysis of an Interdepartmental Course," in *The Changing Medical Curriculum*, ed. V. W. Lippard and E. Purcell (New York: Josiah Macy Jr. Foundation, 1972), pp. 117–129. See also chapter 2, on the Curriculum.

4. S. W. Bloom, "The medical school as a social system. A case study of faculty-student relations," *Milbank Mem Fund Q* 49 (1971): 1–196.

5. S. W. Bloom, "Role of the teaching hospital in the community," *Lancet* 1 (1973): 709–11.

6. R. F. Badgley and S. W. Bloom, "Behavioral sciences and medical education: the case of sociology," *Soc Sci Med* 7 (1973): 927–41.

7. S. W. Bloom, "The medical school as a social organization: the sources of resistance to change," *Med Educ* 23 (1989): 228–41.

8. S. W. Bloom, *The Word as Scalpel: A History of Medical Sociology* (New York: Oxford University Press, 2002).

9. See note 3 and chapter 2, on the Curriculum.

10. A. L. Silver and D. N. Rose, "Kurt W. Deuschle and community medicine: clinical care, statistical compassion, community empowerment," *Mt Sinai J Med* 59 (1992): 439–43.

11. L. A. Johnson, *The People of East Harlem* (New York: Commonwealth Fund, 1974).

12. K. W. Deuschle, Introduction to L. A. Johnson, *The People of East Harlem,* as quoted in S. W. Bloom, J. Moss, R. Belville, et al., "Community medicine: its contribution to the social science of medicine," *Mt Sinai J Med* 59 (1992): 461–68.

13. Report on Primary Care in East Harlem, 1979. Internal document, Department of Community and Preventive Medicine, The Mount Sinai Archives.

14. S. W. Bloom, J. Moss, R. Belville, et al., "Community medicine: its contribution to the social science of medicine," *Mt Sinai J Med* 59 (1992): 461–68.

15. S. J. Bosch, H. E. Bass, H. M. Gold, and H. D. Banta, "Medical student roles in prepaid group practice," *J Med Educ* 48 (1973) (Supplement): 144–53.

16. S. J. Bosch, H. D. Banta, R. Watkins, et al., "The Mount Sinai-HIP joint program," *Health Serv Rep* 89 (1974): 219–24.

17. L. Eisenberg, "Science in medicine: too much or too little and too limited in scope?" *Am J Med* 84 (1988): 483–91.

18. S. S. Goldwater, as quoted in J. Hirsch and B. Doherty, *The First Hundred Years of The Mount Sinai Hospital of New York,* (New York: Random House, 1952), p. 139.

19. S. J. Bosch and K. W. Deuschle, "Social work: an important component of community medicine at Mount Sinai School of Medicine," *Mt Sinai J Med* 56 (1989): 459–67.

20. Website of the National Association of Social Workers Foundation: http://www.naswfoundation.org/pioneers/s/siegel, accessed September 19, 2002.

21. H. Rehr and B. Gordon, "Aging ward patients and the hospital social service department," *J Am Geriatr Soc* 15 (1967): 1153–62.

22. B. Berkman, H. Rehr, D. Siegel, et al., "Utilization of in-patient services by the elderly," *J Am Geriatr Soc* 19 (1971): 933–46.

23. B. Berkman and H. Rehr, "The search for early indicators of social service need among elderly hospital patients," *J Am Geriatr Soc* 22 (1974): 416–21.

24. H. Rehr and B. G. Berkman, "Social service casefinding in the hospital-its influence on the utilization of social services," *Am J Public Health* 63 (1973): 857–62.

25. H. Rehr, "Medical care organization and the social service connection," *Health Soc Work* 10 (1985): 245–57.

26. B. Berkman and H. Rehr, "Social work undertakes its own audit," *Soc Work Health Care* 3 (1978): 273–86.

27. B. J. Morrison, H. Rehr, G. Rosenberg, and S. Davis, "Consumer opinion surveys: a hospital quality assurance measurement," *QRB Qual Rev Bull* 8 (1982): 19–24.

28. H. Rehr, S. Blumenfield, A. T. Young, and G. Rosenberg, "Social work accountability: a key to high-quality patient care and services," *Mt Sinai J Med* 60 (1993): 368–73.

29. H. Rehr, G. Rosenberg, and S. Blumenfield, *Creative Social Work in Health Care: Clients, the Community, and Your Organization* (New York: Springer, 1998).

30. B. Berkman, H. Rehr, and G. Rosenberg, "A social work department develops and tests a screening mechanism to identify high social risk situations," *Soc Work Health Care* 5 (1980): 373–85.

31. http://naswfoundation.org/pioneers/r/rehr, [accessed September 19, 2002].

32. J. S. Brook, E. B. Balka, and M. Whiteman, "The risks for late adolescence of early adolescent marijuana use," *Am J Public Health* 89 (1999): 1549–54.

33. J. S. Brook, D. W. Brook, and M. Whiteman, "The influence of maternal smoking during pregnancy on the toddler's negativity," *Arch Pediatr Adolesc Med* 154 (2000): 381–85.

34. J. G. Johnson, P. Cohen, D. S. Pine, et al., "Association between cigarette smoking and anxiety disorders during adolescence and early adulthood," *JAMA* 284 (2000): 2348–51.

35. D. W. Brook, J. S. Brook, L. Richter, et al., "Coping strategies of HIV-positive and HIV-negative female injection drug users: a longitudinal study," *AIDS Educ Prev* 11 (1999): 373–88.

36. J. S. Brook, P. Cohen, and D. W. Brook, "Longitudinal study of co-occurring psychiatric disorders and substance use," *J Am Acad Child Adolesc Psychiatry* 37 (1998): 322–30.

37. S. J.Bosch, *The Charles G. Bluhdorn Professorship of International Community Medicine: Period October 1, 1983–July 31, 1991*, Report for the Department of Community and Preventive Medicine, The Mount Sinai School of Medicine, The Mount Sinai Archives.

38. S. J. Bosch, R. Merino, E. Fischer, and K. W. Deuschle, "The international role of a U.S. medical school in developing health services," *Am J Prev Med* 1 (1985): 44–49.

39. S. J. Bosch and A. Silver, "Community-oriented health services education: a strategy for international cooperation," *J Community Health* 19 (1985): 207–14.

40. E. H. Robitzek and I. J. Selikoff, "Hydrazine derivatives of isonicotinic acid (rimifon, marsalid) in the treatment of active progressive caseous-pneumonic tuberculosis; a preliminary report," *Am Rev Tuber* 65 (1952): 402–28. See also A. H. Aufses, Jr. and B. Niss, *This House of Noble Deeds: The Mount Sinai Hospital, 1852–2002* (New York: New York University Press, 2002), pp. 116–21.

41. I. J. Selikoff, J. Churg, and E. C. Hammond, "Asbestos exposure and neoplasia," *JAMA* 188 (1964): 22–26.

42. I. J. Selikoff, E. C. Hammond, and J. Churg, "Asbestos exposure, smoking and neoplasia," *JAMA* 204 (1968): 106–12.

43. R. Ehrlich, R. Lilis, E. Chan, et al., "Long term radiological effects of short term exposure to amosite asbestos among factory workers," *Br J Ind Med* 49 (1992): 268–75.

44. P. J. Landrigan, S. B. Markowitz, S. J. Bosch, et al., *Occupational Disease in New York State: Proposal for a Statewide Network of Occupational Disease Diagnosis and Prevention Changes. Report to the New York State Legislature.* New York: Division of Environmental and Ocupational Medicine, Department of Community Medicine, Mount Sinai School of Medicine (CUNY), February 1987, The Mount Sinai Archives.

45. M. S. Wolff, P. G. Toniolo, E. W. Lee, et al., "Blood levels of organochlorine residues and risk of breast cancer," *J Natl Cancer Inst* 85 (1993): 648–52.

46. M. S. Wolff, G. S. Berkowitz, S. Brower, et al., "Organochlorine exposures and breast cancer risk in New York City women," *Environ Res* 84 (2000): 151–61.

47. M. S. Wolff, J. A. Britton, and V. P. Wilson, "Environmental risk factors for breast cancer among African-American women," *Cancer* 97 (2003): 289–310.

48. A. Silver and K. W. Deuschle, "Graduate training in community medicine at Mount Sinai: the development of the Master of Science in Community Medicine," *Mt Sinai J Med* 57 (1990): 362–67.

49. C. Stine, F. P. Kohrs, D. N. Little, et al., "Integrating prevention education into the medical school curriculum: the role of departments of family medicine," *Acad Med* 75 (2000): S55–59.

50. Mission Statement, The Mount Sinai Medical Center.

NOTES TO CHAPTER 7

1. For more on Hirschhorn's accomplishments as Chairman of the Department of Pediatrics, see A. H. Aufses, Jr. and B. Niss, *This House of Noble Deeds: The Mount Sinai Hospital, 1852–2002* (New York: New York University Press, 2002), chapter 27.

2. D. F. Bishop, R. Kornreich, and R. J. Desnick, "Structural organization of the human alpha-galactosidase A gene: further evidence for the absence of a 3′ untranslated region," *Proc Natl Acad Sci USA* 85 (1988): 3903–7.

3. L. E. Quintern, E. H. Schuchman, O. Levran, et al., "Isolation of cDNA clones encoding human acid sphingomyelinase: occurrence of alternative processed transcripts," *EMBO J* 8 (1989): 2469–73.

4. A. M. Wang, D. F. Bishop, and R. J. Desnick, "Human alpha-N-acetyl-galactosaminidase-molecular cloning, nucleotide sequence, and expression of a full-length cDNA. Homology with human alpha-galactosidase A suggests evolution from a common ancestral gene," *J Biol Chem* 265 (1990): 21859–66.

5. E. H. Schuchman, M. Suchi, T. Takahashi, et al., "Human acid sphingomyelinase. Isolation, nucleotide sequence and expression of the full-length and alternatively spliced cDNAs," *J Biol Chem* 266 (1991): 8531–39.

6. A. M. Wang and R. J. Desnick, "Structural organization and complete sequence of the human alpha-*N*-acetylgalactosaminidase gene: homology with the alpha-galactosidase A gene provides evidence for evolution from a common ancestral gene," *Genomics* 10 (1991): 133–42.

7. E. H. Schuchman, O. Levran, L. V. Pereira, and R. J. Desnick, "Structural organization and complete nucleotide sequence of the gene encoding human acid sphingomyelinase (SMPD1)," *Genomics* 12 (1992): 197–205.

8. J. G. Wetmur, D. F. Bishop, C. Cantelmo, and R. J. Desnick, "Human delta-aminolevulinate dehydratase: nucleotide sequence of a full-length cDNA clone," *Proc Natl Acad Sci USA* 83 (1986): 7703–7.

9. S. F. Tsai, D. F. Bishop, and R. J. Desnick, "Human uroporphyrinogen III synthase: molecular cloning, nucleotide sequence, and expression of a full-length cDNA," *Proc Natl Acad Sci USA* 85 (1988): 7049–53.

10. D. F. Bishop, "Two different genes encode delta-aminolevulinate synthase in humans: nucleotide sequences of cDNAs for the housekeeping and erythroid genes," *Nucleic Acids Res* 18 (1990): 7187–88.

11. H. S. Bernstein, D. F. Bishop, K. H. Astrin, et al., "Fabry disease: six gene rearrangements and an exonic point mutation in the alpha-galactosidase gene," *J Clin Invest* 83 (1989): 1390–99.

12. O. Levran, R. J. Desnick, and E. H. Schuchman, "Niemann-Pick disease: a frequent missense mutation in the acid sphingomyelinase gene of Ashkenazi Jewish type A and B patients," *Proc Natl Acad Sci U S A* 88 (1991): 3748–52.

13. T. Takahashi, R. J. Desnick, G. Takada, and E. H. Schuchman, "Identification of a missense mutation (S436R) in the acid sphingomyelinase gene from a Japanese patient with type B Niemann-Pick disease," *Hum Mutat* 1 (1992): 70–71.

14. C. A. Warner, H. W. Yoo, A. G. Roberts, and R. J. Desnick, "Congenital erythropoietic porphyria: identification and expression of exonic mutations in the uroporphyrinogen III synthase gene," *J Clin Invest* 89 (1992): 693–700.

15. C. M. Eng, C. Schechter, J. Robinowitz, et al., "Prenatal genetic carrier testing using triple disease screening," *JAMA* 278 (1997): 1268–72.

16. J. Ekstein and H. Katzenstein, "The Dor Yeshorim Story: Community-based Carrier Screening for Tay-Sachs Disease," in *Tay-Sachs Disease*, ed. R. J. Desnick and M. M. Kaback (San Diego: Academic Press, 2001), pp. 297–310.

17. G. A. Grabowski, "Treatment of Gaucher's disease," *N Engl J Med* 328 (1993): 1565.

18. See chapter 1, on the History of the School during the Kase years.

19. J. G. Wetmur and N. Davidson, "Kinetics of renaturation of DNA," *J Mol Biol* 31 (1968): 349–70. (In *Current Contents, Life Sciences*, no. 3 (1983): 17.)

20. D. F. Bishop, D. H. Calhoun, H. S. Bernstein, et al., "Human alpha-galactosidase A: nucleotide sequence of a cDNA clone encoding the mature enzyme," *Proc Natl Acad Sci U S A* 83 (1986): 4859–63.

21. C. M. Li, J. H. Park, X. He, et al., "The human acid ceramidase gene (ASAH): structure, chromosomal location, mutation analysis, and expression," *Genomics* 62 (1999): 223–31.

22. See note 3.

23. J. P. Davies, B. Levy, and Y. A. Ioannou, "Evidence for a Niemann-Pick C (NPC) gene family: identification and characterization of NPC1L1," *Genomics* 65 (2000): 137–45.

24. See note 4.

25. E. H. Schuchman, C. E. Jackson, and R. J. Desnick, "Human arylsulfatase B: MOPAC cloning, nucleotide sequence of a full-length cDNA, and regions of amino acid identity with arylsulfatases A and C," *Genomics* 6 (1990): 149–58.

26. See note 8.

27. J. G. Wetmur, D. F. Bishop, L. Ostasiewicz, and R. J. Desnick, "Molecular cloning of a cDNA for human delta-aminolevulinate dehydratase," *Gene* 43 (1986): 123–30.

28. See note 9.

29. G. Aizencang, C. Solis, D. F. Bishop, et al., "Human uroporphyrinogen-III synthase: genomic organization, alternative promoters, and erythroid-specific expression," *Genomics* 70 (2000): 223–31.

30. H. W. Yoo, C. A. Warner, C. H. Chen, and R. J. Desnick, "Hydroxymethylbilane synthase: complete genomic sequence and amplifiable polymorphisms in the human gene," *Genomics* 15 (1993): 21–29.

31. M. Mendez, L. Sorkin, M. V. Rossetti, et al., "Familial porphyria cutanea tarda: characterization of seven novel uroporphyrinogen decarboxylase mutations and frequency of common hemochromatosis alleles," *Am J Hum Genet* 63 (1998): 1363–75.

32. See note 10.

33. D. Forrest, L. C. Erway, L. Ng, et al., "Thyroid hormone receptor beta is essential for development of auditory function," *Nat Genet* 13 (1996): 354–57.

34. D. Forrest, E. Hanebuth, R. J. Smeyne, et al., "Recessive resistance to thyroid hormone in mice lacking thyroid hormone receptor beta: evidence for tissue-specific modulation of receptor function," *EMBO J* 15 (1996): 3006–15.

35. S. Gothe, Z. Wang, L. Ng, et al., "Mice devoid of all known thyroid hormone receptors are viable but exhibit disorders of the pituitary-thyroid axis, growth, and bone maturation," *Genes Dev* 13 (1999): 1329–41.

36. L. Ng, J. B. Hurley, B. Dierks, et al., "A thyroid hormone receptor that is required for the development of green cone photoreceptors," *Nat Genet* 27 (2001): 94–98.

37. L. Ng, A. Rusch, L. L. Amma, et al., "Suppression of the deafness and thyroid dysfunction in Thrb-null mice by an independent mutation in the Thra thyroid hormone receptor alpha gene," *Hum Mol Genet* 10 (2001): 2701–8.

38. C. Linnevers, S. P. Smeekens, and D. Bromme, "Human cathepsin W, a putative cysteine protease predominantly expressed in CD8+ T-lymphocytes," *FEBS Lett* 405 (1997): 253–59.

39. M. E. McGrath, J. L. Klaus, M. G. Barnes, and D. Bromme, "Crystal structure of human cathepsin K complexed with a potent inhibitor," *Nat Struct Biol* 4 (1997): 105–9.

40. B. D. Wang, G. P. Shi, P. M. Yao, et al., "Human cathepsin F. Molecular cloning, functional expression, tissue localization, and enzymatic characterization," *J Biol Chem* 273 (1998): 32000–8.

41. D. Bromme, Z. Li, M. Barnes, and E. Mehler, "Human cathepsin V functional expression, tissue distribution, electrostatic surface potential, enzymatic characterization, and chromosomal localization," *Biochemistry* 38 (1999): 2377–85.

42. T. Wex, B. Levy, H. Wex, and D. Bromme, "Human cathepsins F and W: A new subgroup of cathepsins," *Biochem Biophys Res Commun* 259 (1999): 401–7.

43. B. D. Gelb, G. P. Shi, H. A. Chapman, and R. J. Desnick, "Pycnodysostosis, a lysosomal disease caused by cathepsin K deficiency," *Science* 273 (1996): 1236–38.

44. M. Tartaglia, E. L. Mehler, R. Goldberg, et al., "Mutations in PTPN11, encoding the protein tyrosine phosphatase SHP-2, cause Noonan syndrome," *Nat Genet* 29 (2001): 465–68.

45. P. A. Hernandez, R. J. Gorlin, J. N. Lukens, et al., "Mutations in the chemokine receptor gene CXCR4 are associated with WHIM syndrome, a combined immunodeficiency disease," *Nat Genet* 34 (2003): 70–74.

46. J. A. Martignetti, K. E. Heath, J. Harris, et al., "The gene for May-Hegglin anomaly localizes to a <1-Mb region on chromosome 22q12.3-13.1," *Am J Hum Genet* 66 (2000): 1449–54.

47. M. Seri, R. Cusano, S. Gangarossa, et al., "Mutations in MYH9 result in the May-Hegglin anomaly, and Fechtner and Sebastian syndromes, The May-Hegglin/Fechtner Syndrome Consortium," *Nat Genet* 26 (2000): 103–5.

48. J. A. Martignetti, A. A. Aqeel, W. A. Sewairi, et al., "Mutation of the matrix metalloproteinase 2 gene (MMP2) causes a multicentric osteolysis and arthritis syndrome," *Nat Genet* 28 (2001): 261–65.

49. M. Satoda, M. E. Pierpont, G. A. Diaz, et al., "Char syndrome, an inherited disorder with patent ductus arteriosus, maps to chromosome 6p12-p21," *Circulation* 99 (1999): 3036–42.

50. M. Satoda, F. Zhao, G. A. Diaz, et al., "Mutations in TFAP2B cause Char syndrome, a familial form of patent ductus arteriosus," *Nat Genet* 25 (2000): 42–46.

51. G. A. Diaz, M. Banikazemi, K. Oishi, et al., "Mutations in a new gene encoding a thiamine transporter cause thiamine-responsive megaloblastic anaemia syndrome," *Nature Genet* 22 (1999): 309–12.

52. G. A. Diaz, K. T. Khan, and B. D. Gelb, "The autosomal recessive Kenny-Caffey syndrome locus maps to chromosome 1q42-q43," *Genomics* 54 (1998): 13–18.

53. G. A. Diaz, B. D. Gelb, F. Ali, et al., "Sanjad-Sakati and autosomal recessive Kenny-Caffey syndromes are allelic: evidence for an ancestral founder mutation and locus refinement," *Am J Med Genet* 85 (1999): 48–52.

54. R. Parvari, E. Hershkovitz, N. Grossman, et al., "Mutation of TBCE causes hypoparathyroidism-retardation-dysmorphism and autosomal recessive Kenny-Caffey syndrome," *Nat Genet* 32 (2002): 448–52.

55. J. P. Davies, F. W. Chen, and Y. A. Ioannou, "Transmembrane molecular pump activity of Niemann-Pick C1 protein," *Science* 290 (2000): 2295–98.

56. See note 23.

57. See notes 33–36.

58. A. M. Wang, Y. A. Ioannou, E. M. Zeidner, et al., "Generation of a mouse model with alpha-galactosidase A deficiency," *Am J Hum Genet* 59 (1996): A208.

59. A. M. Wang, C. L. Stewart, and R. J. Desnick, "Schindler disease: generation of a murine model by targeted disruption of the alpha-N-acetylgalactosaminidase gene," *Pediatr Res* 35 (1994): 155A.

60. K. Horinouchi, T. Sakiyama, L. Pereira, et al., "Mouse models of Niemann-Pick disease: mutation analysis and chromosomal mapping rule out the type A and B forms," *Genomics* 18 (1993): 450–51.

61. S. Marathe, S. R. Miranda, C. Devlin, et al., "Creation of a mouse model for non-neurological (type B) Niemann-Pick disease by stable, low level expression of lysosomal sphingomyelinase in the absence of secretory sphingomyelinase: relationship between brain intra-lysosomal enzyme activity and central nervous system function," *Hum Mol Genet* 9 (2000): 1967–76.

62. C. M. Li, J. H. Park, C. M. Simonaro, et al., "Insertional mutagenesis of the mouse acid ceramidase gene leads to early embryonic lethality in homozygotes and progressive lipid storage disease in heterozygotes," *Genomics* 79 (2002): 218–24.

63. K. Oishi, S. Hofmann, G. A. Diaz, et al., "Targeted disruption of Slc19a2, the gene encoding the high-affinity thiamin transporter Thtr-1, causes diabetes mellitus, sensorineural deafness and megaloblastosis in mice," *Hum Mol Genet* 11 (2002): 2951–60.

64. See note 20.

65. See note 2.

66. Y. A. Ioannou, D. F. Bishop, and R. J. Desnick, "Overexpression of human alpha-galactosidase A results in its intracellular aggregation, crystallization in lysosomes, and selective secretion," *J Cell Biol* 119 (1992): 1137–50.

67. See note 58.

68. Y. A. Ioannou, K. M. Zeidner, R. E. Gordon, and R. J. Desnick, "Fabry disease: preclinical studies demonstrate the effectiveness of alpha-galactosidase A replacement in enzyme-deficient mice," *Am J Hum Genet* 68 (2001): 14–25.

69. C. M. Eng, M. Banikazemi, R. Gordon, et al., "A phase 1/2 clinical trial of enzyme replacement in Fabry disease: pharmacokinetic, substrate clearance, and safety studies," *Am J Hum Genet* 68 (2001): 711–22.

70. C. M. Eng, N. Guffon, W. R. Wilcox, et al., "Safety and efficacy of recombinant human alpha-galactosidase A-replacement therapy in Fabry's disease," *N Engl J Med* 345 (2001): 9–16.

71. See note 5.

72. See note 60.

73. S. R. Miranda, X. He, C. M. Simonaro, et al., "Infusion of recombinant human acid sphingomyelinase into Niemann-Pick disease mice leads to visceral, but not neurological, correction of the pathophysiology," *FASEB J* 14 (2000): 1988–95.

74. J. Q. Fan, S. Ishii, N. Asano, and Y. Suzuki, "Accelerated transport and maturation of lysosomal alpha-galactosidase A in Fabry lymphoblasts by an enzyme inhibitor," *Nat Med* 5 (1999): 112–15.

75. A. Frustaci, C. Chimenti, R. Ricci, et al., "Improvement in cardiac function in the cardiac variant of Fabry's disease with galactose-infusion therapy," *N Engl J Med* 345 (2001): 25–32.

NOTES TO CHAPTER 8

1. Letter of agreement, John W. Rowe, M.D., President and Chief Executive Officer, Mount Sinai Medical Center, to Mark R. Chassin, M.D., March 28, 1995. Personal communication, M. Chassin to A. H. Aufses Jr., March 2001.

2. M. R. Chassin, "Improving quality of care with practice guidelines," *Front Health Serv Manage* 10 (1993): 40–44.

3. M. R. Chassin, "The missing ingredient in health reform. Quality of care," *JAMA* 270 (1993): 377–78.

4. M. R. Chassin, "Quality is the key to controlling costs and increasing access," Interview by C. Burns Roehrig, *Internist* 34 (1993): 20–23.

5. M. R. Chassin, "Quality of health care. Part 3: improving the quality of care," *N Engl J Med* 335 (1996): 1060–63.

6. M. R. Chassin, "Improving the quality of health care: what strategy works?" *Bull N Y Acad Med* 73 (1996): 81–91.

7. M. R. Chassin, "Assessing strategies for quality improvement," *Health Aff* (Millwood) 16 (1997): 151–61.

8. M. R. Chassin, "Quality improvement nearing the 21st century: prospects and perils," *Am J Med Qual* 11 (1996): S4–7.

9. E. L. Hannan, H. Kilburn Jr., M. Racz, et al., "Improving the outcomes of coronary artery bypass surgery in New York State," *JAMA* 271 (1994): 761–66.

10. E. L. Hannan, A. L. Siu, D. Kumar, et al., "The decline in coronary artery bypass graft surgery mortality in New York State. The role of surgeon volume," *JAMA* 273 (1995) 209–13.

11. M. R. Chassin, "Achieving and sustaining improved quality: lessons from New York State and cardiac surgery," *Health Affairs* 21 (2002): 40–51.

12. M. R. Chassin and A. L. Siu, "Academic quality improvement: new medicine in old bottles," *Qual Manag Health Care* 4 (1996): 40–46.

13. Institute of Medicine (U.S.), Division of Health Care Services, *Medicare: A Strategy for Quality Assurance/Committee to Design a Strategy for Quality Review and Assurance in Medicare,* ed. K. N. Lohr, vol. I (Washington, D.C.: National Academy Press, 1990), p. 21.

14. M. R. Chassin, "Quality of care. Time to act," *JAMA* 266 (1991): 3472–73.

15. E. L. Hannan, J. Magaziner, J. J. Wang, et al., "Mortality and locomotion 6 months after hospitalization for hip fracture: risk factors and risk-adjusted hospital outcomes," *JAMA* 285 (2001): 2736–42.

16. E. A. Eastwood, J. Magaziner, J. Wang, et al., "Patients with hip fracture: subgroups and their outcomes," *J Am Geriatr Soc* 50 (2002): 1240–49.

17. G. M. Orosz, E. L. Hannan, J. Magaziner, et al., "Hip fracture in the older patient: reasons for delay in hospitalization and timing of surgical repair," *J Am Geriatr Soc* 50 (2002): 1336–40.

18. N. A. Bickell, A. H. Aufses, Jr., and M. R. Chassin, "Engaging clinicians in a quality improvement strategy for early-stage breast cancer treatment," *Qual Manag Health Care* 6 (1998): 63–68.

19. N. A. Bickell, A. H. Aufses, Jr., and M. R. Chassin, "The quality of early-stage breast cancer care," *Ann Surg* 232 (2000): 220–24.

20. N. A. Bickell and G. J. Young, "Coordination of care for early-stage breast cancer patients," *J Gen Intern Med* 16 (2001): 737–42.

21. N. A. Bickell and A. L. Siu, "Why do delays in treatment occur? Lessons learned from ruptured appendicitis," *Health Serv Res* 36 (2001): (1 Pt. 1) 1–5.

22. N. A. Bickell, "Race, ethnicity, and disparities in breast cancer: victories and challenges," *Womens Health Issues* 12 (2002): 238–51.

23. C. R. Horowitz, M. H. Davis, A. G. Palermo, and B. C. Vladeck, "Approaches to eliminating sociocultural disparities in health," *Health Care Financ Rev* 21 (2000): 57–74.

24. C. R. Horowitz and M. R. Chassin, "Improving the quality of pneumonia care that patients experience," *Am J Med* 113 (2002): 379–83.

25. E. A. Halm, M. J. Fine, W. N. Kapoor, et al., "Instability on hospital discharge and the risk of adverse outcomes in patients with pneumonia," *Arch Intern Med* 10 (2002): 1278–84.

26. E. A. Halm and A. S. Teirstein, "Clinical practice. Management of community-acquired pneumonia," *New Engl J Med* 374 (2002): 2039–45.

27. E. L. Hannan, A. J. Popp, P. Feustel, et al., "Association of surgical specialty and processes of care with patient outcomes for carotid endarterectomy," *Stroke* 32 (2001): 2890–97.

28. E. A. Halm, M. R. Chassin, S. Tuhrim, et al., "Revisiting the appropriateness of carotid endarterectomy," *Stroke* 34 (2003): 1464–71.

29. E. A. Halm, C. Lee, and M. R. Chassin, "Is volume related to outcome in health care? A systematic review and methodologic critique of the literature," *Ann Int Med* 137 (2002): 511–20.

30. E. C. Becher and N. A. Christakis, "Firearm injury prevention counseling: are We missing the mark?" *Pediatrics* 104 (1999): 530–35.

31. E. C. Becher, C. K. Cassel, and E. A. Nelson, "Physician firearm ownership as a predictor of firearm injury prevention practice," *Am J Public Health* 90 (2000): 1626–28.

32. E. C. Becher and M. R. Chassin, "Improving quality, minimizing error: making it happen," *Health Affairs* 20 (2001): 68–81.

33. M. R. Chassin and E. C. Becher, "The wrong patient," *Ann Int Med* 136 (2002): 826–33.

34. E. C. Becher and M. R. Chassin, "Improving the quality of health care: who will lead?" *Health Affairs* 20 (2001): 164–79.

35. E. C. Becher and M. R. Chassin, "Taking health care back: the physician's role in quality improvement," *Acad Med* 77 (2002): 953–62.

36. V. H. Nguyen and M. A. McLaughlin, "Coronary artery disease in women: a review of emerging cardiovascular risk factors," *Mt Sinai J Med* 69 (2002): 338–49.

37. J. E. Sisk, W. Whang, J. C. Butler, et al., "Cost effectiveness of vaccination against invasive pneumococcal disease among people 50 through 64 years of age: role of comorbid conditions and race," *Ann Int Med* 138 (2003): 960–68.

38. M. R. Chassin, "Is health care ready for Six Sigma quality?" *Milbank Q* 76 (1998): 565–91, 510.

39. "Meet the Green Belts," *Inside Mount Sinai* (newsletter), July 25–August 4, 2002, 1–3, The Mount Sinai Archives.

NOTES TO CHAPTER 9

1. The Jews' Hospital. *Articles of Incorporation, January 15, 1855*, The Mount Sinai Archives.

2. Minutes of the Board of Directors Meeting, The Jews' Hospital, December 5, 1855, The Mount Sinai Archives.

3. The Mount Sinai Hospital, *Annual Report for 1871*, p. 6.

4. Minutes of the Board of Directors Meeting, The Mount Sinai Hospital, February 2, 1873, The Mount Sinai Archives.

5. I. M. Rutkow, "John Wyeth (1845–1922) and the postgraduate education and training of America's surgeons," *Arch Surg* 137 (2002): 748–49.

6. Minutes of the Board of Directors Meeting, The Mount Sinai Hospital, January 8, 1882, The Mount Sinai Archives.

7. A. Meyer, "Recollections of old Mount Sinai days," *J Mt Sinai Hosp* 3 (1937): 295–307.

8. The Levy Library of the Mount Sinai School of Medicine has an extensive collection of reprints of publications by the staff, the earliest published in 1874. These were collected by requesting copies from the authors, and the collection is known to be incomplete. Nevertheless, from the years 1874–1966, the year that the collection ends, there are more than 13,000 reprints.

9. The Mount Sinai Hospital, *Annual Report for 1899*, p. 30.

10. J. Hirsh and B. Doherty, *The First Hundred Years of The Mount Sinai Hospital of New York* (New York: Random House, 1952), p. 107.

11. Ibid., p. 108.

12. The Mount Sinai Hospital, *Annual Report for 1905*, p. 49.

13. The Mount Sinai Hospital, *Annual Report for 1906*, pp. 42–43.

14. A. Flexner, *Medical Education in the United States and Canada: A Report to the Carnegie Foundation for the Advancement of Teaching* (New York: Arno Press, 1972), p. 278. Original edition issued as Bulletin no. 4 of the Carnegie Foundation for the Advancement of Teaching.

15. The Mount Sinai Hospital, *Annual Report for 1911*, p. 27.

16. The Mount Sinai Hospital, *Constitution and By-laws*, Article 2, Objectives, Paragraph 3, Section 3, 1918.

17. The Mount Sinai Hospital, *Annual Report for 1934*, p. 10.

18. G. W. Stephenson, *American College of Surgeons at 75* (Chicago: American College of Surgeons, 1990).

19. B. Stimmel, "House staff demographic analyses: 2001–2002," *Mount Sinai School of Medicine Consortium for Graduate Medical Education*, January 2002. Internal document.

20. See chapter 2, on the Curriculum, for more on the Morchand Center.

21. The Mount Sinai Hospital of New York, Mount Sinai School of Medicine of New York, *Annual Report for 1968*, p. 60.

22. See chapter 2, on the Curriculum, for more on Swartz.

23. M. Swartz, "CME, telemedicine, educational technology for CME," Presentation to a meeting of the Dean and Chairmen, March 10, 2003.

NOTES TO CHAPTER 10

1. K. Ludmerer, *Time to Heal: American Medical Education from the Turn of the Century to the Era of Managed Care* (New York: Oxford University Press, 1999).

2. A full-time physician devotes his or her entire effort to one institution. Although the physician may consult at other hospitals, he or she cannot care for patients outside the "home institution." As defined here, "full" full-time physicians receive their entire income in the form of an institutional salary without regard to practice income. If they are allowed to practice, all fees are turned over to the institution. "Geographic" full-time physicians receive salaries and supplement their income with money derived from patient fees. Although there were full full-time physicians on the faculty early on, by these definitions, all members of the current FPA are "geographic." The allocation of fee income is usually based upon an institutional formula. Mount Sinai's formula is discussed later in this chapter.

3. See A. H. Aufses, Jr. and B. Niss, *This House of Noble Deeds: The Mount Sinai Hospital, 1852–2002* (New York: New York University Press, 2002), for a detailed description of the development of the Departments of the Hospital.

4. See chapter 1, on the history of the school, for more on the Clinical Excellence Committee.

5. B. J. Niss and N. G. Kase, "An overview of the history of the Mount Sinai School of Medicine of The City University of New York, 1963–1988," *Mt Sinai J Med* 56 (1989): 356–366. See also chapter 1, on the History of the School.

6. S. J. Bosch and K. W. Deuschle, "HMO development in an academic medical center: the rise and fall of a prepaid health program in New York City," *J Comm Health* 18 (1993): 183–200. See also chapter 6, on the Department of Community and Preventive Medicine.

7. Ibid., pp. 188–89.

8. State of New York, Department of Health, "HMO and IPA provider contract guidelines," available at http://www.health.state.ny.us/nysdoh/mancare/hmoipa/appendix.htm, accessed December 12, 2002.

9. FPA By-laws, December 1, 1998.

10. Minutes, FPA Board of Governors meeting, July 2002.

NOTES TO CHAPTER 11

1. H. Popper, "Know Your School," *Spectrum* (Spring 1973), n.3.

2. *100 Years . . . History of the Mount Sinai Alumni* (New York: Mount Sinai Alumni, 1996), p. 19. This volume is a complete history of the Alumni written on the occasion of its centennial in 1996.

3. *Ibid.*, p.27.

4. *Alumni News*, June 1969.

5. *Ibid.*, Spring 1993.

6. *100 Years . . .*, p. 142.

Index

Page numbers in italic refer to illustrations.

Aaronson, Stuart, 173–75, 176, 178, 181
Abdenur, Jose, 223
Abramovitz, Max, 16, 29, 334
Academic Computing division, 51, 63
academic convocations, 59, 72
Academic Council, 40, 60, 290
Academic Educational Network, 249
Acs, George, 146–47
Ad Hoc Committee to Review the Curriculum, 84, 86
Adler, Jack, 267
Administrative Medicine, Department of, 37, 50
Advanced Cardiac Life Support (ACLS), 64, 65
Advisory Committee, 13, 15–16, 18, 20
Aggarwal, Aneel, 131
aging: Alzheimer's disease, 154, 167, 169; Brookdale Medical Conference on Aging, 252; Geriatrics and Adult Development Department, 50, 84, 167, 233, 235; Kastor Neurobiology of Aging Laboratories, 114, 168–69
AIDS/HIV, 190, 204, 302–5, 311
AIDS Volunteer Liaison Project, 200
Alberini, Cristina, 129
Albert Einstein College of Medicine, 8, 64
Alberts, Bruce, 328
Alpha Omega Alpha (AOA), 36–37, 81, 136
alumni. *See* Mount Sinai Alumni
Alumni Day, 272
Alumni Library Fund, 271
Alumni News, 271
Alvarez, Melissa, on student life at Mount Sinai, 294, 299, 303, 308, 310
Alzheimer's disease, 154, 167, 169
Ambassador Program, 238
Ambrosone, Christine, 179
American Association of Anatomists, 122
American College of Surgeons (ACS), 245–46
American Medical Association (AMA), 6–7, 8, 13
American Medical Student Association, 85
Anatomic Radiology, 123
Anatomy, Department of, 115–23; Barka as Chair of, 34, 115; becomes Department of Cell Biology and Anatomy, 118; at opening of Mount Sinai School of Medicine, 113, 115; subjects taught by, 76, 115

Anatomy and Functional Morphology, Center for, 120–23
Anderson, Paul J., 52, 115
Annenberg, Mrs. Moses, 30
Annenberg, Walter, 325
Annenberg Building, 30–31; under construction, *36*; delays in development of, 26; design of, 30, 45; development of plan for, 17, 30; and Faculty Practice Plan, 259; interest from building fund of, 44; laboratories of, *45*; Levy Library, 39; in Mount Sinai Alumni logo, *270*, 272; Mount Sinai campus in 2003, *73*; opening of, 44–46
anti-Semitism, 5
Appointments and Promotions (A&P) Committee, 32, 70, 289
Arnstein, Leo, 329
Aron, Jane and Jack, 200, *202*
Aron, Jane B., Residence Hall, 59, 276
Aronson, Harris, 329
"Art and Science of Medicine" course, 89, 203, 308
Asnes, Marvin, 53
Association of American Medical Colleges (AAMC), 6–7, 8, 84, 88, 135
Atran Laboratory, 3
Atweh, George, 223
Aufses, Arthur H., Jr., 48, 49, *59*, 78, 260, 265, 328
Aufses, Arthur H., Sr., 271
Aufses, Beatrice, 271
Auxiliary Board (Social Service Auxiliary), 199–200

Bader, Mortimer, 315
Bader, Richard, 315
Baehr, George, 13, 14–15, *19*, 257, 325
Baerwald, Edith J., Professor of Community Medicine (Social Work), 200, *202*
Baine, Kenneth, 100
Baltimore, David, 327
Bancroft, F. Carter, 130, 132
Bane Report, 7, 9
Banikazemi, Maryam, 224, 228
Barka, Tibor, 34, 75, 115–16, *116*, 117, 118, 276
Baron, Margaret, 106

393

Holland, James, 173
Hollier, Larry H., 331
homeless health, 303
honorary degree recipients, 325–28
Honors Research Track, 59, 86–87
Hood, Leroy E., 326
Horowitz, Carol R., 234
Horowitz, Frances Degan, 94
Horowitz, Saul, Jr., 29, 51
Horowitz, Saul, Jr., Memorial Award, 51–52, 323–24
Horvath, Curt, 188–89
hospital administration, Department of Administrative Medicine and, 37
Hospital for Joint Diseases and Medical Center, 35, 79
house staff, 240, 241, 242
Howell, Elizabeth, 234
Human Genetics, Department of, 217–30; Center for Jewish Genetic Diseases, 220–21; clinical programs of, 224–25; core facilities of, 229; departmental status for, 157–58; educational programs of, 225–26; establishment of, 58, 221–22; faculty recruitment, 222–24; Familial Cancer Registry, 181; philanthropic support for, 229; research highlights of, 226–28; as spanning School and Hospital, 114; training grants for, 96
Human Genome Project, 219, 227
humanities, 47, 61, 74, 305–8
Humanities and Medicine Program, 164
HUMED project, 47
Huntley, George, 170–71
Hybridoma Core Facility, 141

Icahn, Carl C., 182, 333
Icahn, Carl C., Institute for Gene Therapy and Molecular Medicine, 58, 114, 182–87, 222
Immunization Registry Project, 214
Immunobiology, Center for, 58, 114, 187–91
immunology, 58, 99, 134, 139–41, 157
inflammatory bowel disease (IBD), 187–88
influenza virus, 134, 135, 136–38, 141
Ingelfinger, Franz J., 325
Inherited Metabolic Disease Program, 224
institutes. *See* centers and institutes
Institutional Compliance Agreement (ICA), 267
integrated systems teaching, 78–79
"Integrative Core" course, 89
Interdisciplinary Graduate Training Program in Neurosciences, 171
interdisciplinary training: in Graduate School of Biological Sciences, 97, 99–100; in "Introduction to Medicine" course, 78; Laitman as advocate of, 121; Mount Sinai history of, 64. *See also* multidisciplinary research centers
International Community Medicine program, 205–7
International Conference of Social Work in Health and Mental Health, 203–4
Internet, 64, 71, 90, 252
interns, 5, 246–47
Interns and Residents, Committee of, 277

"Introduction to Clinical Medicine" course, 74, 76
"Introduction to Medicine" course, 76–78, 82, 89, 193, 203, 280, 308
Ioannou, Yiannis, 226, 228
Itzkowitz, Steven, 323
Iyengar, Ravi, 148, 152–53, 154, 156

Jacob, Francois, 325
Jacobi, Abraham, 240, 241, 242–43
Jacobi Library, 17, 38, 39
Jacobi Medallion, 137, 146, 147, 271, 272
Jamaica Hospital Medical Center, 213
James, George, 21; as Acting Director of Medical Education Department, 80; at administrative meeting, *40*; and community medicine, 22, 74–75, 192–93, 194, 200; and community service by medical students, 300; on curriculum, 74–75; as Dean of School of Medicine, 21–22, 330; death of, 41; at Dedication Day, *19*, *35*; and Faculty Practice Plan, 258, 259; at first Black Post-Graduate School of Medicine course, *251*; at first commencement, *38*; first students greeted by, 31; on medicine as lifelong pursuit, 43; open-door policy of, 32; and pass/fail system, 82; and Pomrinse, 40, 41; as President of Medical Center, 26, *26*, *27*, 330; and student representation, 288–89; talking to students, *33*; and university affiliation, 24, 25
Javits, Jacob K., 325
Jersey City Medical Center, 213
Jewish Genetic Diseases, Center for, 220–21, 229
Jewish Genetic Diseases Screening Program, 224
Jewish Home and Hospital for Aged, 44, 203
Johnson, Louise, 195
Johnson, Robert Wood, Foundation, 197
Joint Committee on Medical Education, 15
Joint Committee on Medical School Planning, 12
Jones, Lewis P., 333
Journal of The Mount Sinai Hospital, 244–45

Kandel, Eric R., 326
Kane, Richard, *70*
Kaplan, Helene L., 328
Kaplan, Joel, 265
Kaplan, Stephen A., 324
Kaposi's sarcoma, 189
Kardon, Nataline, 222
Kark, Allan E., 3, 19, 258–59
Kase, Nathan G., 55–66; and basic sciences, 58, 98, 221; centers and institutes established under, 58, 157, 158; as Chairman of Obstetrics and Gynecology, 50; at convocation of 2002, 72; as Dean of the School of Medicine, 55, 330; on the East Building, 65; honorary degree for, 328; as Interim Dean and Interim President, 71; at meeting with Department Chairmen, *59*; portrait of, *56*; presentation to Castle and Wood, 13; as President of Mount Sinai Hospital, 330; as President of Mount Sinai Medical Center, 330; retires as Dean, 65–66; and U.S. Medical Licensing Examinations, 59, 87

About the Authors

BARBARA J. NISS, M.A., is the Mount Sinai Archivist.

ARTHUR H. AUFSES JR., M.D., has been at Mount Sinai since 1954 and was Chairman of the Department of Surgery from 1974 to 1996.

They previously co-authored This House of Noble Deeds: The Mount Sinai Hospital, 1852–2002, also published by NYU Press.